Environment and Energy 2018

AA006000

Transportation Research Record: Journal of the Transportation Research Board

Volume 2672, Issue 24

ISBN: 978-1-5108-7828-0

Printed from e-media with permission by:

Curran Associates, Inc.
57 Morehouse Lane
Red Hook, NY 12571

Some format issues inherent in the e-media version may also appear in this print version.

Published by Sage Publications USA

ISBN (Print) 978-1-5108-7828-0
ISBN 2018 Printed Set (All Issues) 978-1-5108-7735-1

Printed by Curran Associates, Inc. (2019)

Additional copies of this publication are available from:

Curran Associates, Inc.
57 Morehouse Lane
Red Hook, NY 12571 USA
Phone: 845-758-0400
Fax: 845-758-2633
Email: curran@proceedings.com
Web: www.proceedings.com

Volume 2672 Number 24 December 2018

TRR
JOURNAL OF THE TRANSPORTATION RESEARCH BOARD

Contents

Environment and Energy

Articles

TRB

TRANSPORTATION RESEARCH BOARD

CORE SPONSORS

Transportation Departments of the 50 States and the District of Columbia

Federal Government

U.S. Department of Transportation
- Federal Aviation Administration (FAA)
- Federal Highway Administration
- Federal Motor Carrier Safety Administration (FMCSA)
- Federal Railroad Administration (FRA)
- Federal Transit Administration
- National Highway Traffic Safety Administration (NHTSA)
- Office of the Assistant Secretary of Transportation for Research and Technology (OST-R)

U.S. Air Force Civil Engineer Center (AFCEC)
U.S. Coast Guard (USCG)
U.S. Department of Energy
U.S. Department of the Interior (DOI)
U.S. Environmental Protection Agency (EPA)

Nongovernmental Organizations

American Public Transportation Association (APTA)
Association of American Railroads (AAR)
California Air Resources Board

The National Academies of
SCIENCES · ENGINEERING · MEDICINE

The **National Academy of Sciences** was established in 1863 by an Act of Congress, signed by President Lincoln, as a private, nongovernmental institution to advise the nation on issues related to science and technology. Members are elected by their peers for outstanding contributions to research. Dr. Marcia McNutt is president.

The **National Academy of Engineering** was established in 1964 under the charter of the National Academy of Sciences to bring the practices of engineering to advising the nation. Members are elected by their peers for extraordinary contributions to engineering. Dr. C. D. Mote, Jr., is president.

The **National Academy of Medicine** (formerly the Institute of Medicine) was established in 1970 under the charter of the National Academy of Sciences to advise the nation on medical and health issues. Members are elected by their peers for distinguished contributions to medicine and health. Dr. Victor J. Dzau is president.

The three Academies work together as the National Academies of Sciences, Engineering, and Medicine to provide independent, objective analysis and advice to the nation and conduct other activities to solve complex problems and inform public policy decisions. The Academies also encourage education and research, recognize outstanding contributions to knowledge, and increase public understanding in matters of science, engineering, and medicine.

Learn more about the National Academies of Sciences, Engineering, and Medicine at **www.national-academies.org.**

The **Transportation Research Board** is one of seven major programs of the National Academies of Sciences, Engineering, and Medicine. The mission of the Transportation Research Board is to increase the benefits that transportation contributes to society by providing leadership in transportation innovation and progress through research and information exchange, conducted within a setting that is objective, interdisciplinary, and multimodal. The Board's varied activities annually engage about 7,000 engineers, scientists, and other transportation researchers and practitioners from the public and private sectors and academia, all of whom contribute their expertise in the public interest. The program is supported by state transportation departments, federal agencies including the component administrations of the U.S. Department of Transportation, and other organizations and individuals interested in the development of transportation.

Learn more about the Transportation Research Board at **www.TRB.org**.

Article

TRR
JOURNAL OF THE TRANSPORTATION RESEARCH BOARD

Transportation Research Record
2018, Vol. 2672(24) 1–10

DOI: 10.1177/0361198118758649
journals.sagepub.com/home/trr
SAGE

Most Likely Bridges as Roosting Habitat for Bats: Study for Iowa

Basak Aldemir Bektas[1], Zachary Hans[1], Brent Phares[1], Emmanuel Nketah[2], Joe Carey[3], Mary Kay Solberg[4], and Kraig McPeek[5]

Abstract

Bats play an important role in the natural balance of many ecosystems. There has been a growing concern about the bat population in the United States, mainly because of white-nose syndrome (WNS). The primary objective of this work was to better understand what types of bridges are the most likely to be used by bats as roosting locations. In one of the most comprehensive studies in the United States to date, 517 structures in the state of Iowa were inspected for evidence of bat roosting. Logistic regression models were fitted to identify structure, land cover distribution, and predicted bat species distribution characteristics that increase the probability of bat roosting. The final model indicated that probability of bat roosting on bridges increases under the following conditions: structures are prestressed concrete continuous, prestressed concrete or steel continuous; increased superstructure height above ground; increased superstructure depth; increased wetland coverage within a 0.1-mile radius of the structure; and increased number of potential bat species present at the location. The findings show that bridge characteristics, combined with land cover and bat species distribution data, are significant for higher probabilities of bat roosting. This information can be useful to transportation agencies as they plan bridge maintenance and renewal and can also help conservation efforts targeted toward bats. It is thought that the integration of objective, geospatial land cover data with potential bat presence data, and estimation of quantitative and relative influence of variables on probability of bat roosting are unique to this study.

Bats play an important role in the natural balance of many ecosystems. There has been a growing concern about the bat population in the United States, mainly because of white-nose syndrome (WNS). WNS was identified in New York in 2006, and is estimated to have killed more than 5.7 million bats in eastern North America since (*1*). Current research suggests that bat fatalities are also caused by collisions with turbine blades. It is reported that at some sites in the Midwest and in the Eastern United States, species that are already battered by WNS can account for up to 60% of wind energy fatalities (*2*). Concern over bat population has also arisen because habitats used by bats for roosting and foraging have been disturbed, altered, or reduced. In Iowa, at least one federal endangered bat (Indiana bat) is known to exist and thought to be potentially affected by habitat influences.

Conservation efforts targeted toward bats can be hampered by a lack of information on their habitats and usage. As a result, many questions may go unanswered as to how to provide suitable environments, especially at critical roosting times (e.g., maternity season). Fortunately, many bat species are able to adapt to a variety of roosting locations. These can include natural "structures" (e.g., caves, trees, rocks, etc.) and man-made structures (e.g., bridges, buildings, etc.). Further complicating the situation is that some locations may be used only during brief time durations and sometimes for specific usages (hibernation, maternity season, etc.). Although it is widely accepted that bats use bridges as roosting sites, little attention has been given to understanding the characteristics associated with their use as day, night, or maternity season roosting sites. Therefore, it is important to

[1]Institute for Transportation (InTrans), Iowa State University, Ames, IA
[2]CBB, Saint Louis, MO
[3]Ankeny High School, Ankeny, IA
[4]Iowa Department of Transportation Office of Location and Environment, Ames, IA
[5]US Fish and Wildlife Service, Illinois and Iowa ES Field Office, Moline, IL

Corresponding Author:
Address correspondence to Basak Aldemir Bektas: basak@iastate.edu

Table 1. Studies Related to Bridge Characteristics and Bat Roosting

Study	Variables considered	Significant or influential variables
Keeley and Tuttle (1999)	Structure type, crevice width and depth, roost height, traffic	Structure type, crevice width and depth, roost height, traffic
Feldhamer et al. (2003)	Structure type, land cover	Structure type
Gore and Studenroth (2005)	Structure type, age, length, height, traffic	Structure type*, age, traffic*
Hendricks et al. (2005)	Structure type, riparian corridors, land cover	Structure type, riparian corridors, land cover
Cleveland and Jackson (2013)	Structure type, riparian corridors, land cover, crevice type	Structure type, riparian corridors, land cover, crevice type

*Statistical significance was reported.

investigate how, why, and when bats use bridges as roosting sites.

A major goal for this study was to better understand when bridge repair or renewal projects have the potential for "taking" (i.e., harassing, injuring, or killing) federally threatened or endangered bat species. The primary objective of this work was to better understand what types of bridges (based on bridge characteristics including local topography and habitat availability) are the most likely to be used by bats as roosting locations. The secondary objective was to document the means and methods developed and followed to conduct the project such that the evaluation protocol can be exported to other states/regions. The objectives were achieved by

- Establishing the general types of landscape characteristics and "structure" characteristics bats generally prefer based on the literature review;
- Identifying, using geographical information systems (GIS) data, bridges in Iowa that are in bat-friendly landscapes;
- Conducting field surveys of selected bridges to identify
  - Evidence of bats using the bridges as roosting locations
  - Structural characteristics of all bridges in the study sample;
- Analyzing information collected to identify structural and environmental conditions that are most likely to attract roosting bats; and
- Documenting processes and procedures for determining bridge likelihood as a bat roosting habitat.

Literature Review

Bridges provide shelter and protection to bats and serve as different roosting sites; findings on bats' use of bridges as roosting habitat have been presented in the literature (*3–16*). The studies vary in terms of coverage (national, statewide, or structure-specific) and study focus (bridge characteristics, monitoring techniques, bat types). Some studies looked into the impact of a particular factor on roosting (*14*) and others focused on field studies that tried to identify the bridge characteristics that made bridges more likely to be roosting locations for bats (*5, 7–9, 11*). The latter group of studies that investigated bridge characteristics as they relate to bat roosting on bridges are summarized in Table 1. Major findings from these studies are also summarized in this section.

The research by Keeley and Tuttle (*11*) has been the most comprehensive and cited study on bat roosting on U.S. bridges. A total of 2421 bridges in 25 southern and western states were surveyed for bat roosting. Bats were found in 211 structures (8.7%). Ideal bridge characteristics for crevice-dwelling bat species (in descending order of priority) were reported as: relatively warm location (primarily in the southern half of the United States); constructed out of concrete; vertical crevices of 0.5 to 1.25 in.; vertical crevices of 12 in. or greater in depth; roost height of 10 ft or more above the ground; rain sealed at the top; full sun exposure; not situated over busy roadways.

In a Florida study, bats were found in 16 (5.4%) of the 299 randomly visited bridges (*8*). Prestressed concrete bridges with multiple I-beams, older bridges, and bridges with less than average daily traffic were found more likely to be home to a bat.

In a Montana study, evidence of bat use was found at 78 of 130 highway structures examined (*9*). Roosts were found in all highway system categories, but relatively more were in the locally maintained or state-maintained category; maternity season colonies were in all but the interstate category. Use of bridges for roosting, and intensity of use at night roosts, were generally unrelated to the landscape within 1.86 miles (3 km) of the structure. The only factor that was significantly greater around day roost structures was the mean percent of forest cover, but substantial overlap among unused, night roost, and day roost categories indicated that this pattern was a trend and not the major influence on structure use by bats. All day roosts were found within 5 miles (8 km) of riparian corridors. Bats used 75.9% of concrete structures, 37.5% of steel structures, and 31.6% of wooden ones. Slab bridges were the least preferred concrete spans because they provided few if any protected sites for roosting bats on the underside of the deck.

Another study in Georgia (*5*) aimed to determine roost selection preferences of bats, specifically identifying those

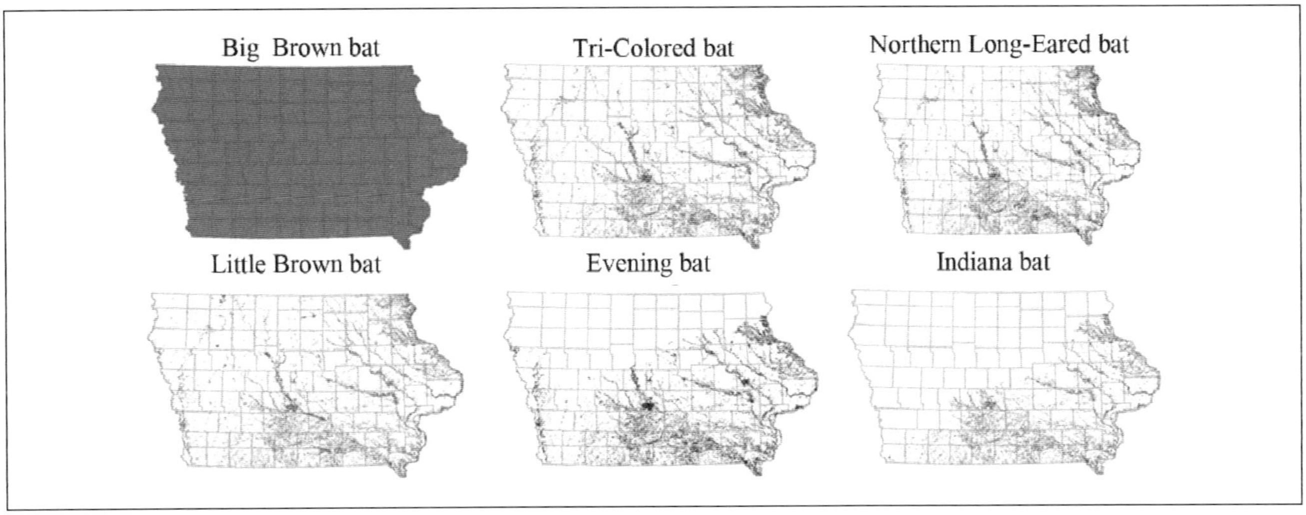

Figure 1. Screenshots of various bat species' coverage in Iowa.

structures being utilized as bat roosts as well as the characteristics that make a bridge a suitable roost site. During the period from August 2003 to April 2005, 540 randomly selected bridges located in Georgia were surveyed. Within this sample, 55 bridges were identified as currently or previously occupied by roosting bats. The data from this study suggested that bats prefer to roost in bridges primarily constructed of concrete materials with open crevices. Roost bridges were most frequently surrounded by woodland/riparian habitat, although some were also found surrounded by residential dwellings, commercial areas, open farms, and ranch lands.

In Southern Illinois, 232 bridges were surveyed in nine Southern Illinois counties for the presence of roosting bats during July 2001, and from June to August 2002 (*7*). Fifteen bridges (6.5%) had bats roosting on them at the time they were surveyed. Bats occurred in four of the five types of bridge designs surveyed; flat slab bridges were never occupied by bats. No relationships between bat presence and habitat features around bridges could be determined.

In a recent study in New England, Civjan et al. (*16*) monitored three regions and did rapid visual screenings of 191 bridges with a more detailed investigation of 18 selected bridges. The main objective of this study was to evaluate monitoring technologies, including acoustic methods, infrared imaging, borescope inspection, and visual inspection. Fourteen bridges (five monitored in project, nine notified by the New England Department of Transportation) have been positively identified as bat roosting sites through this project, with possible roosting at several other sites. This study focused on M. septentrionalis and other nationally or regionally listed threatened species and developed a supplement to the FHWA/FRA (*17*) form for bridge and structure inspection.

The findings from the literature suggest that bats are more likely to use bridges as roosting habitat when the structures are: concrete, typically girder structures that provide vertical crevices for shelter; close to or on riparian corridors; on roads with low traffic volumes; and close to woodland.

Data

Data Sources

Primary sources of information for this investigation were data provided by: (1) the Iowa Department of Transportation (DOT) Geographic Information Management System (GIMS) database; (2) the Iowa Natural Resources GIS Library; (3) the National Gap Analysis Program (GAP); and the National Bridge Inventory (NBI). State-level inquiry and integration processes of each source as of its most recent available data year were implemented using the Esri ArcMap software program. From the GIMS database, base record road and structure information pertaining to all public roads in Iowa as of 2014 were of interest. To that end, surrounding land coverage and GAP-predicted bat species distribution or range were also attained for the most recent available year (2002). The structure data included the data items of interest among 116 NBI data items for all NBI structures in the state of Iowa. However, for the purpose of this study, those structures serving railroads, or owned by other private organizations, were excluded, leaving 24,486 structures of interest.

Predicted distributions of bat species for the state of Iowa were obtained from the Iowa Gap Analysis Program (IAGAP) at the Iowa Department of Natural Resources (DNR). Distributions were presented at a spatial resolution of 30 m and were based on habitat models derived from IAGAP land cover data, hexagon range maps, survey data, and expert review. Similar data are also available nationally through the United States Geological Survey National GAP (*18*).

Predicted distributions for each bat species were spatially integrated with sample bridge locations, and the total number of possible species at each sample bridge derived. Figure 1

Table 2. Distribution and Range of Bat Species in Iowa

#	Bat common name	Complete data		Location	Season
		Distribution	Range		
1.	Big Brown bat (*Eptesicus fuscus*)	–	Yes	Statewide	All
2.	Tri-Colored bat (*Perimyotis subflavus*)	Yes	Yes	Statewide	All
3.	Eastern Red bat (*Lasiurus borealis*)	Yes	Yes	Statewide	Summer
4.	Evening bat (*Nycticeius humeralis*)	Yes	Yes	Southern (2–3)	All
5.	Hoary bat (*Lasiurus cinereus*)	Yes	Yes	Statewide	Summer
6.	Indiana bat (*Myotis sodalis*)	Yes	Yes	Southeast	Summer
7.	Little Brown bat (*Myotis lucifugus*)	Yes	Yes	Statewide	All
8.	Northern Long-Eared bat (*Myotis septentrionalis*)	Yes	Yes	Statewide	All
9.	Silver-Haired bat (*Lasionycteris noctivagans*)	Yes	Yes	Statewide	All

Note: – = Missing data.

Table 3. Sample Distribution by Structure Type

Structure type (material)	Sample structures		All Iowa structures	
	Count	%	Count	%
Concrete	49	9.5%	2,620	10.7%
Continuous concrete	63	12.2%	6,794	27.7%
Steel	70	13.5%	6,227	25.4%
Continuous steel	61	11.8%	1,910	7.8%
Prestressed concrete	77	14.9%	4,459	18.2%
Prestressed continuous concrete	52	10.1%	318	1.3%
Timber	100	19.3%	2,076	8.5%
Masonry	45	8.7%	82	0.3%
Total	517	100.0%	24,486	100.0%

shows the coverage of some bat species found in Iowa. By use of spatial proximity and selection tools in ArcGIS, bat species shapefiles were integrated with structures of interest. Throughout the project, IAGAP-based predicted bat distribution was referred to as "potential bat presence."

As exemplified, some species are more common across Iowa than others; some cover a wider range or have a wider distribution because of their adaptability. According to the National GAP Species data, species distributions were defined as the spatial arrangement of environments suitable for occupation by a species. In other words, species distributions were created using deductive models to predict areas suitable for occupation within a species range (*19*). On the other hand, species ranges were defined as a coarse representation of the total area extent of a species or the geographic limits within which a species can be found. Table 2 shows distribution and range of bat species found in Iowa, combined from Iowa and National GAP Species data.

Data Sampling

Initially, a random sample of 570 structures was selected from the database, which integrated the aforementioned data resources. During field inspections, some bridges could not be inspected because of traffic or time limitation. As a result, a total of 517 structures were sampled and inspected. Table 3 shows a comparison of the inspected sample versus the statewide distribution of structure type (material). Although the sample of structures was selected randomly, the distribution of the sample by location and structure type was checked to make sure the sample was reasonably representative of the population. Another concern during sampling was to make sure a sufficient number of locations were selected with higher potential bat presence to properly compare roosting sites with non-roosting sites. Figure 2 shows a map of all inspected structures. Green dots in Figure 2 indicate the bridge locations with bat presence. The numbers within the dots indicate the number of potential bat species at that location according to IAGAP-based predicted bat distributions.

By use of Network Analyst tools in ArcMap, weekly routes were optimized to maximize travel time efficiency and increase sample size. In this process, supplemental bridges could, and were, inspected in addition to the initial assigned sample for each week. Each team was provided with GIS-based maps of non-Interstate bridges surrounding the required sample sites to better facilitate identification of proximate bridges for possible inspection.

Land Cover

Land cover (land use) classifications for the state of Iowa were obtained from the Iowa DNR. Classifications were derived from satellite imagery collected between May 2002 and May 2003 and presented at a spatial resolution 15 m. The 17 possible classifications included:

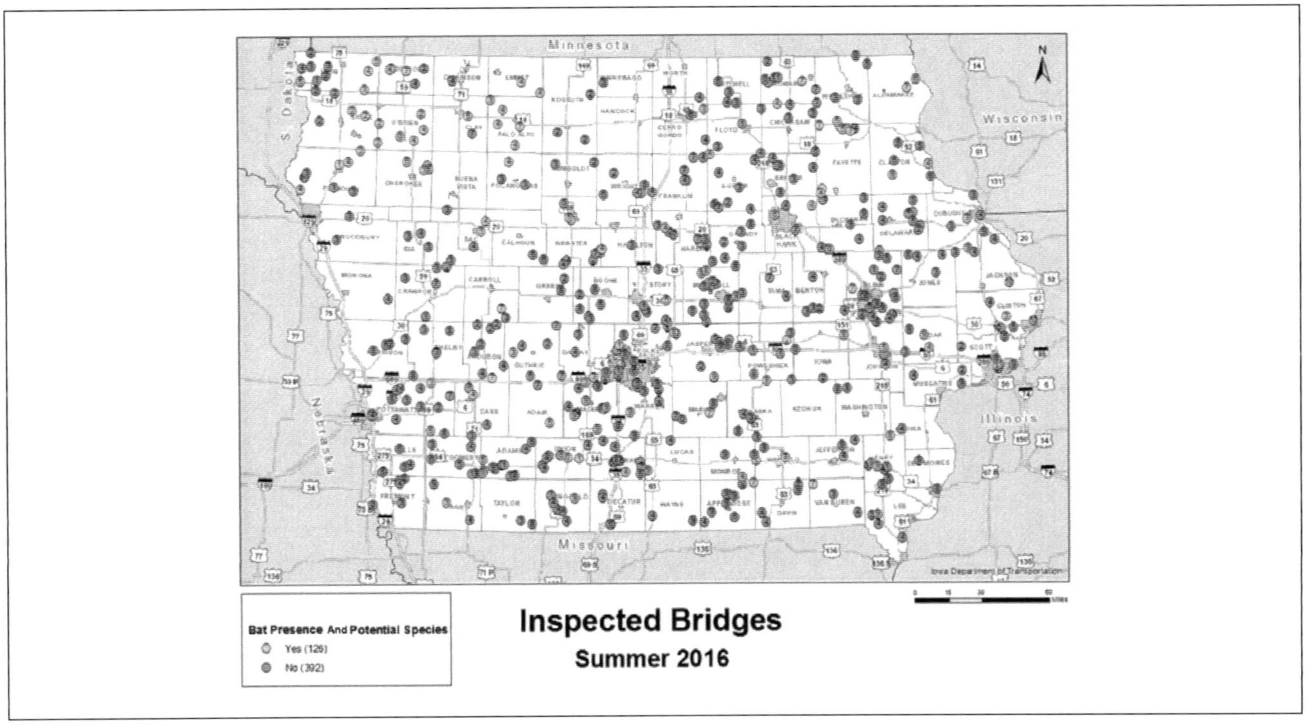

Figure 2. Map of inspected structures.

- Barren;
- Clouds/shadow/no data;
- Bottomland forest;
- Coniferous forest;
- Deciduous forest;
- Alfalfa/hay;
- Planted grassland;
- Grazed grassland;
- Ungrazed grassland;
- Roads;
- Corn;
- Soybeans;
- Other row crop;
- Water;
- Wetland;
- Residential;
- Commercial industrial;
- Unclassified.

Land cover data were spatially integrated with sample bridge locations, which were previously integrated with other sources of data, at five incremental buffer distances: 0.1, 0.25, 0.50, 0.75, and 1.0 miles. These distances were utilized based on relevant literature and recommendations from the project monitor. Land classifications were aggregated for each bridge and buffer combination, and the percentage of each classification was estimated.

For comparison purposes, high-resolution land cover classifications for selected sample bridges within the state of Iowa were obtained from the Iowa DNR. Classifications were based on three dates of aerial imagery, with a target interpretation year of 2009, as well as elevation data derived from aerial lidar. Fifteen different classes were derived and presented at a spatial resolution of 1 m.

Because of the high-resolution nature of the classification data, a statewide dataset was not available. Data were available only at the county level. This, in conjunction with the number of classification records associated with each county, created some challenges in data processing and analyses. Therefore, six sample bridges with somewhat diverse surrounding land cover were identified for comparison to the lower-resolution (and older) land cover dataset. The high-resolution land cover data were spatially integrated with the six sample bridge locations at five incremental buffer distances: 0.1, 0.25, 0.50, 0.75, and 1.0 miles. Land classifications were aggregated for each bridge and buffer combination, and the percentage of each classification estimated. As an example of the increased fidelity of the higher-resolution data, the number of land classifications within 1.0 mile of a bridge increased by approximately 225 times, from approximately 36,000 to 8,136,000 records. An attempt was made to establish consistent classes between the low- and high-resolution datasets, and group classes as appropriate. A comparison of the land classification between high- and low-resolution data showed that the dominant land cover class for the bridges did not change, although there were differences between the percentages of each class. Considerable effort was required to gather

Figure 3. Examples of tools used during field inspections.

high-resolution land cover classification for each structure in the sample and there was a time limitation. It was also found that the dominant land cover for the six selected bridges did not change for low- and high-resolution classification. As a result of these three factors, high-resolution land cover data have not been gathered for the entire sample. However, in comparison to other studies on this topic, the GIS-based collection of land cover data in this study presents a unique approach and provides quantitative land use data.

Field Inspections

In the summer months (May, June, and July) of 2016, sample bridges were inspected for bat presence and the aim of acquiring detailed evidence (*17*) of bat roosting at bridges was achieved. This evidence could include bat droppings, visual sighting of flying bats, or presence of roosting bats. Supplementing each inspected bridge structure included photographic and documented indications of the existence or inexistence of bats. The following are examples of some of the detailed attributes obtained from field reconnaissance:

- Bat presence/evidence (i.e., bat droppings, observed stationary or flying, roosting);
- Roost type (crevice, plugged drain, swallows' nests, etc.);
- Roost dimensions (i.e., height, depth, length);
- Condition surrounding roost (roadway, water, vegetation, etc.);
- Surrounding habitat (woodland, grassland, residential, etc.).

Figure 3 shows example of tools used during inspections of structures, including a Pettersson ultrasound detector D240x, THruNite NC36 UT flashlights (7,300 lumens), a FLIRone infrared smartphone camera, and a GoPro camera with a 24-ft access pole.

To collect the required information about bat presence and structural characteristics of the bridges, four field investigation personnel with varying backgrounds were employed. Team member backgrounds included engineering, bat ecology, and environmental science. Prior to conducting the data collection, each of the team members received two days of training that included being given information on use of inspection tools, signs of bat presence, indication of bat species, and bridge engineering terminology. In addition, each team member was asked to evaluate and document two bridges for the presence of bats. These two bridges had previously been identified as one having and one not having a bat presence. Data from each of the teams were collectively evaluated and a detailed discussion followed. During the data collection period the data collectors worked in groups of two. To ensure that there was no team bias, the groups of two were mixed throughout the data collection period. In addition, the various combinations of teams were asked to document the same bridges blindly. Following the duplicate inspections, a follow-up session was held with each team to discuss any minor differences in data collected. In parallel, senior members of the research team conducted independent reviews of approximately 10% of the inspected bridges as part of the overall quality control/quality assurance process. Table 4 lists the additional data items collected in the field and Figure 4 presents pictures from field inspections.

Analysis

Methodology

In this study, logistic regression models were used to estimate the probability of bat presence based on bridge characteristics, potential bat presence, land cover, and data items collected in the field. Within the logistic regression equation, the natural algorithm (LN) of the odds represents a logit transformation, in which the logit function can be given as (*20*)

$$\begin{aligned} Y_i = \text{logit}(P_i) &= \text{LN}\left(\frac{P_i}{1-P_i}\right) \\ &= \beta_0 + \beta_1 X_{1,i} + \beta_2 X_{2,i} + \beta_3 X_{3,i} + \ldots + \beta_{K,i} X_{K,i} \end{aligned} \quad (1)$$

where β_0 is the model constant (intercept) and $\beta_1, \ldots \beta_K$ are unknown parameters corresponding to the explanatory variables (X_K, k=1,…,K). In Equation 1, the unknown binomial probabilities are a function of the explanatory variables. For this study, the unknown parameters in the models were estimated, as is usual, using maximum likelihood estimation.

Table 4. Data Items Collected in the Field

Inspection identification	Bridge components	Conditions beneath bridge (percentage of each)
Bridge coordinates	Superstructure type	Bare ground
Bridge number	Height above ground (ft)	Vegetation
Team	Superstructure depth (ft)	Flowing water
Week	Number of spans	Standing water
Day	Substructure type	Railroad
		Roadway
General	**Roost information**	**Traffic information**
Date	Typical roost height (ft)	Number of lanes below
Time	Typical crevice depth (ft)	Roadway material below
Cloud cover	Typical crevice length (ft)	Roadway carried
Temperature	Roost material	Number of lanes
County		Roadway material
Roadway		Estimated traffic (H/M/L)
Bridge number		
Latitude		
Longitude		
Elevation (ft)		
Bat presence	**Surrounding habitat (percentage of each within 5 miles)**	**Data collection records**
Bats present?	Residential	Photographs
Visual?	Agricultural	IR photographs
Droppings?	Commercial	Bat sounds
Infrared indication?	Woodland	
Echo meter	Grassland	
Species	Mixed	
Number of bats		
Number of roosts		

Note: H/M/L = high/medium/low.

Using the estimated parameters, the probability that the outcome takes the value 1 (bats are present) can be estimated using

$$P_i = \frac{\mathrm{EXP}\begin{bmatrix}\beta_0 + \beta_1 X_{1,i} + \beta_2 X_{2,i} \\ +\beta_3 X_{3,i} + \ldots + \beta_{K,i} X_{K,i}\end{bmatrix}}{1 + \mathrm{EXP}\begin{bmatrix}\beta_0 + \beta_1 X_{1,i} + \beta_2 X_{2,i} \\ +\beta_3 X_{3,i} + \ldots + \beta_{K,i} X_{K,i}\end{bmatrix}} \quad (2)$$

When the value of an explanatory variable increases by one unit, and all other variables are held constant, the probability ratio can be represented as

$$\left(\frac{P_i}{1-P_i}\right)^* = \left(\frac{P_i}{1-P_i}\right)\mathrm{EXP}\left[\hat{\beta}_i\right] \quad (3)$$

Here, $\hat{\beta}_i$ is the estimated parameter of the associated variable X_i. Thus, a one unit increase in X_i increases the odds ratio $(P_i/(1-P_i))$ by the factor $\mathrm{EXP}\left[\hat{\beta}_i\right]$.

Variables

As an initial step in the analysis, variables from all specified resources were examined for their statistical distributions, potential outliers, errors, and multicollinearity. A series of logistic regression models that estimate the probability of bat presence for single variables were also fitted. Table 5 presents the most influential variables from these analyses, based on the highest McFadden's pseudo R^2 values. High pseudo R^2 values indicate a good model fit. It should, however, be noted that high R^2 values are rare in categorical models. Regardless, pseudo R^2 is a measure of total uncertainty attributed to the model.

Bat presence, indicated by 1 for presence and 0 for no evidence, was the dependent variable for the logistic regression models. Overall, 124 (24%) of the 517 structures had bat presence based on field inspections.

The most influential variable among bridge characteristics was main structure type, which indicates structure material. Prestressed concrete continuous, prestressed concrete, and steel continuous structures had the highest probability of bat presence, respectively. This finding is consistent with the literature. Structure length and deck width were other influential bridge characteristics. With increased length and deck width, the probability of bat presence increased. This finding is also intuitive because larger structures typically have higher dimensions that could provide better shelter.

Superstructure height above ground and superstructure depth were the most influential variables among the ones collected in the field. Consistent with the literature, higher values for each variable significantly increased probability of bat presence. Superstructure depth was presented with a categorical variable later in the analysis. Values of less than 2 ft (median) were coded as 0, values between 2 ft and 3 ft were coded 1, and values above 3 ft were coded 2.

Wetland coverage explained more of the uncertainty in the data with respect to other land cover variables. Only land cover variables for the smallest radius (0.1 mile) were significantly correlated with probability of bat presence at 95% confidence level. Percentage wetland, forest, and water land cover had all positive parameter estimates, indicating increasing probability of bat presence for higher percentages, whereas percentage crop land cover had a negative parameter estimate.

Potential bat presence (IAGAP variables) also showed a significant relationship. The number of potential bats was a numeric variable with a minimum of zero, maximum of nine, and a median of four, and had a positive parameter estimate. The categorical variable named as "more potential bats" has a value of one when the number of potential bat species at a location is more than or equal to four (median value). This categorical variable explained slightly more variability in the model in comparison to the numeric variable.

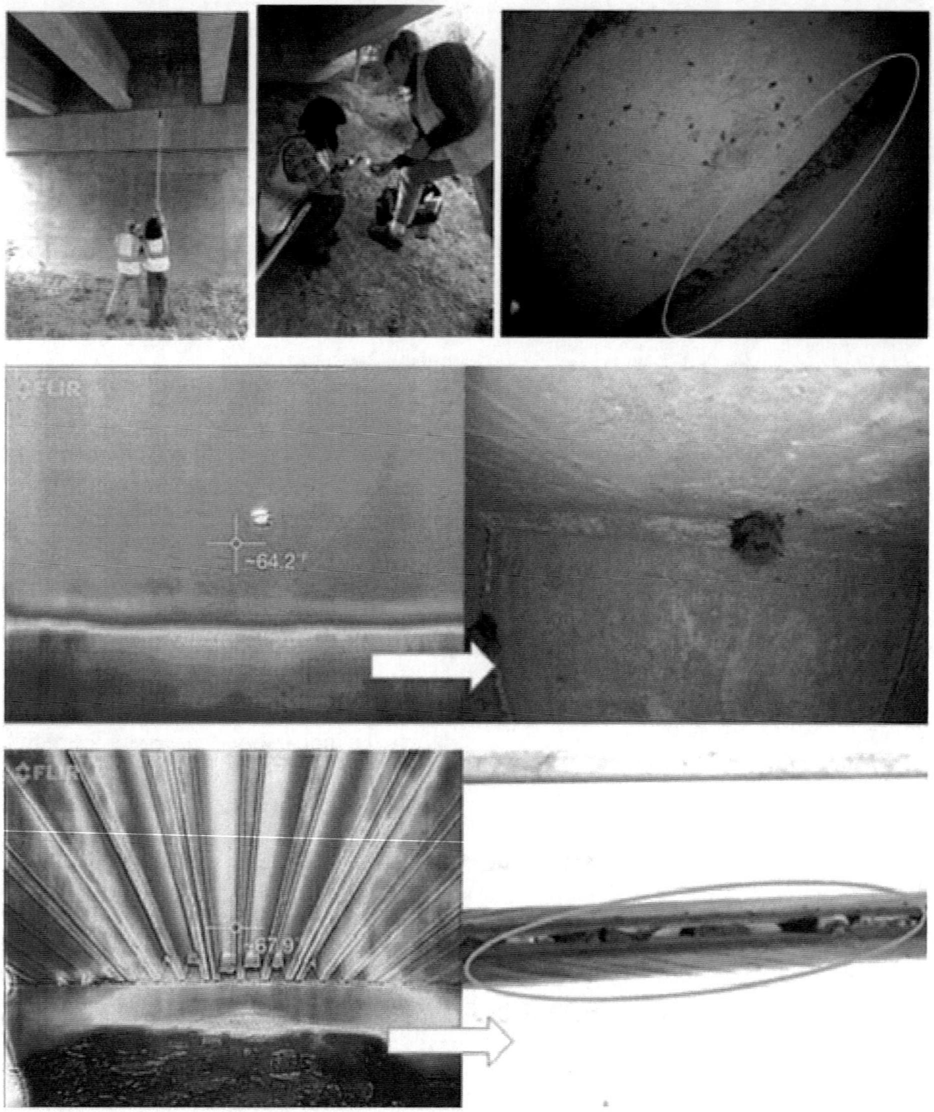

Figure 4. Pictures from field inspections.

Results

After fitting individual models, stepwise logistic regression analysis was done to select the final model that brings together the most relevant and significant variables, and explains the most variability in the data. Table 6 presents the final model that includes three bridge characteristics, one land cover variable, and one variable on potential bat presence (pseudo $R^2 = 17.9\%$).

The parameter notation for the categorical variables in Table 6 indicate the variable levels with respect to parameter estimates. For example, material variable groups prestressed concrete continuous, prestressed concrete, and steel continuous structures together, are coded as 1, and compare them with other structures (coded as −1 in the model). The superstructure depth variable groups together the first two previously discussed classes (0 and 1), and compares them with the final group with the deepest superstructures. The positive parameter estimates in Table 6 indicate higher probability of bat presence for higher values of given parameters.

The parameter estimates for the variables and their values can be entered into Equations 1–3 to calculate predicted probabilities of bat presence for varying structures. For example, given a timber structure with a superstructure height above ground of 10 ft, a superstructure depth of less than 2.5 ft (class 1), 10% wetland land cover, and several potential bat species out of five (class 1), the probability of bat presence is estimated to be 27%. The probability increases to 56% if this structure is prestressed concrete. Looking at a scenario in which the same structure is prestressed concrete

Table 5. Most Influential Variables for Bat Presence

Variable	Type	Source	Pseudo R^2
Material	Categorical	NBI	13.56%
Superstructure type	Categorical	NBI	12.96%
Structure length (ft)	Numeric	NBI	9.58%
Superstructure height above ground (ft)	Numeric	Field	7.53%
Superstructure depth (0–2)	Categorical	Field	5.00%
Number of superstructure spans	Numeric	NBI	5.53%
Percentage wetland within 0.1-mile radius	Numeric	Land cover	4.84%
Deck width (ft)	Numeric	NBI	4.67%
Wetland within 0.25-mile radius (0,1)	Categorical	Land cover	3.14%
Substructure type	Categorical	NBI	3.61%
Road material	Categorical	Field	3.57%
Approach roadway width (ft)	Numeric	NBI	2.73%
Age	Numeric	NBI	2.16%
More potential bats (0,1)	Categorical	IAGAP	2.25%
Number of potential bats	Numeric	IAGAP	2.08%
Number of lanes	Numeric	NBI	1.63%
Percentage forest within 0.1-mile radius	Numeric	Land cover	1.38%
Percentage water within 0.1-mile radius	Numeric	Land cover	1.58%
Average daily traffic	Numeric	NBI	1.57%
Percentage crop within 0.1-mile radius	Numeric	Land cover	1.18%

Table 6. Parameter Estimates of the Final Model

Parameter	Values	Estimate	Standard error	Chi square	Probability > chi square
Intercept		−2.869	0.383	56.18	<.0001*
Material (PS, PSCont., StCont., others)	(1,−1)	0.608	0.124	24.09	<.0001*
Superstructure height above ground	(0–60)	0.101	0.027	14.17	0.0002*
Superstructure depth (0 and 1–2)	(0,1)	0.508	0.245	4.29	0.038**
PCT wetland (0.1 miles)	(0–31)	0.042	0.024	3.03	0.081***
More potential bats	(0,1)	0.548	0.271	4.09	0.043**

Note: PS = prestressed concrete; PSCont. = prestressed concrete continuous; StCont. = steel continuous; PCT = percentage. *Significant at 99% confidence level; **Significant at 95% confidence level; ***Significant at 90% confidence level.

but percentage wetland land cover is 5%, predicted probability of bat presence drops down to 50%.

In summary, the final model indicates that probability of bat presence on bridges increases significantly when the following conditions are met: prestressed concrete continuous, prestressed concrete, or steel continuous structures; increased superstructure height above ground; increased superstructure depth; increased wetland coverage within a 0.1-mile radius of the structure; increased number of potential bat species at the structure location.

Conclusions

In one of the most comprehensive studies in the United States based on the number of bridges inspected, 517 structures were investigated for evidence of bat roosting. Logistic regression models were fitted to identify structure, land cover distribution, and predicted bat species distribution characteristics that increase the probability of bat roosting. The final model indicated that probability of bat roosting on bridges increases significantly when the following conditions are met: prestressed concrete continuous, prestressed concrete or steel continuous structures; increased superstructure height above ground; increased superstructure depth; increased wetland coverage within a 0.1-mile radius of the structure; increased number of potential bat species at the location. The findings show that bridge characteristics, combined with land cover and bat species distribution data, can help identify locations with higher probabilities of bat roosting. This information can be useful to transportation agencies as they plan bridge replacement and repair projects and can help conservation efforts targeted toward bats. It is thought that the integration of objective, geospatial land cover data with potential bat presence data, and estimation of quantitative and relative influence of variables on probability of bat roosting are unique to this study.

Acknowledgments

The research team acknowledges the Iowa DOT for sponsoring this research and the FHWA for state planning and research funds used for this project.

References

1. Gillies, K. Bats in Decline: Today's Fight Against White Nose Syndrome. Presented at 96th Annual Meeting of the Transportation Research Board, Washington, DC, 2017.
2. Bat Conservation International. Bats and Wind Energy, 2017. http://www.batcon.org/our-work/regions/usa-canada/.
3. Benedict, R. A., Benedict, S. K., and Howell, D. L., *Use of Buildings by Indiana Bats (Myotis sodalis) and Other Bats in South-central Iowa*, The American Midland Naturalist, Vol. 178, Issue 1, pages 29-35 (2017).
4. Carden, R., T. Aughney, C. Kelleher, and N. Roche. *Irish Bat Monitoring Schemes*. BATLAS Republic of Ireland, Report for 2008–2009. Bat Conservation Ireland, Virginia, Ireland, 2010.
5. Cleveland, A. G., and J. G. Jackson. Environmental Factors Influencing the Status and Management of Bats Under Georgia (USA) Bridges. *Proc., 2013 International Conference on Ecology and Transportation*, June 23-27, 2013 in Scottsdale, Arizona, USA, 2013.
6. Erickson, G. *Bat and Bridges Technical Bulletin (Hitch Hikers Guide to Bat Roosts)*. California Department of Transportation, Sacramento, Calif., 2002.
7. Feldhamer, G. A., T. C. Carter, A. T. Morzillo, and E. H. Nicholson. Use of Bridges as Day Roosts by Bats in Southern Illinois. *Transactions of the Illinois State Academy of Science*, Vol. 96, No. 2, pp. 107-112, 2003,
8. Gore, J. A., and K. R. Studenroth Jr. *Status and management of bats roosting in bridges in Florida*. Final report for FDOT Research project #BD433. 64 pages. 2005.
9. Hendricks, P., S. Lenard, C. Currier, and J. Johnson. *Bat use of highway bridges in south-central Montana*. 2005. https://doi.org/10.5962/bhl.title.54934
10. Keeley, B. *Bats and Bridges: An Evaluation of Selected Bridges in Laois and Offaly*. Laois County Council and Offaly County Council, 2007.
11. Keeley, B. W., and M. D. Tuttle. Resource publication No. 4. *Bats in American Bridges*. Bat Conservation International Inc., 1999.
12. Perlmeter, S. Bats and Bridges: Patterns of Night Roost Activity in the Willamette National Forest. *Proc., Bats and Forests Symposium*, October 19-21, 1995, Victoria, British Columbia, Canada, 1995. Pages 132-150.
13. Shiel, C. *Bridge Usage by Bats in County Leitrim and County Sligo*. The Heritage Council, Kilkenny, Ireland, 1999.
14. Smith, H. J., and J. S. Stevenson. The Thermal Environment of a Concrete Bridge and its Influence on Roost Site Selection by Bats *(Mammalia chiroptera). Proc., 2013 International Conference on Ecology and Transportation*, Scottsdale Arizona, United States, North Carolina State University, Raleigh Center for Transportation and the Environment Raleigh, NC. pages 1-13. 2013.
15. Timpone, J. C., J. G. Boyles, K. L. Murray, D. P. Aubrey, and L. W. Robbins. Overlap in Roosting Habits of Indiana Bats (*Myotis sodalis*) and Northern Bats (*Myotis septentrionalis*). *The American Midland Naturalist*, Vol. 163, No. 1, 2010, pp. 115–123.
16. Civjan, S. A., A. Berthaume, A. Bennett, and E. Dumont. Bat Roosting in Bridges: Pros and Cons of Assessment Methods from a New England Regional Study. *Transportation Research Record: Journal of the Transportation Research Board*, 2017. 2628: 120-128.
17. FRA, FHWA, and U.S. Fish and Wildlife Service. *Range-Wide Biological Assessment for Transportation Projects for Indiana Bat and Northern Long-Eared Bat*. Final Report. U.S. Department of Transportation, 2015.
18. U.S. Geological Survey. National Gap Analysis Project. USGS Gap Analysis Program Species Ranges. U.S. Department of the Interior, 2013, https://gapanalysis.usgs.gov/wp-content/uploads/2013/09/GAPSpeciesRangemetadata.pdf.
19. U.S. Geological Survey. National Gap Analysis Project. USGS Gap Analysis Program Species Distribution Models. U.S. Department of the Interior, https://gapanalysis.usgs.gov/wp-content/uploads/2013/09/GAPSpeciesDistributionModelmetadata.pdf, 2013.
20. Washington, S. P., M. G. Karlaftis, and F. L. Mannering. *Statistical and Econometric Methods for Transportation Data Analysis*. CRC Press, Chapman and Hall/CRC, 544 Pages, ISBN 9781420082852 - CAT# C285X 2011.

The Standing Committee on Ecology and Transportation (ADC30) peer-reviewed this paper (18-06026).

The opinions and conclusions expressed or implied in the paper are those of the researchers and do not necessarily reflect those of the Iowa DOT, the FHWA, or the U.S. Fish and Wildlife Service.

Article

TRR
JOURNAL OF THE TRANSPORTATION RESEARCH BOARD

Transportation Research Record
2018, Vol. 2672(24) 11–19

DOI: 10.1177/0361198118758691
journals.sagepub.com/home/trr
SAGE

Environmental Benefits of Cold-in-Place Recycling

Angela Pakes[1], Tuncer Edil[1], Morgan Sanger[1], Renee Olley[1], and Tyler Klink[1]

Abstract

The conventional highway resurfacing technique of mill and overlay (M&O) partially removes the existing pavement and replaces it with asphalt derived from some recycled but mostly virgin materials. Cold-in-place recycling (CIR) is an alternative highway resurfacing method that partially mills the existing pavement and uses it beneath a thinner layer of new asphalt. CIR has become widely used for convenience and cost benefits, but the environmental impacts are poorly quantified. The objective of this study was to quantify the environmental life cycle benefits of using CIR for highway resurfacing instead of M&O. Material quantities and equipment used for CIR and what would have been used in M&O for the same project were provided by contractors for nine highway resurfacing projects in Wisconsin. With this information, a life cycle assessment (LCA) tool was used to determine the relative environmental impacts of the two methods, with energy consumption, water usage, and carbon dioxide emissions chosen as the metrics of the LCA. Results show average environmental savings of 23% in energy consumption and carbon dioxide emissions and 20% in water consumption when using CIR instead of M&O for highway resurfacing. Additionally, CIR reduced virgin aggregate consumption by 37%. Environmental savings achieved by using CIR were found to be directly related to the reduction in volume of new hot mix asphalt (HMA) used, and to the reduction in transportation of materials to and from the site. Linear correlations that can be used to estimate savings of future CIR projects were projected.

The United States uses approximately 1.3 billion tons (1.2 billion tonnes) of aggregate every year, 58% of which is for road construction (*1*). Furthermore, 90% of aggregate used in road construction is virgin aggregate (*1*). With the increasing cost of virgin materials and the growing pressure to adopt more sustainable construction, the use of recycled materials in roads is becoming increasingly widespread. The triple bottom line of sustainability requires that a project be economically, socially, and environmentally beneficial relative to conventional methods. Cold-in-place recycling (CIR) is a method for highway resurfacing that has become more widely used in the past decade for its conceived benefits to the triple bottom line.

CIR has the potential to yield economic savings and improve the quality of roads. Surface irregularities are remediated without disturbing the base and subgrade, and traffic disruptions are reduced when using CIR in place of mill and overlay (M&O) (*2*). CIR saves up to 50% in resurfacing costs compared with other methods by: eliminating the need for material disposal through reuse of reclaimed asphalt on site; reducing both the demand for nonrenewable virgin resources, for example, hot mix asphalt (HMA); and by reducing the transportation of materials to and from the site (*3*). Disadvantages of CIR that should be recognized include relatively weak early-life strength and longer curing times; however, in the long term, CIR improves the strength and extends the life of the road without need for reconstruction (*4*).

Literature Review

Despite the understanding of the advantages of CIR, there is insufficient literature that quantifies the environmental benefits of CIR in comparison with the conventional M&O method. One study by Turk et al. compared CIR and conventional construction on one road with an LCA tool, determining that CIR reduced acidification by 18%, reduced fossil fuel consumption by 15%, reduced primary energy consumption by 16%, and reduced global warming potential by 1% when compared with conventional methods (*5*). This study, however, used cement in the process and looked at the use of recycled asphalt pavement (RAP) in the subbase layer, as opposed to using RAP in the surface wearing course layer of the road (*5*). Another study by Thenoux et al. compared asphalt overlay, total reconstruction, and CIR in rural Chile, and found CIR to have the fewest environmental impacts (*6*);

[1]Recycled Materials Resource Center, University of Wisconsin–Madison College of Engineering, Madison, WI

Corresponding Author:
Address correspondence to Angela Pakes: angela.pakes@wisc.edu

however, this study is not directly applicable to Wisconsin because of differences in construction techniques. It was demonstrated by these studies that hauling distance to the nearest asphalt plant plays a significant role in environmental savings associated with CIR, but the other relevant impact factors are not discussed (*4–6*).

Studies by Robinette et al., Giani et al., Alkins et al., and Cross et al. demonstrated that CIR had fewer environmental impacts than conventional road construction methods; however, these studies evaluated equivalent, hypothetical 1-kilometer or 1-mile sections that are not representative of typical project lengths and do not encompass variability in actual construction (*7–10*). Alkins et al. only produced cumulative environmental savings and did not delineate the results by life cycle stage (*9*). Cross et al. assumed a 25-mile hauling distance to the HMA plant and a 100-mile hauling distance to an asphalt emulsion plant, when industry data shows much shorter hauling distances are common (*10*). Other studies by Shatec Engineering Consultants and Chan et al. stated environmental savings when using CIR in place of M&O, but provided little evidence of this (*11, 12*). In reviewing these studies, it was determined that there is a gap in understanding the life cycle components attributed to cumulative environmental savings. Additionally, there is a lack of validation case studies to provide information relevant to construction.

Methodology

To address the research gaps identified in the literature review, the Recycled Materials Resource Center (RMRC) at the University of Wisconsin–Madison has worked closely with the Wisconsin Department of Transportation (WisDOT) to quantify the relative environmental impacts of CIR and M&O. For this report, case studies of nine highway projects across Wisconsin that utilized CIR have been analyzed and compared with conventional M&O using LCA. The nine project locations are shown in Figure 1.

M&O and CIR Processes

The first step in the M&O process, which is also called mill and fill, is to mill the existing pavement to a specified depth dependent on distress of the roadway; for the projects in this study, the milling depth was between 2 in. and 5 in. (5 to 12 cm). The milled material is then hauled to the nearest asphalt plant to be recycled, and 4 to 4.5 in. (10 to 11 cm) of new HMA produced from virgin (80%) and recycled (20%) materials is paved on top of the milled surface (*13*).

Like M&O, the first step in the CIR process is to mill the existing roadway. In the nine cases studied here, and for most other cases, milling depth is 2 to 4 inches (5 to 10 cm) (*2*). Depending on the distress of the roadway, however, some pre-milling may be necessary for a project. Generally, all the RAP generated during the milling of the existing road is used for reconstruction (*2*). After milling, the material is crushed and graded, a stabilizing agent (e.g., asphalt emulsion) is added, and the mixture is paved on to the roadway using a traditional asphalt paver. The new stabilized base is compacted, and the CIR mixture is left to cure. Curing periods for CIR can take anything from a few hours up to several weeks, depending on conditions. The most common curing period is 2 to 3 days (*3*). Traffic can drive on the CIR compacted base during the curing period. After curing, new HMA is paved as a wearing course layer; the wearing course required for CIR construction is thinner, and therefore requires fewer virgin materials than M&O. A side-by-side road profile comparison of the M&O and CIR processes is illustrated in Figure 2.

It is also important to note that the CIR surface post-compaction may exhibit a series of ripples in the pavement surface perpendicular to traffic, known as corrugations (*14*). These can disturb the ride quality and may require an additional layer of HMA overlay as a leveling layer to smooth the surface before paving the wearing course layer. Of the nine projects evaluated in this study, only STH 64 required a leveling layer, and environmental savings were still achieved. Although the CIR method has a more involved construction process, it requires less new HMA and reduces transportation of materials to and from the HMA plant.

There are presently three methods of CIR construction: single-unit recycling train, two-unit recycling train, and multi-unit recycling train. The single-unit recycling train accomplishes the CIR process at one fell swoop. The milling machine, the crushing and sizing machine, and the pugmill machine are all combined into one unit that mills the roadway using a down-cutting rotor, grades the milled material, and adds the stabilizing agents in the cutting chamber (*2*). A paver then relays the modified RAP and compaction rollers stabilize the base. After the curing period, the road is ready for the HMA overlay. Only one project analyzed in this study used a single-unit recycling train: STH 27. Similarly, a two-unit recycling train consists of a milling machine and a mix paver, in which the mix paver acts as both a pugmill machine to add the stabilizing agent and a paver. No projects evaluated in this report utilized a two-unit recycling train.

Multi-unit recycling trains involve different machines for each of the different processes. A typical multi-unit recycling train consists of a milling machine to mill the existing roadway, a screening and crushing machine to grade the milled material, a pugmill machine to add the stabilizing agent, and a paver to relay the modified RAP mixture (*2*). A compaction roller then finishes the job and the stabilized base is left to cure until it is ready for the HMA overlay. A multi-unit recycling train was used in all the case studies presented in this report, with the exception of STH 27.

Environmental Impacts Analysis

To determine the environmental benefits associated with the implementation of the CIR process as effectively as possible, an LCA of both the CIR and M&O processes was performed.

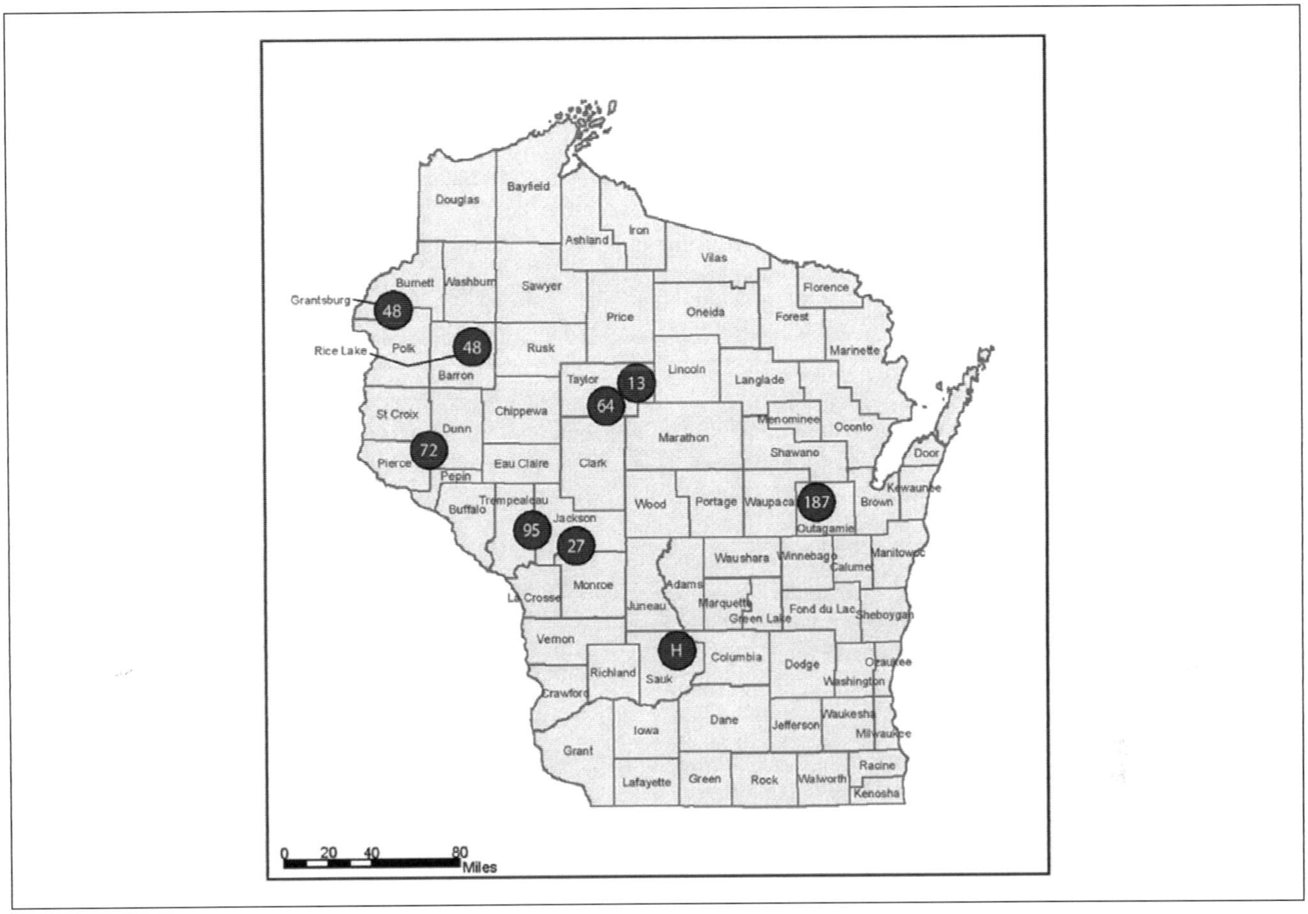

Figure 1. CIR projects in Wisconsin, listed by highway name and closest municipalities:
CTH H (Reedsburg to Wisconsin Dells);
STH 13 (Medford to Westboro);
STH 27 (Sparta to Black River Falls);
STH 48 (Grantsburg to Frederic);
STH 48 (Rice Lake to Birchwood);
STH 64 (Gilman to Medford);
STH 72 (Ellsworth to Elmwood);
STH 95 (Blair to Merrillan); and
STH 187 (Shiocton to North County Line).
Note: 1 mi = 1.61 km; CTH = county highway; STH = state highway.

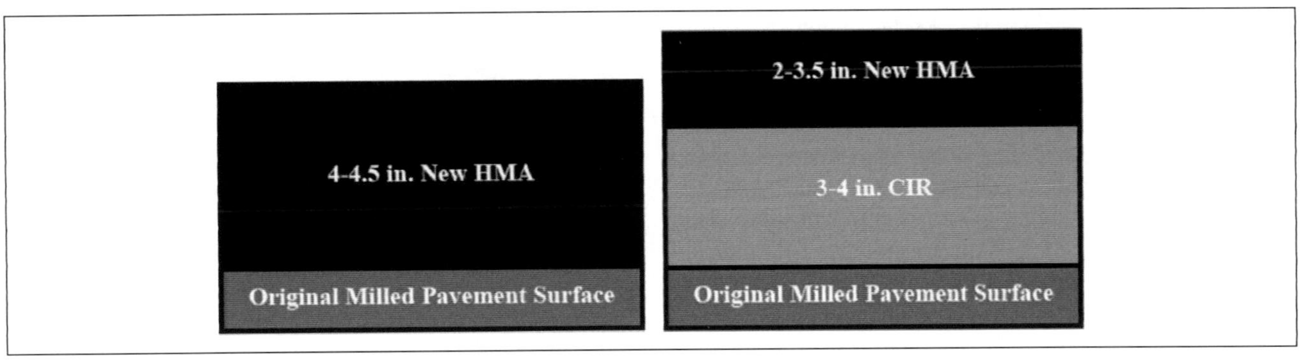

Figure 2. M&O and CIR road profiles.

LCA refers to the systematic evaluation of a process or product in which the environmental impacts associated with all stages of the process are considered. LCAs can assist in gaining a better understanding of the environmental impacts of materials and processes throughout the product life cycle, also known as a cradle-to-grave analysis, and provide relevant data to make informed decisions. To achieve this, the LCA tool PaLATE (pavement life cycle assessment tool for environmental and economic effects) was chosen. PaLATE is a spreadsheet LCA program that was developed by the Consortium on Green Design and Manufacturing from the University of California–Berkeley under the sponsorship of RMRC to assess the environmental and economic effects of pavement and road construction (*15*). The program follows the production of materials and their transportation, use in construction, maintenance, and end-of-life processes. Many of the PaLATE outputs are based on the volumes or weights of materials used and the parameters of specific equipment utilized. The environmental outputs of PaLATE include: energy consumption (kg), water consumption (kg), CO_2 emissions (kg), NO_x emissions (kg), PM_{10} emissions (kg), SO_2 emissions (kg), CO emissions (kg), leachate information (mercury, lead), and hazardous waste generated (*15*).

The first step in executing an LCA is to define the functional scope of the project. Here, energy use, water consumption, and carbon dioxide emissions were the chosen environmental factors for impact analysis. The scope of this project included the benefits associated with using CIR in place of M&O; thus, the benefits of utilizing recycled materials within the HMA in either process were not specifically investigated.

Next, a complete inventory of each component of the construction process is taken within the defined scope of the project. To determine the equipment and materials used during the CIR process, the RMRC research team worked closely with WisDOT and contractors. The nine projects were all constructed using CIR, with materials and equipment tracked by the contractors on site. Additionally, contractors were asked to provide hypothetical material quantities and equipment specifications for the nine projects if M&O were to be used. Productivity and fuel consumption data for the equipment were obtained from the manufacturers (*16, 17*). For each project, two PaLATE scenarios were performed: one for the actual CIR construction, and another for the hypothetical M&O construction. For the eight projects that utilized a multi-unit recycling train, two PaLATE spreadsheets were needed to accommodate the equipment inputs and, therefore, the total CIR environmental impacts for the multi-unit recycling trains were the sum of the outputs of the two spreadsheets. Information used to perform LCAs included amount of HMA and tack coat, and the surface area of milling for the CIR process and the hypothetical M&O one and, in addition, amount and type of asphalt stabilizing agent and surface area of the CIR layer for the CIR process. CIR thickness and HMA thickness varied by project to meet the design requirements of the road. Hauling distances from the asphalt plant to the project site were found using site locations provided by the contractors and were calculated to the midpoint of each project using Google Maps.

With all inputs compiled, each assessment was run in the PaLATE spreadsheet according to the standard procedures described in the program manual (*18*). For this study, the impact assessment results for energy use, water consumption, and carbon dioxide emissions only were compared for both CIR and M&O. Conclusions were drawn such that the results of this study can help future contractors and departments of transportation to estimate the savings associated with using CIR instead of M&O for their highway construction projects.

Virgin Aggregate Reduction

M&O construction requires more virgin aggregate than CIR. Reduction in virgin aggregate consumption was also considered as a benefit and was evaluated. To find reduction in virgin aggregate consumption by using CIR in place of M&O, a simple volume reduction calculation was used:

$$VA_{reduction} = VA_{M\&O} - VA_{CIR} \quad (1)$$

where VA = Volumne of Virgin Aggregate (yd^3).

Assumptions

This research used the PaLATE database program with some updates, and used the following assumptions:

- All M&O projects were assumed to have depths of 4 to 5 in. and HMA overlay of 4 to 4.5 in.
- Mix design was assumed to be the same for both the M&O process and the CIR process for a given project; however, the HMA mix design varied between each project based on asphalt binder percentages provided from the job mix formulas (*19*).
- Material quantities were assumed to be those found in the Statewide Transportation Improvement Program specific to each project.
- Hauling distances were assumed to be from the midpoint of each project to the closest HMA plant provided by each contractor.
- Hauling distance was assumed to be the same for material hauled to the project site and material hauled away from it.
- Material densities were assumed to be the listed densities in PaLATE.
- Water trucks were not included in the analyses because they were used in both the CIR process and the M&O alternative.
- Manufacturers' fuel consumption and productivity specifications were unavailable for some of the older equipment used. Thus, research on comparable equipment was conducted to choose an equivalent piece of machinery that had the most similar fuel consumption and productivity specifications. This allowed the use of the same equipment for each multi-unit recycling train project.
- Initial construction was not considered because each of the projects was completed on existing road.

Table 1. Summary of Project Information

Project	M&O HMA (in)	CIR base (in)	CIR HMA (in)	Road width (ft)	Project length (mi)	Hauling distance (mi)	Excess RAP hauled away (tons)[a]	Recycling train
CTH H	4.5	4	3.5	30	9.5[b]	5.3	0	multi
STH 13	4	4	2.25	30	5.64	11.6	5,811	multi
STH 27	4	4	2.25	30	8.99	8.7	9,206	single
STH 48 RL	4	3	2	30	8.10	10.3	8,898	multi
STH 48 GB	4	4	2.25	24	12.5	4.3	10,382	multi
STH 64	4	4	3	30	4.46[c]	3.7	5,426	multi
STH 72	4	4	2.25	30	4.63	18.3	0	multi
STH 95	4	4	2.5	30	4.42	24.4	0	multi
STH 187	4	3	2.5	30	9.84	21.3	5,575	multi

Note: RL = Rice Lake; GB = Grantsburg; 1 in = 2.54 cm; 1 ft = 0.305 m; 1 mi = 1.61 km; 1 ton = 0.907 tonnes.
[a]If the asphaltic surface was too distressed to use CIR, excess RAP was hauled to the HMA plant.
[b]Originally a 12.3-mile project, 2.8 miles were constructed using a single-unit recycling train and the remaining 9.5 miles using a multi-unit recycling train. The project was looked at as a 9.5-mile multi-unit one and the quantities were adjusted.
[c]This is a 13.3-mile project for which 4.5 miles were constructed using a multi-unit recycling train and the remaining 8.8 miles using M&O because of inclement weather.

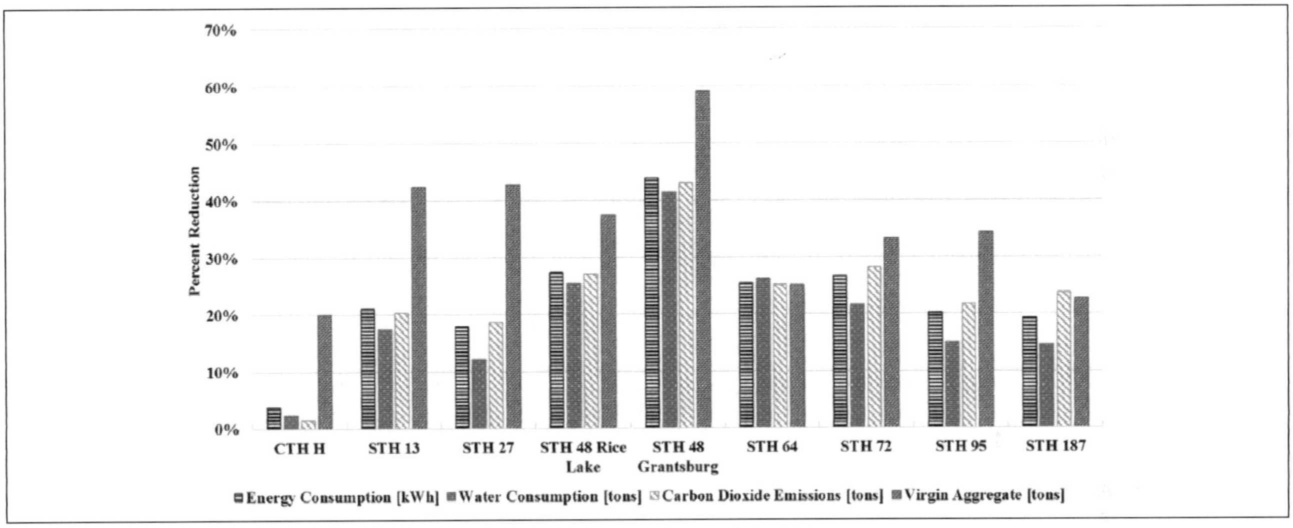

Figure 3. Percent reductions achieved using CIR in place of M&O for each project.
Note: 1 kWh = 3.6 MJ; 1 ton = 0.907 tonnes.

Maintenance materials, transportation, and construction were analyzed.

- It is noted that PaLATE calculates emission factors from national averages from 1996 to 2002.

Results

The variables that were subject to change with every project are listed in Table 1 for all the nine projects. Thickness of HMA for M&O and CIR, road width, and project length all affect the quantities of materials needed for construction, as well as determining the amount of hauling trips needed to transport the materials to and from the site. Distance from the midpoint of the project to the HMA plant, the type of recycling train used, and equipment for M&O all control the transportation- and construction-related environmental impacts.

Environmental parameters were assessed at the material production, transportation, and construction phases and combined as total percent reductions. The percent reductions within each of the environmental output categories when using CIR instead of M&O for each project are illustrated in Figure 3. Percent reductions in environmental outputs are relatively consistent throughout the nine projects. The average reduction in energy consumption and carbon dioxide emissions was 23% and in water usage 20%. The average reduction in virgin aggregate consumption was 37%.

A listing of savings for each project for each environmental parameter considered is provided in Table 2. The nine projects saved a total of 24,341,387 kWh (87,628,993 MJ) of energy, 30 tons (27 tonnes) of water, 5,029 tons (4,562 tonnes) of carbon dioxide emissions, and 81,694 tons of virgin aggregate (74,342 tonnes). The cumulative savings translate to

Table 2. Environmental Savings by Project

Project	Energy consumption (kWh)	Water consumption (tons)	Carbon dioxide emissions (tons)	Virgin aggregate (tons)
CTH H	1,102,742	1.0	209	6,880
STH 13	2,008,621	2.3	411	7,620
STH 27	2,030,254	1.8	395	12,436
STH 48 RL	3,930,466	5.1	820	11,142
STH 48 G	8,394,554	11.0	1,738	23,802
STH 64	3,490,967	5.3	752	4,068
STH 72	1,042,298	1.1	214	4,762
STH 95	1,200,413	1.2	250	5,159
STH 187	1,141,070	1.0	239	5,826
Total	24,341,387	29.7	5,029	81,694

Note: 1 kWh = 3.6 MJ; 1 ton = 0.907 tonnes.

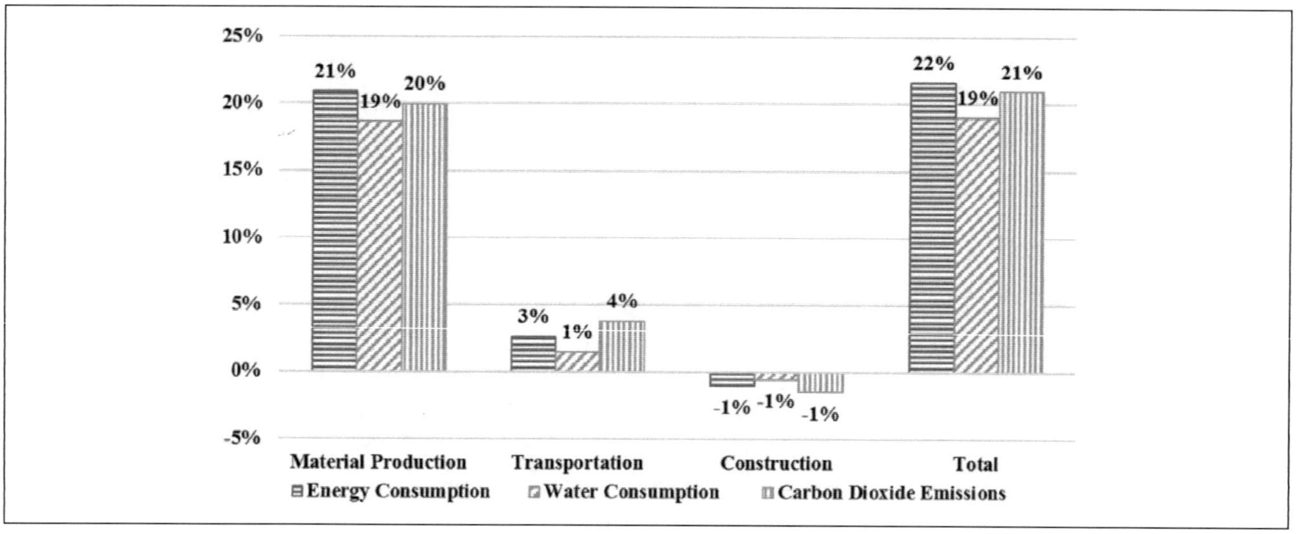

Figure 4. Percent reductions achieved using CIR in place of M&O for each life cycle stage.

energy savings equivalent to the energy consumption of 2,226 U.S. households for a year, a reduction in carbon dioxide emissions equivalent to pulling 971 cars off the road for a year, and water savings equivalent to 158 bathtubs (*20–22*).

By using CIR, there is a significant reduction in emissions related to materials production. The amount of environmental savings achieved through transportation- and construction-related activities is, therefore, only a fraction of the total environmental savings. The substantial environmental savings, then, come from a reduction in virgin materials used in CIR because of the thinner HMA overlay. Figure 4 shows the cumulative percent savings from each life cycle stage of maintenance construction for the nine projects combined.

The CIR process is more demanding in the construction phase because two layers are placed: compacted CIR and the thinner HMA overlay. Other studies that have looked at the environmental impacts of CIR have concluded that hauling distance is the key factor in savings (*4–6*). Figures 5, 6, and 7 show the savings of each project overlaid with a line representing the hauling distance of each one. These figures indicate that there is another key factor in environmental savings when using CIR: HMA saved is the largest influential factor.

Analysis of Data and Observed Trends

To normalize the data and demonstrate the parameters in a project that will determine the savings, Figures 8, 9, and 10 were generated. These graphs represent a framework for the quantity of savings achieved by using CIR in place of M&O by reducing the project specifications to one number: volume of HMA avoided divided by hauling distance. In the figures, this number is labeled as Normalized HMA Reduction on the horizontal axis. This normalization produces an essentially linear trend, which demonstrates that the two key factors resulting in savings when using CIR instead of M&O are the reduction in HMA production and the hauling distance. The

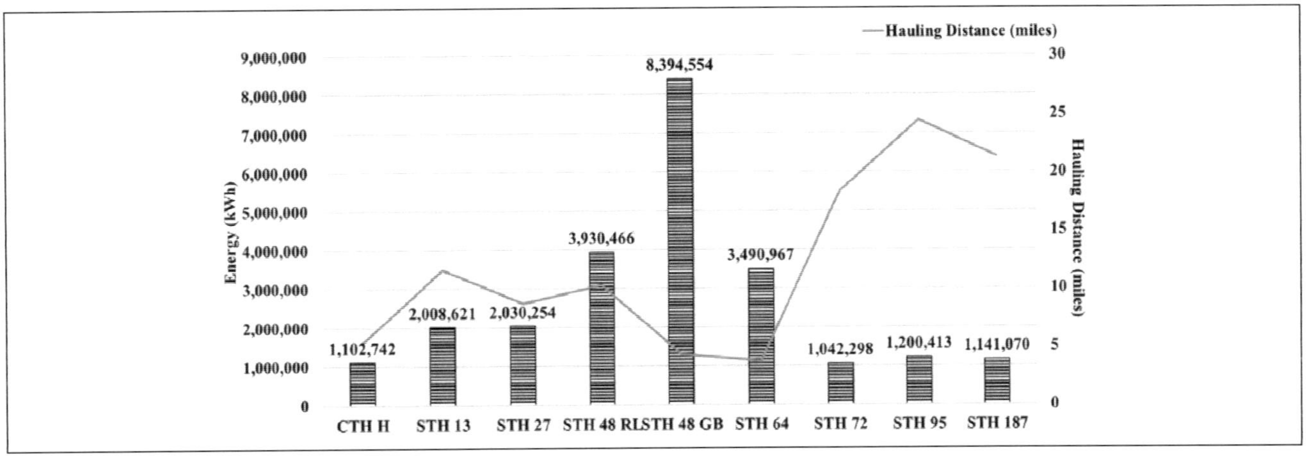

Figure 5. Energy savings achieved per project, plotted with hauling distance.
Note: 1 kWh = 3.6 MJ; 1 mi = 1.61 km; CTH = county highway; STH = state highway; RL = Rice Lake; GB = Grantsburg.

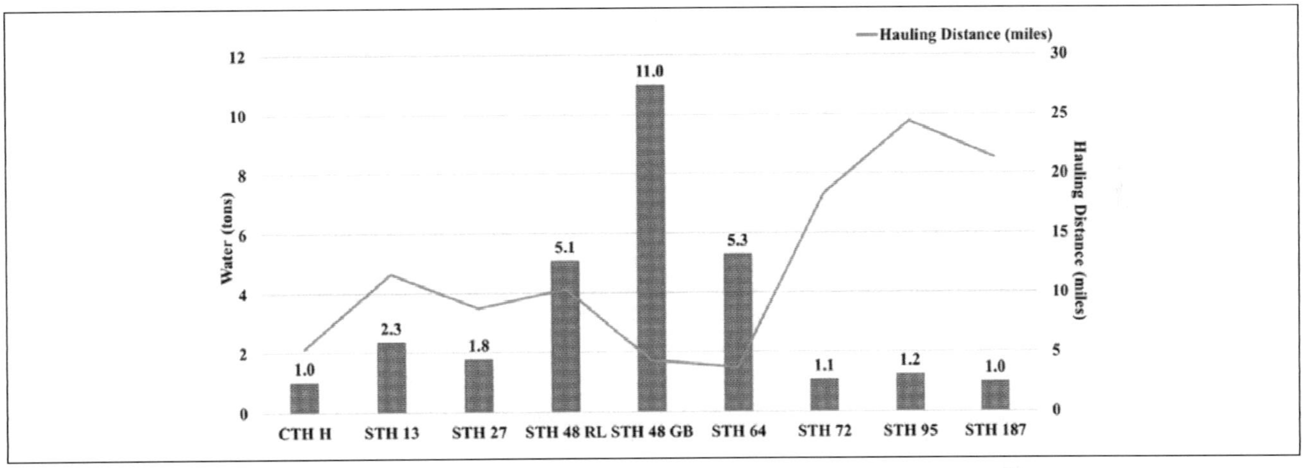

Figure 6. Water savings achieved per project, plotted with hauling distance.
Note: 1 ton = 0.907 tonnes; 1 mi = 1.61 km; CTH = county highway; STH = state highway; RL = Rice Lake; GB = Grantsburg.

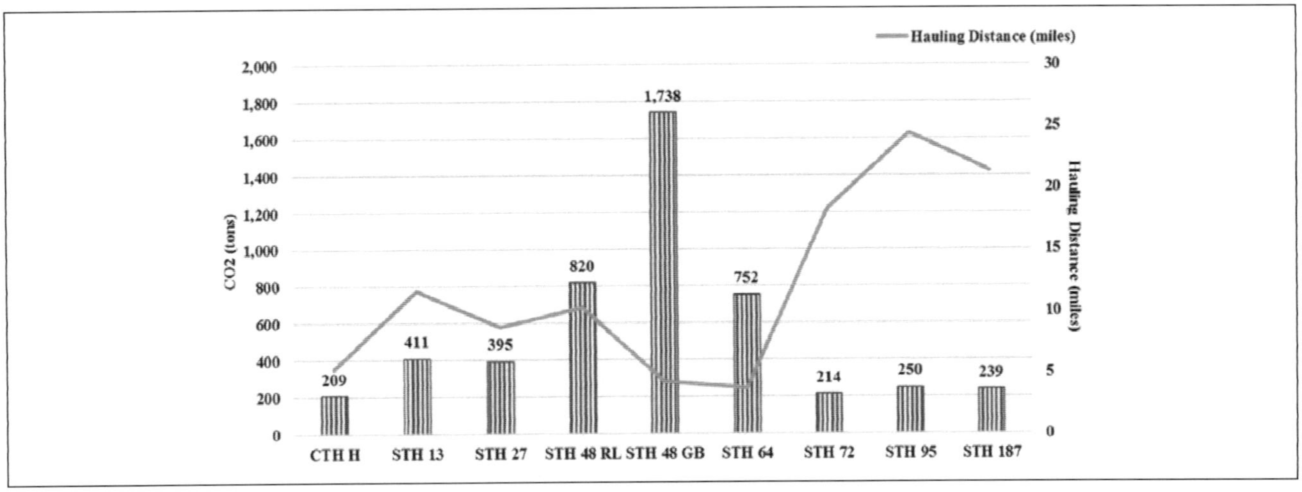

Figure 7. Carbon dioxide emission savings achieved per project, plotted with hauling distance.
Note: 1 ton = 0.907 tonnes; 1 mi = 1.61 km; CTH = county highway; STH = state highway; RL = Rice Lake; GB = Grantsburg.

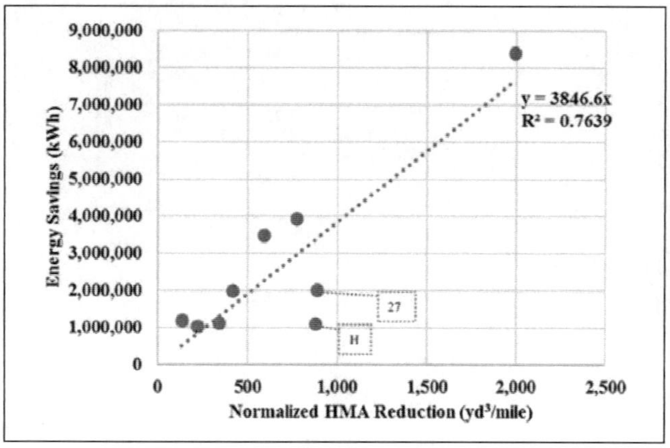

Figure 8. Energy savings predictions.
Note: 1 kWh = 3.6 MJ; 1 yd^3/mile = 0.475 m^3/km.

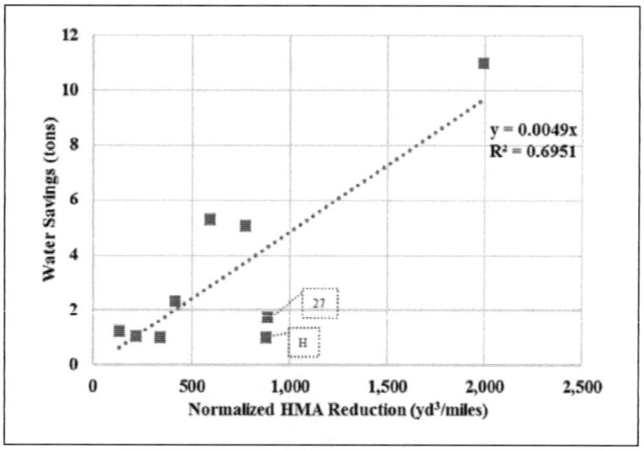

Figure 9. Water savings predictions.
Note: 1 ton = 0.907 tonnes; 1 yd^3/mile = 0.475 m^3/km.

equation of the linear trendlines in Figures 8, 9, and 10 can be used to estimate the respective energy savings, water savings, and carbon dioxide emission savings achieved when using CIR instead of M&O for any highway resurfacing project in Wisconsin. Contractors or the Department of Transportation can make these estimations using only the volume of HMA avoided and the hauling distance. The coefficient of determination, R^2 (given in each of Figures 8, 9, and 10), represents the statistical proximity of the linear trend to the data set, and therefore gives an estimate of confidence in the savings estimations.

It should be noted that when CTH H and the single-unit recycling train project, STH 27, are removed from the data set, the linear correlation improves and the R^2 value increases to around 0.96. For CTH H, the layer of HMA placed over the CIR base is particularly thick. This resulted in only a 1-inch reduction in HMA use when CIR was implemented, relative to traditional M&O, whereas all

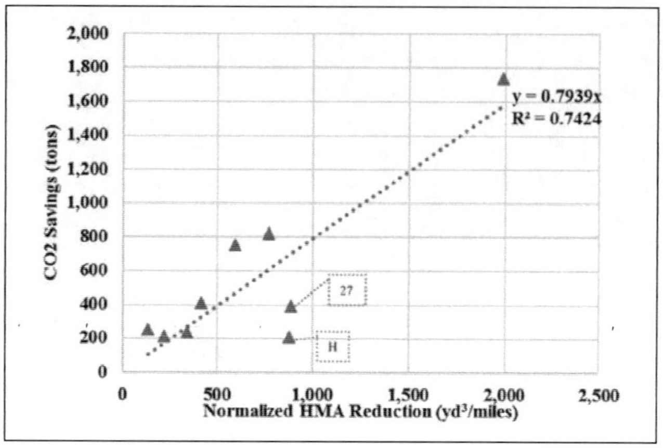

Figure 10. Carbon dioxide savings predictions.
Note: 1 ton = 0.907 tonnes; 1 yd^3/mile = 0.475 m^3/km.

other projects used much less HMA proportionally. The resource-intensive nature of asphalt makes reduction of HMA a key factor in the environmental savings achieved by using CIR instead of M&O. For this reason, the environmental savings achieved in CTH H are less significant than in other projects because there is a smaller reduction in the HMA profile.

Conclusions

The nine projects in summation saved 24,341,387 kWh (87,628,993 MJ) in energy consumption, 5,029 tons (4,562 tonnes) in carbon dioxide emissions, 30 tons (27 tonnes) in water usage, and 81,694 tons (74,112 tonnes) of virgin aggregate. It was determined that the environmental savings achieved by using CIR are directly related to the reduction in volume of HMA used in the thinner hot mix asphalt overlay, and to the reduction in transportation of reclaimed materials to and from site. Linear correlations using volume of HMA avoided and hauling distance estimate the energy consumption, water usage, and carbon dioxide emission savings achieved when using CIR in place of conventional M&O for future highway resurfacing projects. They exhibit the environmental savings potential CIR holds for road refurbishing projects in the future.

Acknowledgments

This research was funded by the Wisconsin Department of Transportation under the oversight of the College of Engineering Construction and Materials Support Center. The authors gratefully acknowledge their support. The authors thank Girum Merine, Barry Paye, and Peter Kemp from WisDOT, Ervin Dukatz of Mathy Construction, Michael Gonnering and Ric Szalewski of Northeast Asphalt, Dustin Albert of American Asphalt, WK, Midstate Reclamation, Gary Whited of the Construction and Materials Support Center, and others who contributed to data collection for this report.

References

1. Carpenter, C., K. H. Gardner, J. Fopiano, C. H. Benson, and T. B. Edil. Life Cycle Based Risk Assessment of Recycled Materials in Roadway Construction. *Waste Management*, Vol. 27, No. 10, 2007, pp. 1458–1464.
2. Asphalt Recycling and Reclaiming Association. *Basic Asphalt Recycling Manual*. U.S. Department of Transportation, 2001, pp. 19–25.
3. *Cold Recycling: The Future in Pavement Maintenance and Rehabilitation*. Asphalt Recycling and Reclaiming Association, Annapolis, Md., 2016, pp. 3–8.
4. Tabakovic, A., C. McNally, and E. Fallon. Specification Development for Cold In Situ Recycling of Asphalt. *Construction and Building Materials*, Vol. 102, No. 1, 2016, pp. 318–328.
5. Turk, J., A. M. Pranjic, A. Mlandenovic, and Z. Cotic. Environmental Comparison of Two Alternative Road Pavement Rehabilitation Techniques: Cold-in-Place Recycling Versus Traditional Reconstruction. *Journal of Cleaner Production*, Vol. 121, 2016, pp. 45–55.
6. Thenoux, G., A. Gonzalez, and R. Dowling. Energy Consumption Comparison for Different Asphalt Pavement Rehabilitation Techniques Used in Chile. *Resources, Conservation and Recycling*, Vol. 49, No. 4, 2007, pp. 325–339.
7. Robinette, C., and J. Epps. Energy, Emissions, Material Conservation, and Prices Associated with Construction, Rehabilitation, and Material Alternatives for Flexible Pavement. *Transportation Research Record: Journal of the Transportation Research Board*, 2010. 2179: 10–22.
8. Giani, M., G. Dotelli, N. Brandini, and L. Zampori. Comparative Life Cycle Assessment of Asphalt Pavements Using Reclaimed Asphalt, Warm Mix Technology and Cold In-Place Recycling. *Resources, Conservation and Recycling*, Vol. 104, 2015, pp. 224–238.
9. Alkins, E., B. Lane, and T. Kazmierowski. Sustainable Pavements: Environmental, Economic, and Social Benefits of In Situ Pavement Recycling. *Transportation Research Record: Journal of the Transportation Research Board*, 2008. 2084: 100–103.
10. Cross, S., W. Chesner, H. Justus, and E. Kearney. Life Cycle Environmental Analysis for Evaluation of Pavement Rehabilitation Options. *Transportation Research Record: Journal of the Transportation Research Board*, 2011. 2227: 43–52.
11. Shatec Engineering Consultants, LLC. *Whitepaper: Cold-in-Place Recycling with Expanded Asphalt Mix Technology*. Graniterock Corporation, 2013. https://www.slideshare.net/DennisMcElroy/shatec-engineeringconsultants-graniterock-dba-fmg-cir-whitepaper. Accessed January 30, 2017.
12. Chan, P., S. Tighe, and S. Chan. Exploring Sustainable Pavement Rehabilitation: Cold-in-Place Recycling with Expanded Asphalt Mix. Presented at 89th Annual Meeting of the Transportation Research Board, Washington D.C., 2010.
13. Mathy Construction. *Single-Unit CIR Train*, 2016.
14. Asphalt Recycling and Reclaiming Association. *Basic Asphalt Recycling Manual*. U.S. Department of Transportation, 2016, p. 40.
15. Consortium on Green Design and Manufacturing, University of California, Berkeley. PaLATE. http://cgdm.berkeley.edu/. Accessed January 30, 2017.
16. CMI Roadbuilding, Inc. Autograde TR-4. http://cmi-roadbuilding.com/concrete-paving/autograde-tr-4. Accessed January 30, 2017.
17. Cummins Engine Company, Inc. *Marine Performance Curve*. http://www.sbmar.com/Engines/PDF/6BT/6BT%20210%20Power%20Curve-%20Nov%2000.pdf. Updates available: www.cummins.com. Accessed January 30, 2017.
18. Nathman, R. K. *PaLATE User Guide, Example Exercise, and Contextual Discussion*, 2008. https://www.ce.udel.edu/UTC/Presentation%2008/rachel_Nathman_Thesis_Spring08.pdf. Accessed April 10, 2017.
19. Atwood Systems, Inc. Highway Quality Management System. http://www.atwoodsystems.com/iibv2/projectmain.cfm. Accessed April 10, 2017.
20. United States Environmental Protection Agency. Transportation, Air Pollution, and Climate Change. https://www.epa.gov/air-pollution-transportation. Accessed April 10, 2017.
21. Portland Water Bureau. *Shower and Bath Fact Sheet*. https://www.portlandoregon.gov/water/article/305153. Accessed April 10, 2017.
22. U.S. Energy Information Administration. How Much Electricity Does an American Home Use? https://www.eia.gov/tools/faqs/faq.cfm?id=97&t=3. Accessed April 10, 2017.

The Standing Committee on Resource Conservation and Recovery (ADC60) peer-reviewed this paper (18-04381).

Article

TRR
JOURNAL OF THE TRANSPORTATION RESEARCH BOARD

Transportation Research Record
2018, Vol. 2672(24) 20–30

DOI: 10.1177/0361198118759015
journals.sagepub.com/home/trr
SAGE

An Investigation of Battery Electric Vehicle Driving and Charging Behaviors Using Vehicle Usage Data Collected in Shanghai, China

Jie Yang[1], Jing Dong[2], Qi Zhang[3], Zhiyuan Liu[3], and Wei Wang[3]

Abstract

This paper investigates the driving and charging behaviors of battery electric vehicle (BEV) drivers observed in Shanghai, China. The summary statistics are compared with the observations from the U.S. EV Project. A machine-learning approach, namely self-organizing feature map (SOM), is adopted as a classifier to analyze BEV drivers' habitual behaviors. The inter-driver heterogeneities are examined in terms of: the distributions of distance traveled per day, the start time of charging, the number of charges per day, distance traveled between consecutive charges, battery state of charge (SOC) before and after charging, and time-of-day electricity demand. It is found that (*a*) BEV drivers demonstrate conservative charging behaviors, leading to short distances between consecutive charging events; (*b*) a significant number of BEV drivers in Shanghai charge during daytime; (*c*) the distributions depicting the driving and charging patterns vary greatly due to the diversity in travel activities among different drivers.

The depletion of fossil fuels, as well as their adverse impact on the environment and human well-being, have raised serious concerns over the sales and usage of fossil-fuel powered vehicles. Consequently, to overcome these challenges, many countries have declared plans to ban the sale of gas and diesel cars in the coming years while the trend and focus of the automotive industry are shifting towards environmentally-friendly vehicles. These initiatives have led to the development and popularity of battery electric vehicles (BEVs) which produce zero tailpipe emissions. Thus, replacing the vehicle fleet with BEVs has the potential to reduce greenhouse gas emissions and other harmful pollutants. In many metropolitan cities of China, where the high concentration of vehicles has caused severe air pollution, there is an increasing interest in large-scale deployment of BEVs in the near future. The Chinese government has set pure electric driving as the primary strategic direction of new energy vehicles (NEVs) development and the auto industry transition (*1*). Correspondingly, a series of financial incentive policies has been implemented to promote the deployment of BEVs (*2*).

To better quantify the societal benefits of BEVs, forecast their market acceptance, design charging infrastructure, and reduce vehicle energy consumption, it is essential to characterize the driving and charging behaviors of BEV users. The EV Project in the U.S., initiated and managed by ECOtality North America, has collected and analyzed vehicle usage data from approximately 8,300 Nissan Leaf BEVs and Chevrolet Volt extended-range electric vehicles (EREVs) since 2010. Similar projects have been developed to analyze the behavioral and mobility patterns of electric vehicles, including the SwitchEV project (*3*), the CABLED project (*4*), the FREVUE project (*5*), the ZeEUS project (*6*) and so forth. These have provided realistic and systemic data that helps researchers to better understand users' real-world driving and charging patterns (*7–11*). Likewise, cities in China, such as Beijing, Shanghai, and Shenzhen, have launched various programs to develop NEV monitoring and service platforms. Several pioneering works, focused on commercial vehicles (i.e., BEV taxis), have been carried out based on the data set collected from these platforms (*12, 13*). In this paper, we use the data collected by Shanghai Electric Vehicles Data Center (SHEVDC) which was developed to remotely monitor electric vehicles driven across the city. Fifty BEVs, used

[1]Development Research Institute of Transportation Governed by Law, School of Law, Southeast University, Nanjing, P. R. China
[2]Department of Civil, Construction and Environmental Engineering, Iowa State University, Ames, IA
[3]Jiangsu Key Laboratory of Urban ITS, Jiangsu Province Collaborative Innovation Center of Modern Urban Traffic Technologies, School of Transportation, Southeast University, Nanjing, P. R. China

Corresponding Author:
Address correspondence to Jie Yang: jieyang@seu.edu.cn

Table 1. Summary Metrics of Driving Behaviors

	Mean					
	Shanghai	U.S. EV Project	Med.	Max.	Min.	St.D.
Avg. distance traveled per day when the vehicle was driven (km)	50.7	43.0	55.4	95.3	17.5	18.8
Avg. distance traveled between charging events (km)	49.5	38.5	43.7	101.8	33.9	14.5

Note: Avg. = average; St.D. = standard deviation.

as personal vehicles and all the same model, the Roewe E50, are extracted from the SHEVDC database.

Empirical observations show that variations exist in BEV owners' travel activities. With respect to such variations, there are many existing studies which focus on modeling the diversity in trip length (*13, 14*), daily vehicle kilometers traveled (VKT) (*15–18*), dwell time (*14*), travel distance between charges (*14, 17*) and so on. The travel pattern of each individual BEV owner is relatively constant, yet contingencies make the travel demand vary from day to day (i.e., intra-driver heterogeneity). To better understand the diversity of travel activities for different drivers (i.e., inter-driver heterogeneity), this paper not only presents the summary metrics of the real-world BEV owners' driving and charging behaviors, but also investigates inter-driver heterogeneity by classifying vehicles according to their distance traveled daily and start of charging time, which helps to eliminate the interference of the intra-driver heterogeneity.

The main contributions of this paper can be summarized as follows: (*a*) compare the driving and charging behaviors of BEV owners in Shanghai with the observations from the EV Project; (*b*) use a machine-learning approach, namely self-organizing feature maps (SOM), to cluster the collected vehicle samples; (*c*) examine inter-driver heterogeneity from the perspective of the distributions of distance traveled per day, start time of charging, number of charging events daily, distance traveled between consecutive charges, battery state of charge (SOC) before and after charging, and time-of-day electricity demand.

The next section describes the observed driving and charging behaviors of Roewe E50 owners in Shanghai, China, followed by vehicle grouping using the SOM method in the third section. After that, results and discussion are presented in the fourth section. Conclusions and caveats of the study are discussed in the final section.

Data Description

Data Collection

This study makes use of a rich database collected from 50 BEVs (all Roewe E50s) over periods of four to 12 months covering a time period from June 2015 to June 2016. The longest recording time lasted for 357 days (i.e., VID 14), while the shortest recording time is 94 days (i.e., VID 43). The average recording time for each vehicle is about 225 days. Developed by SAIC Motor, the Roewe E50 is a pure electric passenger car with a 22.4 kWh battery pack and a claimed range of 170 km (*19*).

Vehicle terminals, such as global positioning system data loggers and instruments to measure voltage and current, are installed on BEVs to record in-use data and vehicle driving information at a rate of about two samples per minute. In-use data includes turn on time, turn off time, total mileage, SOC, voltage, current, etc. Vehicle driving information includes time-stamped location (i.e., longitude and latitude), spot speed, azimuth, etc. After data cleansing and consistency checking procedures to remove invalid and erroneous data from the records, BEV owners' driving and charging information, such as the average distances traveled per day, the average number of charging events per day, distances traveled between charging events, start time for charging, SOC before and after charging, etc., were extracted.

Observed Driving Behavior

Table 1 presents summary metrics describing the driving behaviors of the 50 BEVs considered in this study. The mean values of the daily VKT and the distance traveled between charges are compared with the observations from the EV Project, which collected in-use electronic data from 3,499 Nissan Leafs (*20*). The Nissan Leaf is a BEV with a 24 kWh battery pack, and the manufacturer claims an expected range of 100 to 220 km on a single full charge depending on driving conditions (*21*).

The average daily VKT is adopted to characterize the diversity of different drivers as it indicates drivers' travel intensity (*22–24*). Overall, Roewe E50 owners have a higher average daily VKT than Leaf owners. The maximum average daily VKT is 95.3 km, observed from VID 8. The average daily VKT of VID 17 is the lowest, indicating a low vehicle utilization. All the average daily travel distances are far below the range of the E50. Drivers might have limited their daily driving due to concerns about the range of the vehicle.

Examining distance traveled between consecutive charging events helps to understand better the behavior of drivers with respect to BEV range. Similarly, for both Shanghai and U.S. BEV drivers, the average distance traveled between charging is lower than the average daily VKT. The results reveal that most drivers would charge their vehicles when provided with

Table 2. Summary Metrics of Charging Events

	Mean					
	Shanghai	U.S. EV Project	Med.	Max.	Min.	St.D.
Avg. number of charging events per day when the vehicle was driven	1.1	1.1	1.3	1.6	0.2	0.5
Avg. charging power (kW)	3.8	3.3	3.8	3.9	3.6	0.1
Avg. charging time (hour)	2.6	3.3	2.5	4.4	1.8	0.5
Avg. SOC before charging (%)	45.6	44.0	45.4	61.6	18.5	8.3
Avg. SOC after charging (%)	92.9	89.6	92.3	100	87.4	3.7
Avg. start time of charging events (hh:mm)	14:14	NA	14:03	21:33	9:13	2:06

Note: Avg. = average; St.D. = standard deviation; SOC = battery state of charge.

convenient opportunities, regardless of the remaining range. This conservative behavior helps to avoid the risk of fully depleting their battery before the next charging opportunity. Nevertheless, some drivers prefer to use a large portion of the vehicle's range before charging. For example, the longest average distance traveled between charges is observed from VID 11, while its average daily VKT is 20.9 km, only about one-fifth of the distance traveled between two charges. It could be speculated that the remaining range might play a more important role in deciding whether or not to charge for BEV owners with short daily VKTs. The section "Distribution of Distance Traveled per Day", below, presents more detailed analysis regarding these findings.

Observed Charging Behavior

Table 2 shows summary metrics of charging events. Because of missing records, VID 7 only has the information of the number of charging events and the start time for each charging and thus is not included in the calculation of other metrics. The mean values are compared with the observations from the EV Project (*20*).

Table 2 shows that Shanghai BEV drivers exhibit the same charging frequency as their U.S. counterparts, charging 1.1 times per day on average. This is consistent with the observation described in the preceding section, where distance driven between charging events is slightly less than distance driven per day. The average charging frequency varies from vehicle to vehicle, from 0.2 charges to 1.6 charges per day. Drivers with short daily VKTs choose not to charge more often than is necessary.

The charging power is calculated based on the charging time and the increase in SOC. It can be inferred that almost all the charging events are performed using AC Level 2 electric vehicle supply equipment (EVSE) at home or workplace (*25*). The starting SOC indicates how much of the battery pack's capacity has been depleted prior to charging. There is a wide variation in the starting SOC from vehicle to vehicle. A high starting SOC (e.g., VID 38 with an average value of 61.6%) indicates drivers are likely to charge their vehicles whenever they have charging opportunities. Battery SOC at the end of charging represents how full the battery pack is after charging. The data shows that the majority of charging events result in a fully charged or nearly fully charged battery pack, which is consistent with the findings from the EV Project (*20*). Thus, it can be speculated that the variation in charging time found in this sample is mainly determined by the starting SOC.

The starting time of charging represents when a BEV is connected to a charging unit. Based on charging power and charging demand, we can assess the impact of charging BEVs on the electricity grid. Similar to the daily VKT, each individual BEV owner has a relatively fixed time for charging on different days. The mean and median values of starting time are around 14:00, and 33.9% of the participants plug in their vehicles in the afternoon (from 12:00 to 18:00). Some drivers prefer charging the battery in the morning (e.g., VID 36 is often plugged in at 9:00); while some prefer to connect their vehicles to their residential EVSE in the late evening (e.g., VID 1 is often plugged in after 21:00). The impact of charging demand on the electricity grid is discussed in the section "Distribution of Charged Electric Energy" (below).

Methodology

Self-Organizing Feature Map

To gain better insights into inter-driver heterogeneous behaviors, we apply the self-organizing feature map (SOM), an unsupervised machine-learning approach, to group the 50 BEVs.

The SOM consists of neurons arranged systematically on a two-dimensional surface (known as a "map"). Each neuron has a prototype weight vector that represents the characteristic features in the input space. Such a structure is capable of mapping patterns in the high dimensional input space into a two-dimensional map. According to the unsupervised learning rule, vectors that are similar to each other in the multidimensional space will be clustered in the same neighborhood in the SOM's two-dimensional space, hence it can be adopted as a tool for data classification. Because of its unique structure, users of the SOM do not need to specify

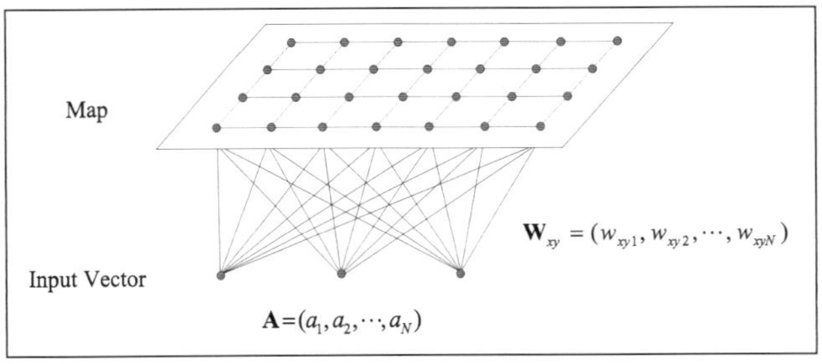

Figure 1. General architecture of a self-organizing feature map *(26)*.

the function between the input features and its output variable. No equation needs to be pre-defined and no parameter calibration is necessary. Therefore, the SOM is considered as a non-parametric approach to measure similarities and dissimilarities among different BEV drivers.

A SOM is a two-dimensional grid, as shown in Figure 1. The map is usually square, but can be of any rectangular or hexagonal shape. Each point on the grid, denoted by its coordinates (x,y), has a neuron and an associated weight vector W_{xy}. The N-dimensional weight vector $W_{xy}=(w_{xy_1}, w_{xy_2}, \cdots, w_{xy_n}, \cdots, w_{xy_N})$ represents the centroid of a data cluster of similar input vectors. The weight vectors are collectively known as the SOM's memory.

When an input vector $A=(a_1, a_2, \cdots, a_n)$ is presented to the SOM, the "distance" between A and each of the weight vectors in the entire SOM is computed. The neuron whose weight vector is "closest" to A will be declared the "winner" and has its output set to 1, while others are set to 0. Mathematically, the output b_{xy} of a neuron located at (x,y) is

$$b_{xy} = \begin{cases} 1 & \text{if } A-W_{xy} = \min\limits_{\forall i,j}\left\{\left\|A-W_{ij}\right\|\right\} \\ 0 & \text{otherwise} \end{cases} \tag{1}$$

where the operator $\| \ \|$ represents the Euclidean distance and i, j are indices of the grid positions in the SOM. The input vectors that are categorized into the same cluster, that is, the same winning neuron, have similar driving and charging patterns.

The training of a SOM is to code all the W_{xy} so that each of them represents the center of a cluster of similar training vectors. Once trained, W_{xy} is known as a prototype vector of the cluster it represents. The SOM training is based on a competitive learning strategy. More details on the SOM training can be found in *(26)*.

Mapping Framework

The SOM is used to classify the vehicles embedded in the input vectors. The average distance traveled per day when the vehicle was driven (denoted as d_{VKT}) and the starting time of charging (denoted as T_{start}) are selected to form the input vectors. That is, $A=(d_{VKT}, T_{start})$. These two components are selected because different drivers may have different magnitudes of d_{VKT} and T_{start} due to their personal preferences. The distribution of d_{VKT} is commonly used to investigate whether the range of a certain type of BEV can satisfy the daily VKT of most drivers *(16, 18)*. The general insights into T_{start} can help assess the impact of BEV charging on the electric grid.

Before the SOM training, each component of the input vector is linearly scaled to [0, 1] between its minimum and maximum values in the data set, that is, $a_n \in [0,1], n=1,2$. The training of the SOM is performed with the MATLAB R2015b Neural Network Toolbox *(29)* running on a personal workstation with 3.50 GHz CPU and 16GB of RAM. After training, the matrix of W_{xy} is formulated representing the characteristic of each group. Once the winning neuron has been identified, the distributions of driving and charging patterns associated with each neuron in the SOM can be plotted and analyzed. The parameters of the patterns of the winning neuron are used to study inter-driver heterogeneity.

The size of the SOM is selected considering the following two factors. First, the grid has to be large enough so that there are sufficient neurons to distinguish the varied patterns among the prototype weight vectors. Since the SOM has two input components and the value of each component may be viewed at two or three levels (e.g., d_{VKT} may be described as long, medium, and short; T_{start} may be described as daytime and nighttime), there will be six possible combinations of input levels. Second, the number of neurons must be small enough such that most, if not all, neurons have sufficient winning frequencies (sample sizes) to observe the distribution of the pattern characteristics. After some trials involving SOMs with different numbers of neurons and with different arrangements (square grid, rectangular grid, and linear) on the map, the SOM is determined to have four neurons arranged in a 2x2 square grid. Accordingly, we name these four groups Cluster A, Cluster B, Cluster C, and Cluster D.

Group Characteristics

After training and clustering, each neuron in the SOM has a winning frequency of 34, 2, 5, and 9, respectively. The trained weight matrix W_{xy} is as follows:

$$W_{xy} = \begin{matrix} \text{Cluster A} \\ \text{Cluster B} \\ \text{Cluster C} \\ \text{Cluster D} \end{matrix} \overset{\begin{matrix} W_{xy_1} & W_{xy_2} \end{matrix}}{\begin{bmatrix} 0.5391 & 0.3762 \\ 0.9684 & 0.6296 \\ 0.1697 & 0.7687 \\ 0.0649 & 0.3128 \end{bmatrix}}. \tag{2}$$

From the trained weight matrix W_{xy}, Cluster A represents vehicles with a modest daily VKT and generally charging during the daytime. Vehicles in Cluster B are associated with long daily VKTs, and often charging in the late afternoon. Clusters C and D capture the vehicles with relatively short daily VKTs. Vehicles in Cluster C are usually connected to

Table 3. Summary Metrics of Input Components for Different Groups

Type	Number of Vehicles	Avg. distance traveled per day when the vehicle was driven (km)			Avg. start time of charging events (hh:mm)		
		Sample size	Mean	St.D.	Sample size	Mean	St.D.
Cluster A	34	4406	58.0	42.0	4778	13:42	5:40
Cluster B	2	240	93.4	44.3	265	17:16	6:41
Cluster C	5	1059	31.7	30.1	443	18:48	5:54
Cluster D	9	1570	23.3	19.0	484	13:10	4:40

Note: Avg. = average; St.D. = standard deviation.

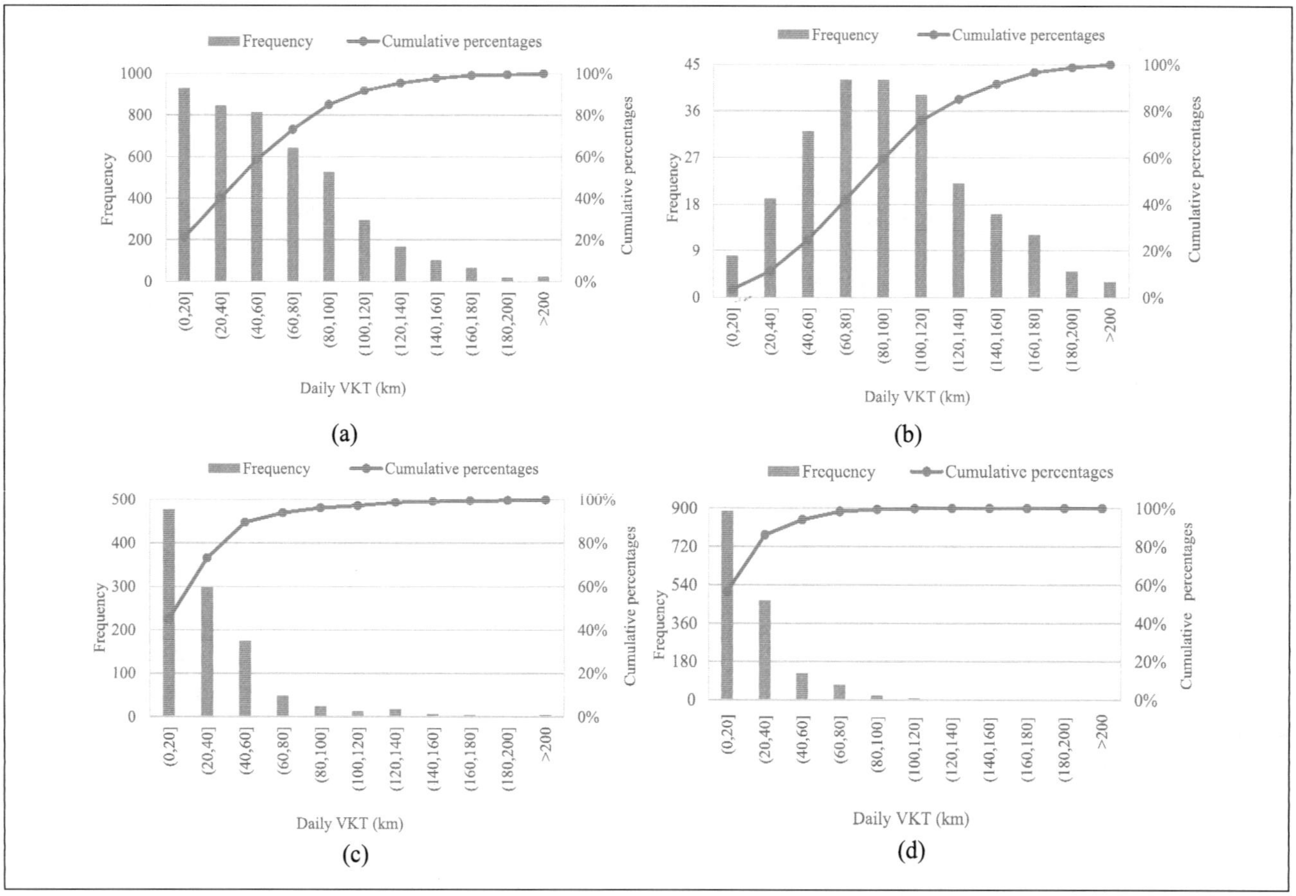

Figure 2. Distribution of distance traveled per day when the vehicle was driven: (a) Cluster A, (b) Cluster B, (c) Cluster C, and (d) Cluster D.

the charger at night while vehicles of Cluster D are usually plugged in during the daytime.

Table 3 presents the statistics of the two input components for different groups. There are 34 vehicles grouped into Cluster A. Although some initial trials have been carried out in an attempt to further divide this group, such as taking the number of daily charging events, or the variance of d_{VKT} and T_{start} as additional input components, no obvious differences are found among these vehicles. The numbers of travel days in Clusters A and B are less than the numbers of charging events, indicating the charging frequency is more than once per day. For Clusters C and D, the numbers of charging events are less than the days when the vehicle was driven. This matches the observations mentioned previously that BEVs with short daily VKTs are usually not charged every day.

Results and Discussion

Distribution of Distance Traveled per Day

Figure 2 plots distributions of daily VKTs in different groups. It can be seen that daily VKT distributions are significantly

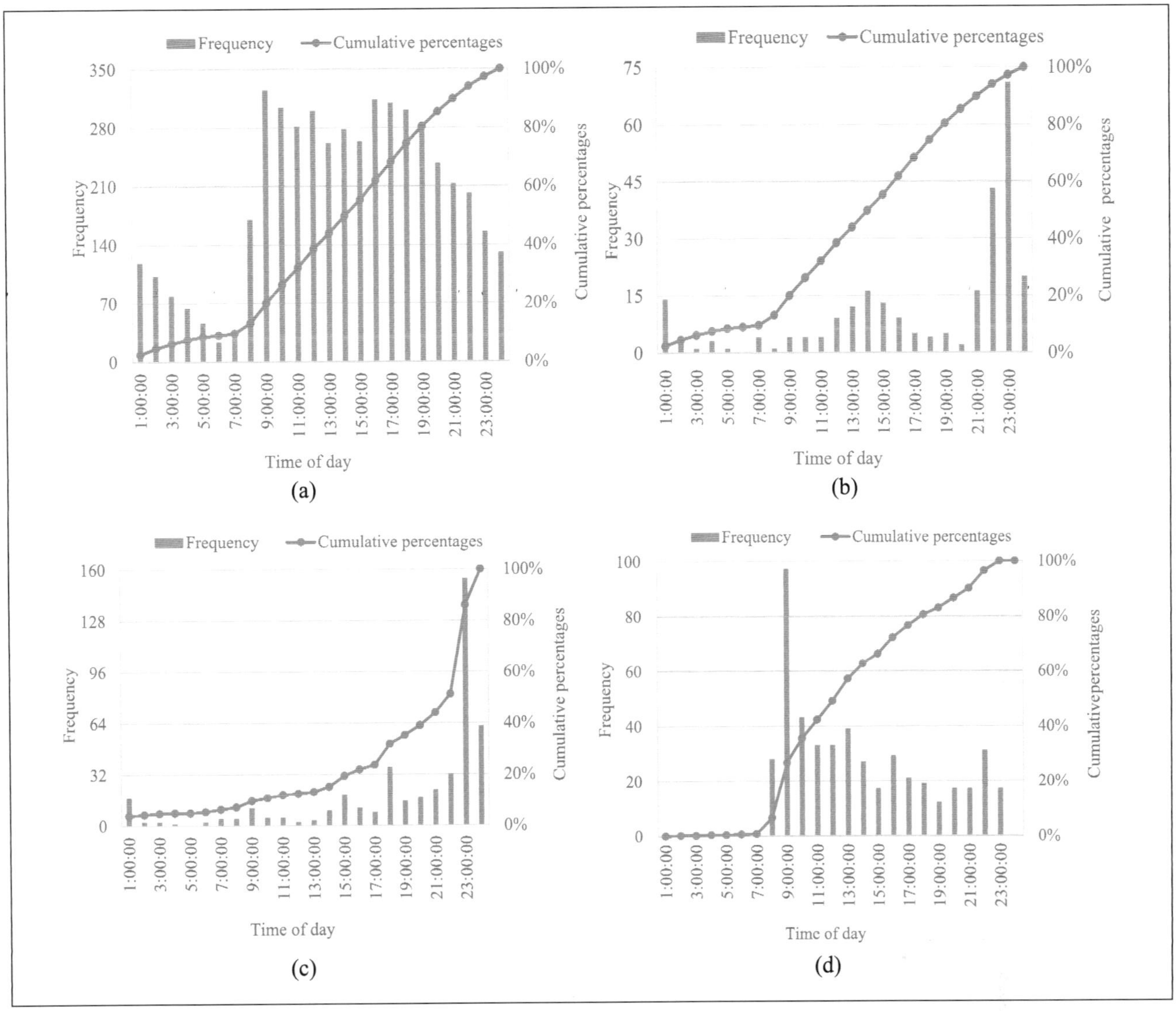

Figure 3. Distribution of start time of charging events: (a) Cluster A, (b) Cluster B, (c) Cluster C, and (d) Cluster D.

different among the four clusters. In the case of Clusters A, C, and D, the distribution interval with the largest probability is 0 to 20 km, and the share declines with the increase in the distance. As compared with Clusters C and D, the decline of Cluster A is much slower, with a similar pattern as that observed in (*7*). Because the average daily VKTs of Clusters C and D are 31.7 km and 23.3 km respectively, more than 70% of daily VKTs occur in the interval of 0 to 40 km. In the literature, a gamma distribution is often used to characterize daily VKT variations (*17, 28, 29*). We find that the distribution of Cluster B satisfies a gamma distribution with the shape of 3.44 and the rate of 0.04.

Distribution of Start Time of Charging Events

Figure 3 shows the distribution of starting time, that is, when drivers plug in their vehicles to EVSEs. For Cluster A, the percentage of BEVs charging begins to increase after 8:00 and maintains a steady rate during the daytime; 65% of the charging events occurred between 8:00 and 18:00. The other three groups have an obvious peak starting time for charging. The peak occurred between 22:00 and 23:00 for Clusters B and C, while it occurred between 8:00 and 9:00 for Cluster D. There is a smaller peak between 13:00 and 15:00 for Cluster B, presumably as individuals with long daily travel demand attempt to avoid depleting the battery and recharge their vehicles to overcome their anxiety over range. There is almost no start of charging from 1:00 to 6:00, as most vehicles have already been plugged in before midnight.

Distribution of Number of Charging Events per Day

The distribution of the number of charging events per day when the vehicle was driven is shown in Figure 4. For

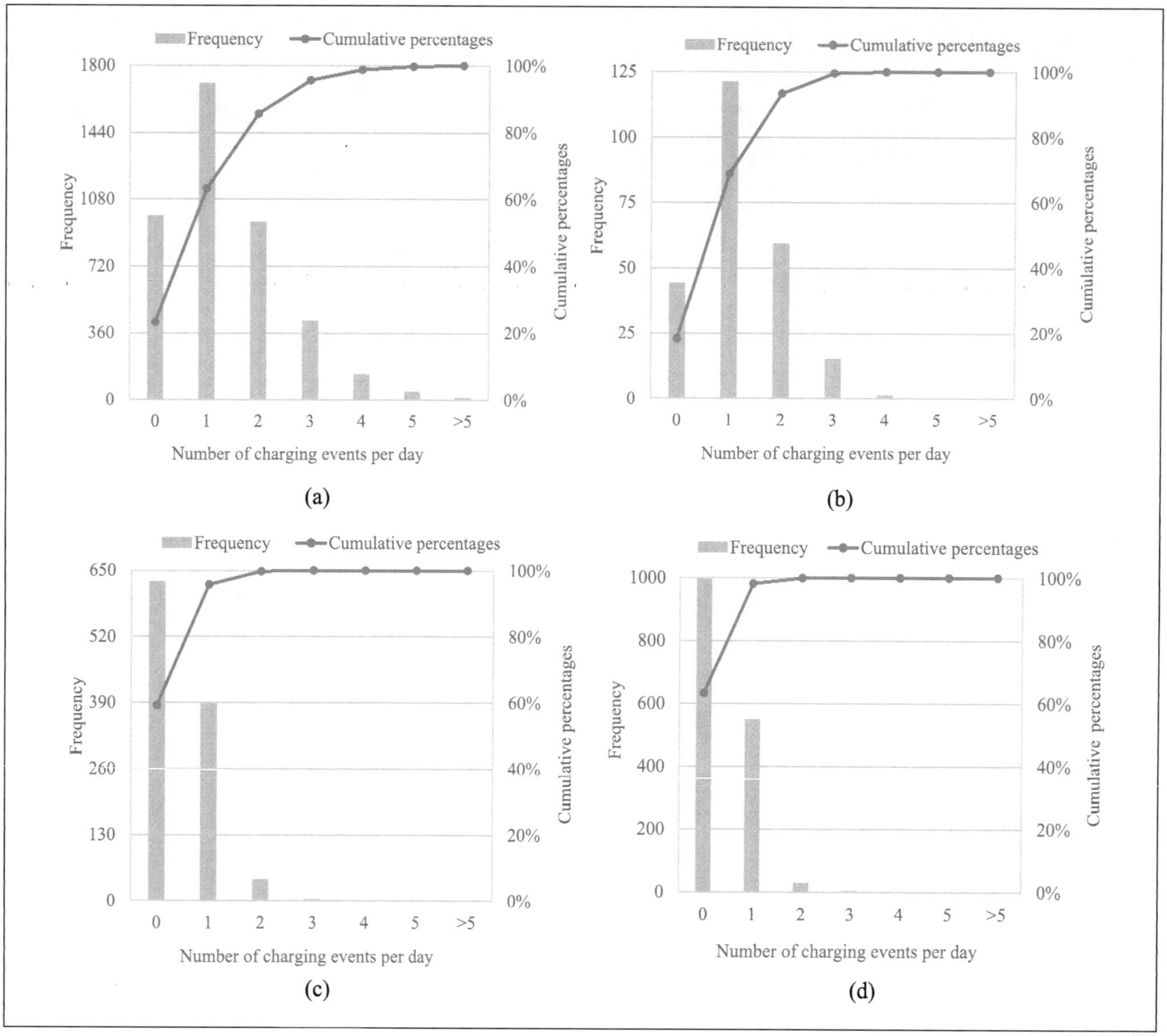

Figure 4. Distribution of number of charging events per day when the vehicle was driven: (a) Cluster A, (b) Cluster B, (c) Cluster C, and (d) Cluster D.

Clusters A and B, charging once per day is the frequency with the largest probability, accounting for 39.9% and 50.4%, respectively. Drivers of Cluster C do not charge the battery on 59.2% of days. This proportion is even larger for Cluster D, accounting for 63.5% of the driving days. Since the average daily VKTs of Clusters C and D are less than one-fifth of the E50's range (i.e., 170 km), it is reasonable for drivers to choose not to charge the vehicle every day.

Distribution of Distance Traveled between Charges

Figure 5 displays the distribution of distance traveled between consecutive charging events. The travel distance between two charges is often assumed to be gamma-distributed with different parameters in previous research (*14, 17*). However, the observations from Figure 5*a* may violate that assumption. That is, some conservative drivers are likely to charge the battery whenever they have the opportunity, regardless the remaining SOC. As shown in Figure 5*b, c*, and *d*, the percentage grows as the distance increases, until it reaches a peak. The peak occurs at the interval of 60 to 80 km for Clusters B and D, while for Cluster C it is observed at the interval of 40 to 60 km, both of which are far less than the range of the E50.

Distribution of Battery SOC Before/After Charging

Figure 6 shows the distribution of battery SOC at the start and the end of charging. For Cluster A, the probabilities

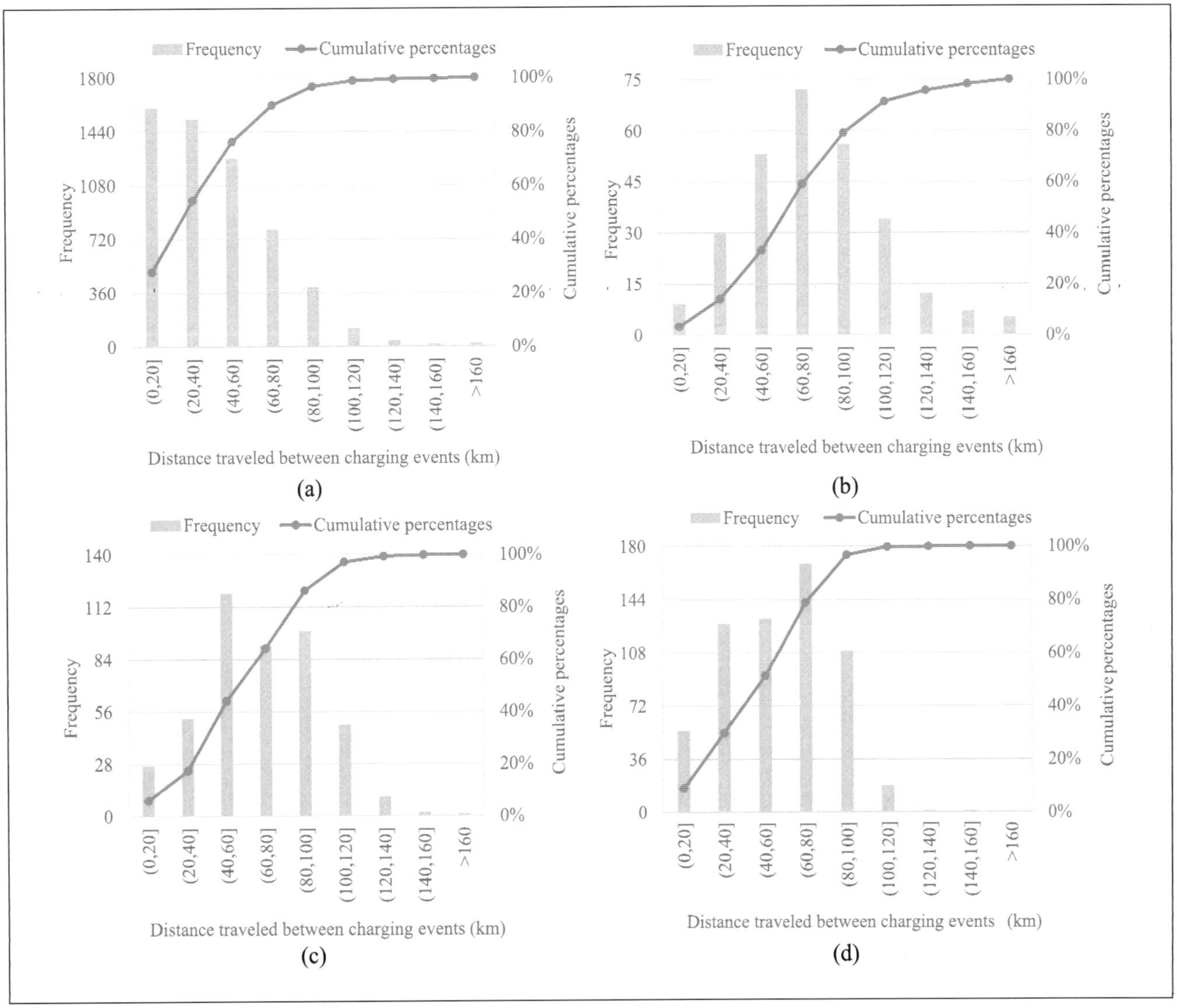

Figure 5. Distribution of distance traveled between charging events: (a) Cluster A, (b) Cluster B, (c) Cluster C, and (d) Cluster D.

of starting SOC are almost uniformly distributed along the range because of drivers' conservative charging behaviors. Due to the long daily VKT, the average starting SOC of Cluster B is lowest. Distributions of Clusters C and D are fairly wide but fall off quickly at each extreme, indicating that drivers are most likely to charge their vehicles when SOC is between 20% and 80%. For all the groups, it is observed that the majority of vehicles are fully or nearly fully charged at the end of each charging event.

Distribution of Charged Electric Energy

Figure 7 plots time-of-day electricity demand for the four groups. To generate these plots, first, the energy obtained from each charging event is estimated for each hour, and then the time-of-day electricity demands of different events are combined into one group. The charging demands of Clusters A and D are primarily centered on the daytime, which may add to peak load on the electricity grid. For Clusters B and C, demand mainly occurs during the off-peak period, increasing in the evening as home charging increases.

Conclusions

Using real-world in-use BEV data collected from Shanghai, China, this paper studies BEV owners' driving and charging behaviors, and compares them with the observations from the EV Project. The SOM is adopted as a classifier to eliminate the interference of intra-driver heterogeneity and capture inter-driver diversity. The key

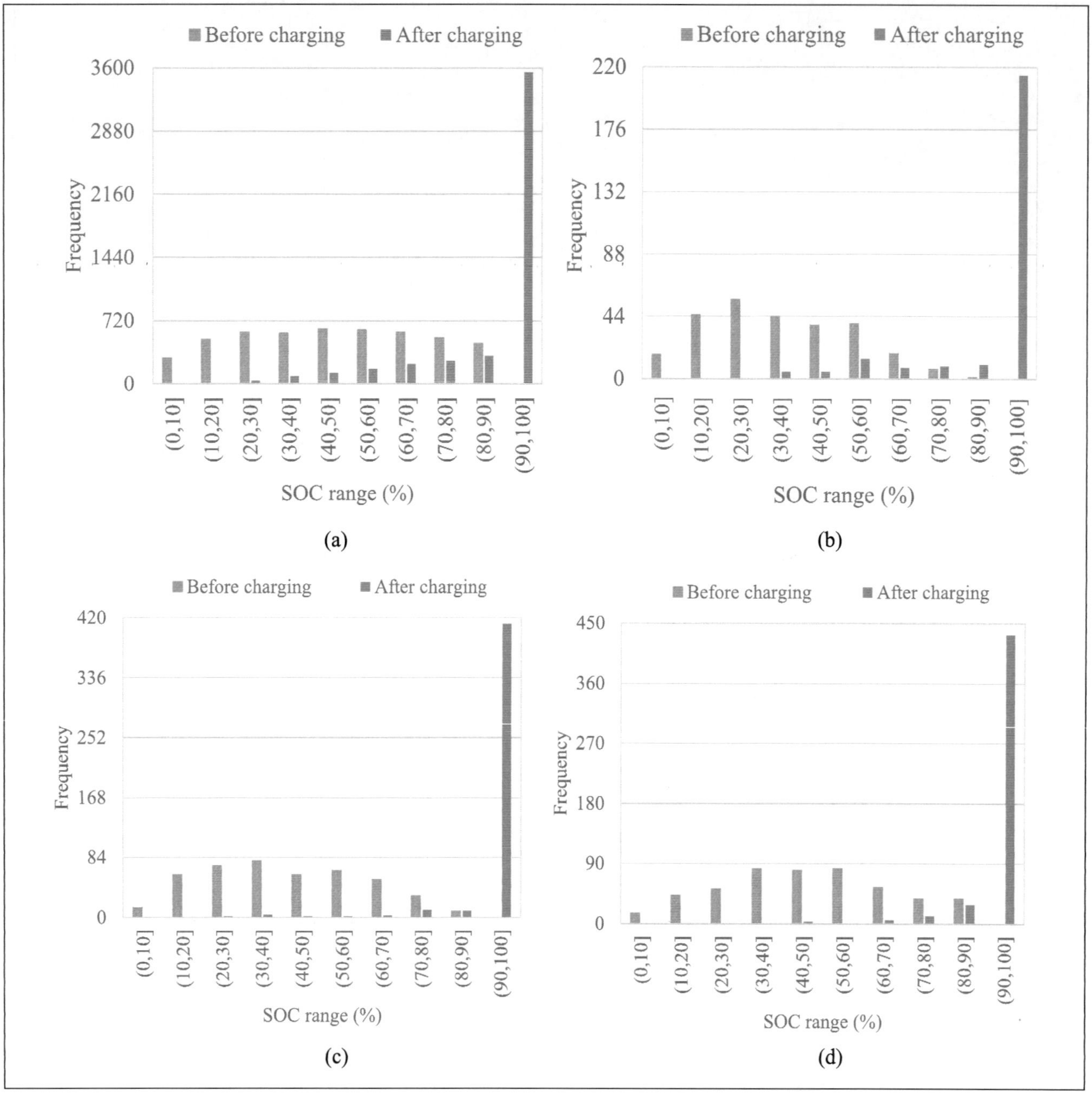

Figure 6. Distribution of battery SOC before/after charging: (a) Cluster A, (b) Cluster B, (c) Cluster C, and (d) Cluster D.

findings from the results include: (*a*) similar to the observations from the EV Project, the Shanghai E50 owners manifested conservative charging behaviors, resulting in short distances between charging events and on average more than one charge per day when the vehicle was driven; (*b*) the charging demand in daytime is significant, potentially overlapping with the peak electricity demand from non-BEV load; (*c*) the distributions depicting the driving and charging patterns vary in the shape and rate parameters because of inter-driver heterogeneity, for example, habitual charging behavior affects the distribution of distance between charges.

The main caveat of this paper lies in the shortage of socioeconomic information on BEV owners. For example, if the home locations were provided, the proportion of away-from-home charging events could be used to explain the high demand for daytime charging. Information on household vehicle ownership and owners' occupations would also be helpful to depict the intended usage of BEVs.

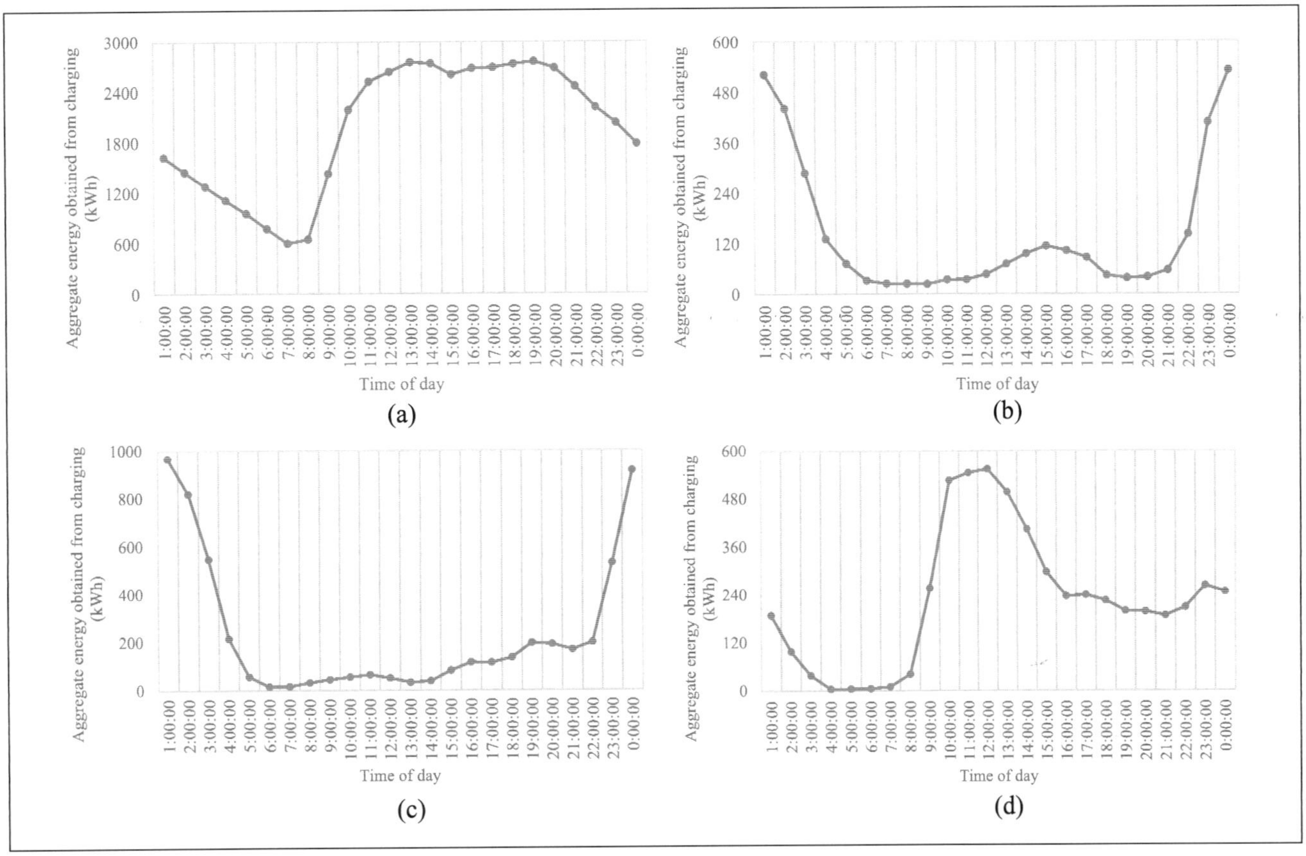

Figure 7. Distribution of charged electric energy: (a) Cluster A: (b) Cluster B, (c) Cluster C, and (d) Cluster D.

Acknowledgments

This work is partially supported by the Major projects of the National Social Science Fund of China (No. 17ZDA131), the Social Science Foundation of Jiangsu Province in China (No. 14FXC001), the Projects of International Cooperation and Exchange (No. 51561135003), the Natural Science Foundation of Jiangsu Province in China (No. BK20150603), and the Projects of National Natural Science Foundation of China (No. 51208101). The second author is supported by Oak Ridge National Lab. Analysis data provided by Shanghai International Automobile City (SIAC) support this study. The authors thank Yutaka Motoaki, John Smart, and Zhenhong Lin for providing data from the U.S. EV Project.

References

1 The State Council, PR China. *The State Council Planning for the Development of the Energy-Saving and New Energy Automobile Industry (2012- 2020)*, 2012. http://www.nea.gov.cn/2012-07/10/c_131705726.htm. Accessed July 18, 2017.

2 The State Council, PR China. *Instruction on Quickening Promotion and Implication of New Energy Vehicles*, 2014. http://www.gov.cn/zhengce/content/2014-07/21/content_8936.htm. Accessed July 18, 2017.

3 Robinson, A. P., P. T. Blythe, M. C. Bell, Y. Hübner, and G. A. Hill. Analysis of Electric Vehicle Driver Recharging Demand Profiles and Subsequent Impacts on the Carbon Content of Electric Vehicle Trips. *Energy Policy*, Vol. 61, 2013, pp. 337–348.

4 Fisher, P., J. Jostins, S. Hilmansen, and K. Kendall. Electronic Integration of Fuel Cell and Battery System in Novel Hybrid Vehicle. *Journal of Power Sources*, Vol. 220, No. 4, 2012, pp. 114–121.

5 FREVUE. *FREVUE: Freight Electric Vehicles in Urban Europe*, 2013. http://frevue.eu. Accessed November 4, 2017.

6. ZeEUS. *ZeEUS: A Flagship Electromobility Project Coordinated by UITP*, 2013. http://zeeus.eu. Accessed November 4, 2017.

7. Smart, J., and S. Schey. Battery Electric Vehicle Driving and Charging Behavior Observed Early in the EV Project. *SAE International Journal of Alternative Powertrains*, Vol. 1, No.1, 2012, pp. 27–33.

8. Schey, S., J. Smart, and D. Scoffield. *A First Look at the Impact of Electric Vehicle Charging on the Electric Grid in the EV Project*. Paper presented at 26th Electric Vehicle Symposium (EVS-26), Los Angeles, 2012.

9. Scoffield, D., J. Smart, S. Salisbury, and M. Kirkpatrick. *Investigating the Usefulness of Grid-Connected Plug-In Electric Vehicles as Controllable Loads*, 2014. https://avt.inl.gov/sites/default/files/pdf/phev/InvestigatingTheUsefullnessOfGridConnectedPHEVasControllableLoads.pdf. Accessed July 18, 2017.

10 Weldon, P., P. Morrissey, J. Brady, and M. O'Mahony. An Investigation into Usage Patterns of Electric Vehicles in Ireland. *Transportation Research Part D Transport and Environment*, Vol. 43, 2016, pp. 207–225.

11. Kim, S., D. Yang, S. Rasouli, and H. Timmermans. Heterogeneous Hazard Model of PEV Users Charging Intervals: Analysis of Four Year Charging Transactions Data. *Transportation Research Part C Emerging Technologies*, Vol. 82, 2017, pp. 248–260.
12. Tian, Z., Y. Wang, C. Tian, and F. Zhang. Understanding Operational and Charging Patterns of Electric Vehicle Taxis Using GPS Records. *Proc., IEEE International Conference on Intelligent Transportation Systems*, Qingdao, China, IEEE, New York, 2014, pp. 2472–2479.
13. Zou, Y., S. Wei, F. Sun, X. Hu, and Y. Shiao. Large-Scale Deployment of Electric Taxis in Beijing: A Real-World Analysis. *Energy*, Vol. 100, 2016, pp. 25–39.
14. Hu, L., J. Dong, Z. Lin, and J. Yang. An Investigation of Taxi Travel Patterns and Implications on the Feasibility of Electric Taxis. Presented at 96th Annual Meeting of the Transportation Research Board, Washington, D.C., 2017.
15. Pearre, N. S., W. Kempton, R. L. Guensler, and V. V. Elango. Electric Vehicles: How Much Range Is Required for a Day's Driving? *Transportation Research Part C: Emerging Technologies*, Vol. 19, No. 6, 2011, pp. 1171–1184.
16. Franke, T., and J. F. Krems. What Drives Range Preferences in Electric Vehicle Users? *Transport Policy*, Vol. 30, No. 3, 2013, pp. 56–62.
17. Dong, J., and Z. Lin. Stochastic Modeling of Battery Electric Vehicle Driver Behavior: Impact of Charging Infrastructure Deployment on the Feasibility of Battery Electric Vehicles. *Transportation Research Record: Journal of the Transportation Research Board*, 2014. 2454: 61–67.
18. Li, Z., S. Jiang, J. Dong, S. Wang, Z. Ming, and L. Li. Battery Capacity Design for Electric Vehicles Considering the Diversity of Daily Vehicles Miles Traveled. *Transportation Research Part C: Emerging Technologies*, Vol. 72, 2016, pp. 272–282.
19. ROEWE. Introduction of Roewe E50. http://www.roewe.com.cn/roeweee50/. Accessed July 18, 2017.
20. The EV Project. *Nissan Leaf Vehicle Summary Report* (October–December, 2013). Idaho National Laboratory, 2013. https://avt.inl.gov/sites/default/files/pdf/EVProj/EVProjectNissanLeafQ42013.pdf. Accessed July 18, 2017.
21. Voelcker, J. *2015 Nissan Leaf Review The Car Connection, 2015*. http://www.thecarconnection.com/overview/nissan_leaf_2015. Accessed July 18, 2017.
22. Liu, Y., C. Kang, S. Gao, Y. Xiao, and Y. Tian. Understanding Intra-Urban Trip Patterns from Taxi Trajectory Data. *Journal of Geographical Systems*, Vol. 14, No. 4, 2012, pp. 463–483.
23. Leng, B., H. Du, J. Wang, L. Li, and Z. Xiong. Analysis of Taxi Drivers' Behaviors within a Battle between Two Taxi Apps. *IEEE Transactions on Intelligent Transportation Systems*, Vol. 17, No. 1, 2015, pp. 296–300.
24. Tang, J., F. Liu, Y. Wang, and H. Wang. Uncovering Urban Human Mobility from Large Scale Taxi GPS Data. *Physica A Statistical Mechanics and Its Applications*, Vol. 438, 2015, pp. 140–153.
25. Kissel, G. EVSE Standards Status. SAE, 2011. https://www.smartgrid.gov/files/Electric_Vehicle_Supply_Equipment_EVSE_Standards_Status_201111.pdf. Accessed April 8, 2018.
26. Engelbrecht, A. P. *Introduction to Computational Intelligence*. John Wiley & Sons, Ltd., Hoboken, 2007.
27. MathWorks. MATLAB Version 8.6 (R2015b). The MathWorks Inc., Natick, Mass., 2015.
28. Greene, D. L. Estimating Daily Vehicle Usage Distribution and the Implications for Limited-Range Vehicles. *Transportation Research Part B: Methodological*, Vol. 19, No. 4, 1985, pp. 347–358.
29. Lin, Z., J. Dong, C. Liu, and D. L. Greene. PHEV Energy Use Estimation: Validating the Gamma Distribution for Representing the Random Daily Driving Distance. *Transportation Research Record: Journal of the Transportation Research Board*, 2012. 2287: 37–43.

The Standing Committee on Alternative Transportation Fuels and Technologies (ADC80) peer-reviewed this paper (18-03896).

Article

TRR
JOURNAL OF THE TRANSPORTATION RESEARCH BOARD

Transportation Research Record
2018, Vol. 2672(24) 31–40

DOI: 10.1177/0361198118756876
journals.sagepub.com/home/trr
SAGE

Cost of Oil and Biomass Supply Shocks under Different Biofuel Supply Chain Configurations

Rocío Uría-Martínez[1], Paul N. Leiby[1], and Maxwell L. Brown[2]

Abstract

This analysis estimates the cost of selected oil and biomass supply shocks for producers and consumers in the light-duty vehicle fuel market under various supply chain configurations using a mathematical programing model, BioTrans. The supply chain configurations differ by whether they include selected flexibility levers: multi-feedstock biorefineries; advanced biomass logistics; and the ability to adjust ethanol content of low-ethanol fuel blends, from E10 to E15 or E05. The simulated scenarios explore market responses to supply shocks including substitution between gasoline and ethanol, substitution between different sources of ethanol supply, biorefinery capacity additions or idling, and price adjustments. Welfare effects for the various market participants represented in BioTrans are summarized into a net shock cost measure. As oil accounts for a larger fraction of fuel by volume, its supply shocks are costlier than biomass supply shocks. Corn availability and the high cost of adding biorefinery capacity limit increases in ethanol use during gasoline price spikes. During shocks that imply sudden decreases in the price of gasoline, the renewable fuel standard (RFS) biofuel blending mandate limits the extent to which flexibility can be exercised to reduce ethanol use. The selected flexibility levers are most useful in response to cellulosic biomass supply shocks.

Blending biofuels with petroleum fuel increases the diversity of the fuel mix used to satisfy U.S. light-duty vehicle (LDV) travel demand. By displacing petroleum barrels, biofuels reduce the exposure of the U.S. economy to petroleum supply shocks. Biofuels can mitigate fuel price volatility if their prices are not too tightly correlated with those of gasoline and diesel (*1*). The biomass feedstocks used to produce biofuel experience their own supply shocks due to adverse weather conditions, pests, and competition from other non-biofuel demand uses. Climate change could affect long-term average corn yield and bioenergy crop yields (*2,3*). A more diverse fuel mix has the potential to be more resilient to supply shocks, depending on, for example, supply chain configuration. The petroleum and corn ethanol portions of the supply chain are already fully developed, but adjustments can still be made to enhance system resilience. Petroleum refinery operations that allow for quick adjustments of ethanol content in fuels, in response to changes in relative prices, can help mitigate price volatility at the pump after a petroleum or biomass supply shock. The supply chain for biomass feedstocks used in producing cellulosic biofuels is still at a nascent stage and its evolution will have implications for system resilience in case of biomass supply shocks. We propose that effectiveness in mitigating supply shocks should be one of the criteria, along with cost and environmental sustainability targets, for biofuel supply chain planning. This analysis explores the costs of selected supply shocks of varying origin (biomass or petroleum), sign (supply increases or decreases), and size under various biofuel supply chain configurations. It provides insight into the effectiveness of different supply chain flexibility levers in mitigating supply shocks.

Methods

The mathematical programming model used for this analysis, BioTrans (*4*), contains a detailed representation of the U.S. farm-to-pump supply chain for biofuels and its competition with the petroleum sector in fulfilling LDV fuel demand. This representation is summarized in Figure 1. BioTrans is a dynamic model that captures intertemporal linkages between periods and solves for the supply chain investment and operation levels that maximize social surplus. The modeling horizon is multi-decadal (2010–2040)

[1]Oak Ridge National Laboratory, Oak Ridge, TN
[2]Colorado School of Mines, Golden, CO

Corresponding Author:
Address correspondence to Rocío Uría-Martínez: uriamartiner@ornl.gov

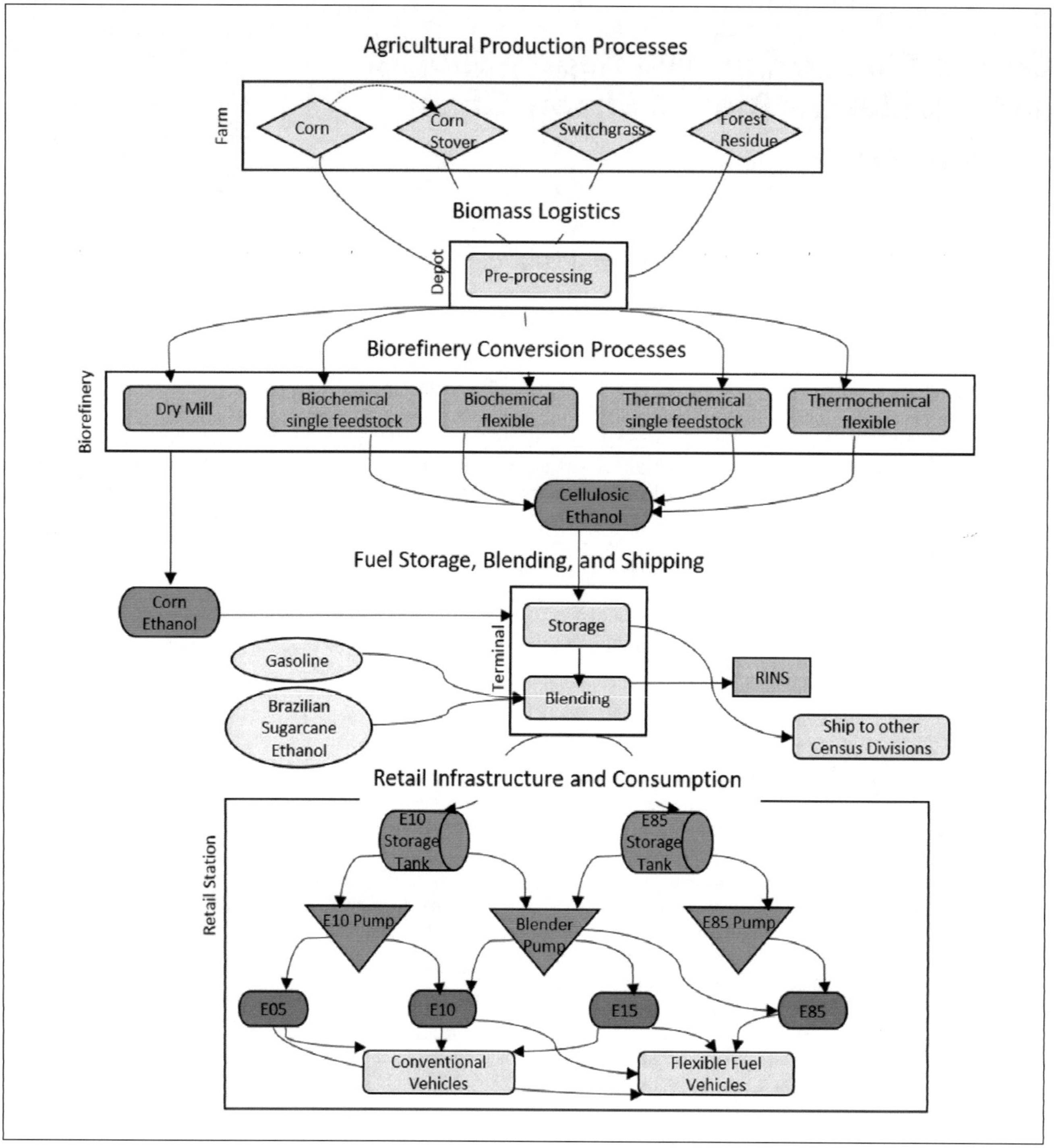

Figure 1. LDV fuel supply chain structure in the BioTrans model.

with annual periods as the unit of temporal disaggregation. BioTrans has national scope and includes regional detail at the census division level. Land allocation decisions and vehicle purchase decisions are the upstream and downstream model boundaries. Neither of them are explicitly modeled. Instead, biomass feedstock supply curves and vehicle stock projections are inputs to BioTrans.

For this study, BioTrans focuses on ethanol, produced from corn or one of three cellulosic feedstocks: corn stover, switchgrass, and forest residues. A biofuel blending constraint representing the renewable fuel standard (RFS) is included in the model. As the 2022 biofuel blending levels originally mandated by EISA 2007 appear too ambitious, a modified RFS target level that equals 60% of the maximum

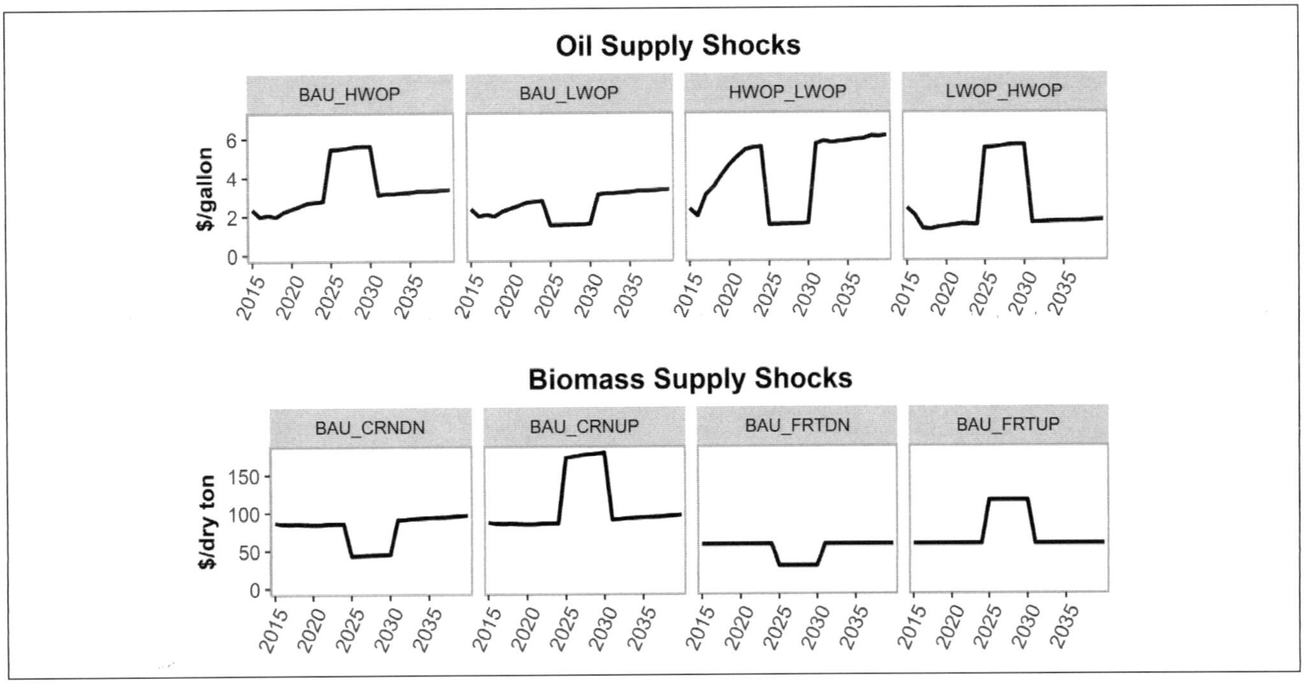

Figure 2. Shocked wholesale gasoline and biomass feedstock reference supply price trajectories.

potential ethanol consumption given the projected LDV fleet is assumed instead. Maximum ethanol consumption would be achieved if flexible fuel vehicles (FFVs) always consumed E85 and other LDVs always used the most ethanol-intensive blend that is compatible with their engines. Based on vehicle stock projections from the AEO2017 reference case, the modeled 2022 blending target is 19 billion gallons of renewable fuel, of which at least four billion gallons must be cellulosic biofuel. The target is kept constant for the remainder of the modeling period (2023–2040). The shadow value of the biofuel blending constraint can be interpreted as the price of a renewable identification number (RIN) which is the marginal cost of compliance with the RFS for obligated parties (petroleum refiners and importers). A positive RIN price indicates that the blending constraint is binding. If ethanol is cheaper than gasoline and it is economical for market participants to use more than what is mandated by the RFS, the RIN price drops to zero.

BioTrans allows depiction and evaluation of two types of strategies used by farmers and biorefiners to manage risks associated with biomass supply shocks: engineering solutions and portfolio diversification (*5*). Particular attention is given here to the value of building flexibility into the supply chain to be able to accommodate changes in the optimal mix of biomass feedstocks or blended fuels depending on the relative prices of gasoline and ethanol.

Selected Supply Shocks

Previous research concerned with representing risk surrounding key parameters for biofuel market participants (e.g., oil price, corn yield, ethanol production) has typically used Monte Carlo analysis based on random draws from the observed probability distributions for those parameters (*6,7*). Instead, the approach used here simulates selected multi-year price excursions under limited foresight and alternative supply chain configurations. Supply curves are shifted upward or downward during the shock years without changing the maximum available quantity. The shocked reference price paths are illustrated in Figure 2.

Figure 2 displays the simulated oil (upper panels) and biomass (lower panels) supply shocks. Outside of the shock period (2025–2030), oil price levels correspond to projections from one of three AEO2017 cases: reference (BAU), high world oil price (HWOP), and low world oil price (LWOP). The oil shocks are each depicted as a deviation from one of those reference price paths to another, for a 6-year duration.

In all the biomass supply shocks, the point of departure is a BAU case based on the 2016 Billion Ton Report (*8*) supply quantities and prices (for a US$60/dry ton biomass payment to farmers). The shocks are introduced as biomass feedstock supply curve shifts that result in supply price changes for the affected feedstocks. In the BAU_CRNUP and BAU_CRNDN shocks, both the corn and corn stover supply curves are shifted by multiplicative factors of 2 and 0.5, respectively. In the BAU_FRTUP and BAU_FRTDN shocks, the same multiplicative factors are used to introduce supply curve shifts for forest residues. BAU_CRNUP could be consistent with an extended period of drought or other event causing a multi-year increase in supply prices; similarly, BAU_CRNDN could correspond to a string of exceptionally large harvests that would result in price reductions. For forest residues, it is more difficult to envision large, multi-year changes in availability at the national level. However, there could be significant increases in the supply price of forest residues faced by biofuel producers due to competition from other uses (e.g., strong world demand for pellets produced in the U.S.).

The eight shocks described above are simulated in a limited foresight implementation of BioTrans. Foresight and expectations are important in a long-run dynamic model with durable capital investments; we adopt an approach between perfect foresight and complete myopia. The model is solved in 6-year overlapping windows and the window is rolled forward one year in each successive model solve. This is consistent with market participants formulating their investment and operation plans with reasonably-accurate near-term information that is updated each year. Shocks, however, are surprises: information about the shocks only becomes available in the 2025–2030 solution period, after they have started.

Selected Flexibility Levers

We describe the flexibility levers and discuss their impact on reference (undisrupted) model outcomes prior to the hypothesized 2025–2030 disruption period.

Ethanol Blending Flexibility (E05–E15). The "rigid" option includes two fuel blends in its choice set: E10 and E85, where the latter can only be used by FFVs. The "flexible" option includes two additional gasoline–ethanol blends: E05, and E15. E15 is assumed to be compatible with gasoline-based LDVs whose model year is 2001 or newer and only can be dispensed through flexible pumps. E05 is compatible with all gasoline-based LDVs and can be retailed through the same pumps used for E10. Adjusting ethanol content downwards from E10 to E05 is modeled as cheaper/easier than adjusting it up from E10 to E15. Petroleum refineries, which would likely have to increase use of aromatics to maintain octane ratings if ethanol content was reduced (*9*), are not explicitly modeled in BioTrans.

In 2024, total LDV fuel consumption ranges between 100 billion and 130 billion gallons of gasoline equivalent (gges) depending on the reference oil price path: BAU, HWOP, or LWOP. E85 consumption is higher under the HWOP oil prices regardless of supply chain configuration. The only noticeable change in the reference (undisrupted) outcomes brought about by the ability to change ethanol content in the E05–E15 range is the consumption of 4 billion gges of E15 consumption in the HWOP case.

Biorefinery Feedstock Flexibility. In this study, two levels of feedstock flexibility at biorefineries were considered: "rigid" and "flexible." In the rigid configuration, four biomass-to-biofuel conversion pathways are included with each dedicated to a single feedstock: dry milling of corn, biochemical conversion of stover to cellulosic ethanol, biochemical conversion of switchgrass to cellulosic ethanol and, thermochemical conversion of forest residue to cellulosic ethanol. In the flexible configuration, thermochemical biorefineries can use any mix of stover, switchgrass, and forest residue. Both configurations lead to similar levels and types of installed biorefinery capacity by 2024 (the year before the start of the simulated shocks) but there are differences in their regional distribution. When thermochemical biorefineries only use forest residue, most are built in the regions with the largest potential supply of that feedstock. With feedstock flexibility, thermochemical biorefinery capacity is concentrated in the regions with the cheapest supply of cellulosic feedstock.

Advanced Logistics Design. The "flexible" option for biomass logistics is the Advanced Logistics Design (*10*). With advanced logistics, cellulosic feedstocks are preprocessed into a more flowable material (e.g., pellets) that is cheaper to transport longer distances by rail and store. The lower transportation cost per unit also allows for taking advantage of economies of scale in biorefinery sizing. Cumulative biorefinery investment cost by 2024 decreases by approximately 9% in the advanced logistics case, relative to the "rigid" conventional logistics design, because of the larger biorefinery sizes. Consistent with the idea that advanced logistics makes biorefineries less dependent on their local supply by creating a larger pool of eligible resources, interregional transportation of biomass feedstocks is allowed in this configuration. The "rigid" option (conventional logistics design) involves transportation of cellulosic feedstocks in bales or raw format by truck to nearer destinations and smaller biorefineries. Therefore, under the "rigid" scenarios, only transportation within the census division in which the feedstock is harvested is allowed. The logistics for corn are assumed to already be mature and optimized but, to explore differences in optimal supply chain configuration and response to shocks with national versus regional biomass feedstock markets, corn is also restricted to only be transported within the census division in which it is harvested in the conventional logistics design.

The largest savings associated with the adoption of advanced biomass logistics accrue to forest residues. Advanced logistics result in a lower forest residue supply price at the roadside because the ability to ship forest residues across census divisions mobilizes cheap supplies from regions where otherwise it would not be collected. Additional savings in transportation and storage costs more than offset the cost of pre-processing forest residues at depots. Simulated forest residue price at the biorefinery gate is US$135/dry ton versus US$106/dry ton under conventional and advanced logistics, respectively.

A transition to advanced biomass logistics, by reducing transportation and storage costs and enabling biorefinery scale economies, would likely impact average distance traveled both by biomass from forest/field to biorefinery and by ethanol from biorefinery to demand center. Figure 3 shows volume-weighted average distances traveled by biomass in

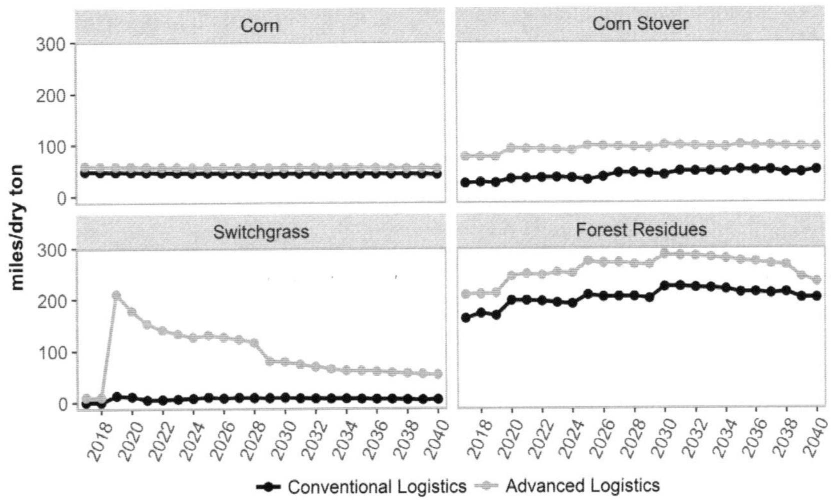

Figure 3. Average distance traveled by biomass under alternative logistics designs (2017–2040).

Table 1. Net Private Cost (or Benefit) of Selected Supply Shocks Under Different Biofuel Supply Chain Configuration (Million 2010 US dollars)

Shock type	All rigid	Biorefinery feedstock flexibility	Advanced logistics design	Ethanol blend flexibility	All flexible
BAU_HWOP	−118,605.8	−118,537.2	−118,481.4	−118,030.5	−117,671.0
BAU_LWOP	70,561.2	70,398.0	70,472.1	70,534.5	70,568.2
LWOP_HWOP	−276,892.5	−275,393.5	−275,849.7	−276,922.4	−275,607.1
HWOP_LWOP	174,717.2	174,546.9	174,603.3	173,257.0	173,166.4
BAU_CRNUP	−23,435.9	−22,727.5	−22,179.4	−23,373.6	−21,999.0
BAU_CRNDN	13,208.9	13,958.3	13,047.8	13,091.8	14,251.1
BAU_FRTUP	−4750.0	−1.6	−4683.8	−4861.2	−739.0
BAU_FRTDN	2617.0	4.2	2851.4	2508.6	2823.8

the conventional versus advanced logistics designs from 2017 to 2040; note that 2017 is the year in which the advanced logistics design is first available in the BioTrans simulations.

Figure 3 shows that corn travels an average of 50 miles from farm to biorefinery and this distance does not change markedly across the two logistics designs. On the other hand, cellulosic feedstocks travel significantly larger distances in the advanced logistics configuration. Switchgrass only becomes available in 2019, which explains the zero distances for years 2017 and 2018. In the 2020s, as switchgrass planted acreage increases, average distance traveled decreases.

Advanced logistics result in cellulosic feedstocks traveling longer distances. On the other hand, transitioning to advanced logistics leads to a substantial reallocation of dry mill capacity within the Midwestern census divisions that reduces the average distance traveled by corn ethanol from 917 to 714 miles. More than half of the corn ethanol produced is transported across census divisions in the advanced logistics scenario. In contrast, most cellulosic ethanol is consumed in the same census division as it is produced, and the advanced logistics case decreases the average shipping distance modestly from 756 to 660 miles.

Results

The set of cases to be discussed in this document includes 40 shocked cases (eight shocks times five supply chain configurations) plus the 15 corresponding unshocked baseline cases (three baseline oil market futures times five supply chain configurations) needed to determine the cost of the shocks. Table 1 summarizes the net private cost of each shock under the various supply chain configurations. Net private cost is a welfare measure that adds up the differences in producer and consumer surplus components along with the differences in transportation and conversion costs between the two solutions.

Each cell in Table 1 represents the cost of a shock (relative to the market outcomes of a model run with the same supply chain configuration and no shock). Reading vertically down a column shows the large variation in shock costs by type and size. Comparing results horizontally within a row serves as an indication of variability in the net welfare effect of a shock under alternative supply chain configurations. Comparing shock costs across configurations is not straightforward because many of the model variables take different base values in 2024 across the various supply chain alternatives. These differences in "starting points" lead to instances in which a more flexible configuration results in a worse net private welfare outcome during a shock. For instance, the LWOP_HWOP shock is slightly more damaging for the *Ethanol Blend Flexibility* than the *All Rigid* case. Ethanol blending flexibility enables the system to better adapt to the market conditions in earlier years, less ethanol consumption during low oil price periods up until 2024, but makes a sudden increase in gasoline prices costlier.

The BAU_HWOP and BAU_CRNDN shocks are the only shocks where the gain, or avoided cost, enabled by the most flexible configuration is larger than the sum of gains from each individual lever. In response to those shocks, the flexibility levers act synergistically. In many of the other shocks, there is a single flexibility lever that is more effective than the others and even more effective than the combination of all three levers. This result shows the importance of evaluating flexibility levers in the context of the full biofuel-petroleum fuel system.

Oil Supply Shocks

For oil supply shocks, the main takeaway is that the flexibility levers, as depicted, are not very effective in mitigating the cost of damaging upward oil supply shocks (BAU_HWOP, LWOP_HWOP) or amplifying the benefits from downward oil supply shocks (BAU_LWOP, HWOP_LWOP). The difference in net private cost across supply chain configurations is less than 1% for those shocks. During supply shocks that increase the cost of ethanol relative to gasoline, there is limited ability to adjust ethanol content down due to the biofuel blending mandate constraint. In contrast, RIN prices drop down to zero at the beginning of the BAU_HWOP shock and for most years of the LWOP_HWOP shock, indicating that it is optimal to use more biofuel than the mandated level during major oil price increases. Increases in ethanol use during those shocks are typically achieved through a combination of increases in sugarcane ethanol imports and investments in additional domestic ethanol production capacity. Availability of additional domestic biofeedstock supplies and E15/E85-compatible retail fuel infrastructure limit the magnitude of the ethanol use increase in response to BAU_HWOP and LWOP_HWOP shocks.

Simulated oil supply shocks result in large changes in the price of wholesale gasoline. The average changes in gasoline prices during the shock period range between –US$3.71/gallon in the HWOP_LWOP shock and US$6.45/gallon in the LWOP_HWOP shock. The average annual fuel expenditure for FFV owners in 2025–2030 across all shocks is 3%–8% lower (depending on supply chain configuration) than for conventional vehicle owners. For conventional vehicle owners, adding flexibility levers reduces average expenditures by 2% relative to the most rigid configuration.

Biomass Supply Shocks

Biomass supply shocks lead to moderate changes in ethanol price. The average changes in ethanol price during the shock period range between –US$0.22/gallon in the BAU_CRNDN shock and US$0.55/gallon in the BAU_CRNUP shock. Biomass supply shocks affecting the feedstock used to produce the majority of biofuel (corn) have a larger impact on ethanol prices than the forest residue supply shocks.

For corn supply shocks (BAU_CRNUP and BAU_CRNDN), the most flexible configuration produces the best outcome from a social welfare perspective. Combining the three flexibility levers leads to an 8% increase in the net benefit of the BAU_CRNDN shock and a 6% decrease in the cost of the BAU_CRNUP shock. The depicted flexibility levers are most consequential in response to supply shocks affecting cellulosic biomass feedstocks. Biorefinery feedstock flexibility mutes the effect from forest residue supply shocks because forest residue is not used in large amounts when biorefiners can choose among multiple feedstocks, *but* the prevailing logistics design is conventional. For BAU_FRTDN, the advanced logistics configuration leads to the highest net benefit.

Variability in farmer revenue from sales of corn or cellulosic feedstocks for ethanol production decreases significantly in the *All Flexible* configuration relative to the *All Rigid* configuration. The coefficient of variation of farmer revenue decreases from 18.4% to 16.9% for corn and from 30.9% to 21.7% for cellulosic feedstocks when comparing the most flexible to the most rigid supply chain configuration. Weighting the probability of all the simulated shocks equally, the highest average farmer revenues correspond to the most rigid supply chain design. These results regarding farmer revenue are due to the very high equilibrium prices of corn and forest residue during the BAU_CRNUP and BAU_FRTUP shocks respectively in the *All Rigid* configuration.

Normalized Shock Costs

Taking into account the differences in scale of biofuel versus gasoline use is important for interpreting the results from Table 1. Average volumetric ethanol content of fuels by 2024 (the year before the start of the simulated shocks) ranges between 13% in cases with LWOP oil prices and 16.8% in cases with HWOP oil prices. The implied shocks to fuel supply price are therefore greater for gasoline than for ethanol (since biomass only represents a fraction of ethanol costs). Shocks on petroleum supply affect a much larger fraction of total light-duty fuel supply and lead to much larger changes,

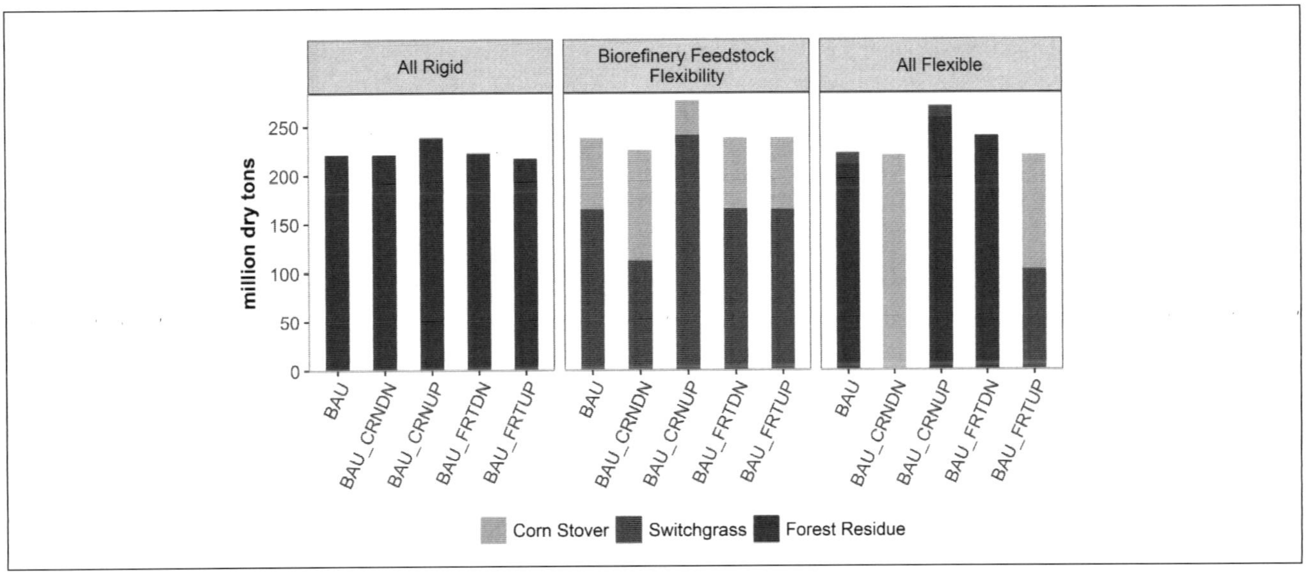

Figure 4. Simulated cellulosic feedstock mix during biomass supply shocks under three alternative supply chain configurations (average, 2025–2030).

in absolute value, in net private welfare than the biomass supply shocks.

Biorefinery Feedstock Flexibility

As our representation of dry mills does not allow the use of any feedstock other than corn, this flexibility lever should primarily be of use in adjusting to supply shocks affecting cellulosic feedstocks (corn stover, switchgrass, and forest residues). Figure 4 shows the annual average mix of cellulosic feedstocks during biomass supply shocks for three alternative supply chain configurations: *All Rigid*, *Biorefinery Feedstock Flexibility*, and *All Flexible*.

Regardless of the biomass supply shock, the *All Rigid* configuration does not allow thermochemical biorefineries to use any other feedstock than forest residue; thus, forest residue is used 100% of the time. With feedstock flexibility, the selected feedstock mix under undisrupted baseline circumstances (BAU case) differs from the *All Rigid* configuration and is made up of corn stover and switchgrass. Large, cost-reducing adjustments take place in response to shocks affecting corn and corn stover: reliance on corn stover increases during the BAU_CRNDN shock. In response to the BAU_CRNUP shock, switchgrass substitutes for much of the stover use. The absence of forest residue from the feedstock mix in the *Biorefinery Feedstock Flexibility* configuration results from transport costs of forest residue being much more expensive than either stover or switchgrass under conventional biomass logistics. The cost of transporting forest residue 50 miles with conventional logistics is US$16.8/dry ton versus US$10.4/dry ton for stover and switchgrass.

In the most flexible configuration, the cost of transporting forest residue becomes comparable to the other cellulosic feedstocks (approximately US$6/dry ton to transport preprocessed material for 50 miles) due to the advanced logistics design. Without shock, the least expensive way to operate the thermochemical biorefineries is by using forest residue almost exclusively. Reliance on forest residue is maintained during the BAU_CRNUP and BAU_FRTDN shocks. However, both unexpected increases in the supply cost of forest residue or decreases in the cost of corn stover result in drastic changes in feedstock mix. In response to the BAU_CRNDN shock, corn stover is used as the unique feedstock in thermochemical biorefineries instead of forest residue. When forest residue supply becomes more expensive, it is completely replaced by a mixture of corn stover and switchgrass (BAU_FRTUP). For perennial energy crops like switchgrass, there is a multi-year lag from planting to harvesting which is not captured in the model. Thus, the surge in switchgrass use during the BAU_FRTUP shock would require a large amount of the crop to have been planted years earlier without being tied to a particular use by a long-term contract. Only with advanced logistics and mature regional or national markets for switchgrass does the existence of those "free" switchgrass supplies appear plausible.

Figure 4 shows that total cellulosic feedstock use increases during the BAU_CRNUP shock regardless of supply chain configuration. During the BAU_CRNUP shock, the cellulosic portion of the RFS blending mandate is not binding and the cellulosic RIN price plunges to zero (even though the overall renewable fuel blending mandate remains binding). In those instances, it is optimal to increase the fraction of total ethanol use satisfied with cellulosic ethanol instead of corn ethanol or sugarcane ethanol imports. A total of 83% of the instances in which the cellulosic RIN price is zero arise in supply chain configurations with biorefinery feedstock flexibility or advanced

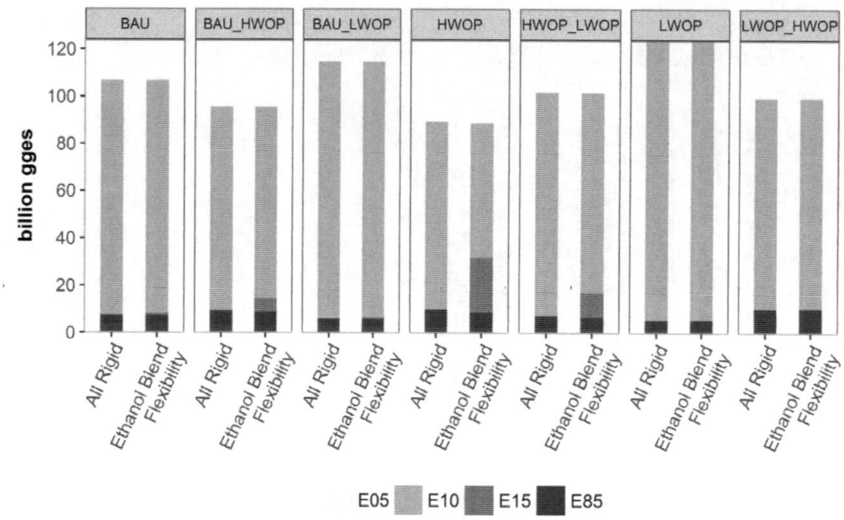

Figure 5. Simulated LDV fuel mix during oil supply shocks under alternative supply chain configurations (average, 2025–2030).

logistics design. Flexibility levers make cellulosic ethanol better able to compete with or complement corn ethanol.

Advanced Logistics

Advanced logistics for biomass feedstocks might help reduce the costs of a biomass supply shock in two ways: 1) by leading to a network of depots and biorefineries able to produce cellulosic ethanol at a lower price than in the conventional logistics case in undisrupted market conditions and 2) by shipping feedstocks between Census Divisions to arbitrage away price differences that arise during the shock period. In our current implementation, biomass supply shocks are applied uniformly across the country which limits the potential for large interregional price differentials opening up during supply shocks. Regional variation in the welfare consequences of a shock for farmers remains when, as it happens in the BAU_FRTUP case, alternative feedstocks with different geographical distribution substitute the one experiencing the shock.

The national average cellulosic ethanol price in 2025–2030 is lower in the *Advanced Logistics Design* configuration than in the *All Rigid* alternative, both with and without biomass supply shocks. In the unshocked BAU case, the simulated cellulosic ethanol price is US$2.38/gallon and US$2.84/gallon under advanced and conventional logistics, respectively. The most flexible supply chain configuration (*All Flexible*) leads to even lower prices, particularly in the BAU_FRTUP shock where advanced logistics yields greater value combined with biorefinery feedstock flexibility.

Advanced logistics leads not only to a lower average cellulosic ethanol price across shock events but also to a less variable price. With the *All Rigid* configuration, cellulosic ethanol prices range between US$2.46-3.56/gallon. The range narrows to US$2.30–3.04/gallon with the *Advanced Logistics* configuration.

Prices for corn ethanol during shocks change much less with supply chain configuration because corn cannot be substituted by other feedstocks at dry mills and the only difference between conventional and advanced logistics is the ability to transport corn across census divisions. For instance, in the BAU_CRNDN shock, the average price of corn ethanol is US$2.07/gallon in the *All Rigid* configuration and US$2.05/gallon with *Advanced Logistics Design*. In the BAU_CRNUP shock, the *All Rigid* average corn ethanol price is US$2.89/gallon and US$2.80/gallon lower with advanced logistics.

Ethanol Blend Flexibility

Ethanol blend flexibility is potentially the most-powerful lever of those analyzed here, allowing adjustment of the ethanol–gasoline mix used to satisfy most light-duty transportation demand. Blending flexibility can play a role in response to both oil and biomass supply shocks. There are limits to this flexibility however, including regulatory constraints establishing a minimum volume of biofuel use and retail infrastructure availability for dispensing E15 and E85. Figure 5 summarizes changes in the mixture of gasoline–ethanol blends consumed during oil supply shocks.

The height of the bars in Figure 5 corresponds to total LDV fuel consumption in billion gges. It changes significantly depending on oil price levels (BAU, HWOP, LWOP) outside of the shock period, as base conditions alter total consumption and blend mix used before the shock starts. The flexibility levers considered here do not substantially modify the total fuel consumption level. The renewable fuel blending mandate is binding except in the first two years of the

BAU_HWOP shock and from 2025 to 2029 in the LWOP_HWOP shock. Under those circumstances, gasoline is sufficiently expensive relative to ethanol such that the optimal level of ethanol use is larger than what is required by the mandate.

For oil price shocks taking place under BAU oil prices, supply chain configuration does not play a significant role. The increase or decrease in biofuel consumption that accompanies the BAU_LWOP and BAU_HWOP shocks, respectively, is mostly attained by adjusting the level of E85 consumption. However, E15 sales are also part of the response during a BAU_HWOP shock.

The ability to adjust ethanol content between E05 and E15 plays a key role in response to the downward HWOP_LWOP oil price shock. In response to a very large decrease in oil prices that unexpectedly interrupts a sequence of high prices (HWOP_LWOP shock), total fuel consumption increases, and total ethanol use decreases through a substitution from E15 to E10. This combination of responses results in a decrease in total ethanol consumption from 18.9 billion gallons in 2024 to 17.9 billion gallons in 2025 under the ethanol blend flexibility configuration.

Prior to the upward LWOP_HWOP shock, 125 billion gges of E10 and 5 billion gges of E85 are estimated to be consumed in 2024. In response to the sudden increase in gasoline prices, E85 consumption roughly doubles during the shock period accounting for approximately 80% of the vehicle-miles traveled (VMT) demand by FFVs for the duration of the shock. Total ethanol use increases by an additional 2.8 billion gallons in 2025 in the *Ethanol Blend Flexibility* configuration in response to the LWOP_HWOP shock. Such an increase in ethanol supply is achieved through a combination of biorefinery capacity additions (both dry mills and cellulosic biorefineries) and an increase in sugarcane ethanol imports. In the LWOP_HWOP shock, the 2025 supply cost of ethanol jumps to US$6.5/gallon. To accommodate the increase in ethanol supply, retail infrastructure investment also adjusts. For instance, during the LWOP_HWOP shock, investment in dedicated E85 pumps increases but E85 consumption is bounded by the number of FFVs (30 million in 2025), which is an exogenous parameter in BioTrans.

Conclusions and Further Work

This analysis has explored the value of three selected flexibility levers for mitigating the impacts of petroleum and biomass supply shocks on the LDV fuel supply. Although not every lever is useful in response to every shock, a more flexible supply chain slightly reduces average fuel expenditures for consumers and variability of revenues for farmers. The reduction in petroleum shock costs enabled by these particular levers is quite limited and merits further investigation.

The flexibility levers affect the undisrupted base supply chain configuration, changing the optimal regional distribution of installed biorefinery capacity, cellulosic feedstock mix, and fuel blend mix. This leads to different pre-disruption conditions on which a given shock is applied. These differences in starting points influence the effectiveness of a given flexibility lever during a shock and complicate the comparison of shock costs across supply chain configurations.

For oil supply shocks, the main takeaway is that the depicted flexibility levers are not very effective in mitigating the cost of damaging supply shocks (BAU_HWOP, LWOP_HWOP) or amplifying the benefits from positive supply shocks (BAU_LWOP, HWOP_LWOP). The RFS constraint is often binding, limiting downward biofuel blend flexibility and occasionally making upward flexibility unnecessary or unattractive unless oil prices increase greatly. Upward ethanol blend flexibility is also limited by short-run surge supply limits, short-run biomass supply elasticity, available logistics, and conversion capacity. The flexibility levers considered are most useful in response to supply shocks affecting cellulosic biomass feedstocks. Biorefinery feedstock flexibility and advanced logistics help decrease the supply cost of cellulosic ethanol so that it becomes more competitive with corn ethanol.

The supply chain representation in BioTrans does not include some response mechanisms that would help increase the effectiveness of flexibility levers during shocks: first, the analysis does not consider short-run flexibility from inventories of fuel or feedstock; second, no adjustment in corn planted acreage in response to shocks that imply large increases in ethanol demand and prices is allowed; third, drop-in biofuels that could substitute for larger shares of gasoline without requiring modifications to vehicle engines or retail infrastructure are not modeled. Including these mechanisms in BioTrans is among the items suggested for future study.

This analysis has explored potential benefits from flexibility levers along the supply chain, but it does not fully account for all the costs associated with implementing that flexibility. It includes estimated costs for flexible fuel pumps and advanced logistics components. Multi-feedstock biorefineries are presumably more expensive to build and operate than those dedicated to a single feedstock and those premia are not yet included in BioTrans. Similarly, *Ethanol Blend Flexibility* would require changes in operations at petroleum refineries whose cost is not depicted here. A more complete accounting of those additional costs could be a useful extension to this analysis.

The set of selected shocks is meant to be diverse enough to ascertain which types of shocks a given flexibility lever is more effective in mitigating. Without a probability distribution attached to those different types, average revenue and expenditure calculations are effectively assuming equal probabilities for all the simulated shocks which is not necessarily realistic. For petroleum and corn supply shocks, historical data can help in assigning probabilities, as long as we assume that the future will follow a similar pattern to the past. For cellulosic feedstocks, the task is much more difficult because historical data are limited.

All the supply shocks we have simulated are nationwide and applied uniformly to all U.S. regions. Additional insight, and measured flexibility benefits, might arise from regional supply shocks, particularly for biomass supply. If only one region is affected by adverse weather or a pest, the combination of biorefinery feedstock flexibility and advanced biomass logistics might significantly dampen the net private welfare cost of the shock by substituting biomass from unaffected regions for the affected crops.

Another reason that shocks' benefits are small is the way benefits are calculated. In this study welfare value has not been ascribed to the reduction of fuel price shock size *per se*. The social surplus measure focuses on deadweight loss, and import costs, to measure U.S. net private welfare loss. A change in price from the shocks we model leads mostly to offsetting gains and losses for producers and consumers, or vice versa. When elasticities are small, the deadweight loss triangles are small. If petroleum imports are replaced by imported ethanol during a shock and elasticities are small, then the social surplus change may be very small.

Acknowledgments

This work has been supported by the U.S. Department of Energy's Bioenergy Technologies Office. Oak Ridge National Laboratory is managed by UT-Battelle, LLC, for DOE under contract DE-AC05-00OR22725.

References

1. Bailis, R. E., B. S. Koeb, and M. W. Sanders. *Reducing Fuel Volatility - An Additional Benefit from Blending Biofuels?* Discussion Paper Series/Tjalling C. Koopmans Research Institute, 2011, 11(01).
2. Schlenker, W., and M. J. Roberts. Nonlinear Temperature Effects Indicate Severe Damages to US Crop Yields Under Climate Change. *Proceedings of the National Academy of Sciences*, Vol. 106, No. 37, 2009, pp. 15594–15598.
3. Efroymson, R. A., M. H. Langholtz, K. E. Johnson, and B. J Stokes. *2016 Billion-Ton Report: Advancing Domestic Resources for a Thriving Bioeconomy, Volume 2: Environmental Sustainability Effects of Select Scenarios from Volume 1*. ORNL/TM-2016/727. Oak Ridge National Laboratory, 2017. http://dx.doi.org/10.2172/1338837.
4. Uría-Martínez, R., P. N. Leiby, and M. L. Brown. *BioTrans Model Documentation*. May 2017. https://www.researchgate.net/publication/318837111_BioTrans_Model_Documentation. Accessed August 1, 2017.
5. Golecha, R., and J. Gan. Cellulosic Biorefinery Portfolio and Diversification: Strategies to Mitigate Cellulosic Biorefinery Risks in US Corn Belt. *Energy Strategy Reviews*, Vol. 13–14, 2016, pp. 147–153.
6. Thompson, W., S. Meyer, and P. Westhoff. How Does Petroleum Price and Corn Yield Volatility Affect Ethanol Markets With and Without an Ethanol Use Mandate? *Energy Policy*, Vol. 37, 2009, pp. 745–749.
7. Christensen, A., and S. Siddiqui. A mixed complementarity model for the US biofuel market with federal policy interventions. *Biofuels, Bioproducts, and Biorefining*, Vol. 9, No. 4, 2015, pp. 397–411.
8. Irwin, S., and D. Good. The Competitive Position of Ethanol as an Octane Enhancer. February 2016. http://farmdocdaily.illinois.edu/2016/02/ethanol-position-as-octane-enhancer.html. Accessed August 1, 2017.
9. Langholtz, M. H., B. J. Stokes, and L. M. Eaton. *2016 Billion-Ton Report: Advancing Domestic Resources for a Thriving Bioeconomy, Volume 1: Economic Availability of Feedstocks*. ORNL/TM-2016/160. Oak Ridge National Laboratory, 2016. http://dx.doi.org/10.2172/1271651.
10. Eranki, P. L., B. D. Bals, and B. E. Dale. Advanced Regional Biomass Processing Depots: A Key to the Logistical Challenges of the Cellulosic Biofuel Industry. *Biofuels, Bioproducts, and Biorefining*, Vol. 5, No. 6, 2011, pp. 621–630.

The Standing Committee on Alternative Transportation Fuels and Technologies (ADC80) peer-reviewed this paper (18-05310).

Article

TRR
JOURNAL OF THE TRANSPORTATION RESEARCH BOARD

Transportation Research Record
2018, Vol. 2672(24) 41–48

DOI: 10.1177/0361198118798286
journals.sagepub.com/home/trr
SAGE

Trip Energy Estimation Methodology and Model Based on Real-World Driving Data for Green-Routing Applications

Jacob Holden[1], Harrison Van Til[1], Eric Wood[1], Lei Zhu[1], Jeffrey Gonder[1], and Matthew Shirk[2]

Abstract

A data-informed model to predict energy use for a proposed vehicle trip has been developed in this paper. The methodology leverages roughly one million miles of real-world driving data to generate the estimation model. Driving is categorized at the sub-trip level by average speed, road gradient, and road network geometry, then aggregated by category. An average energy consumption rate is determined for each category, creating an energy rate look-up table. Proposed vehicle trips are then categorized in the same manner, and estimated energy rates are appended from the look-up table. The methodology is robust and applicable to a wide range of driving data. The model has been trained on vehicle travel profiles from the Transportation Secure Data Center at the National Renewable Energy Laboratory and validated against on-road fuel consumption data from testing in Phoenix, Arizona. When compared against the detailed on-road conventional vehicle fuel consumption test data, the energy estimation model accurately predicted which route would consume less fuel over a dozen different tests. When compared against a larger set of real-world origin–destination pairs, it is estimated that implementing the present methodology should accurately select the route that consumes the least fuel 90% of the time. The model results can be used to inform control strategies in routing tools, such as change in departure time, alternate routing, and alternate destinations to reduce energy consumption. This work provides a highly extensible framework that allows the model to be tuned to a specific driver or vehicle type.

The transportation sector accounts for 28% of total energy consumption in the United States and 72% of petroleum use (*1*), resulting in a third of all greenhouse gas emissions (*2*). To increase the energy efficiency of such a large sector, new technologies and models informed by real-world driving data are necessary. Optimal energy use varies by powertrain and level of vehicle automation. A method to accurately predict the energy required for individual trips is a fundamental necessity to identify optimal solutions for vehicle, trip planning, and routing decisions (*3*). Energy estimation for proposed routes has direct application in the green-routing and mobility planning areas. Accurate trip energy estimation can also be applied to regional- or national-level transportation energy analyses, where trips (i.e., origin–destination pairs) are known, but real and detailed driving data are unavailable.

Rapid global urbanization has led to increased mobility demands, emissions production, and energy consumption being concentrated in urban cores (*4*). The fundamental step to making progress on transportation energy consumption is to accurately understand energy demand and to design optimal mobility options. Energy modeling has limitations in its traditional structure. Typically energy models are considered to be macroscopic, mesoscopic, or microscopic in scale (*5*). Some models specialize in a particular scope, and others, such as the U.S. Environmental Protection Agency's MOtor Vehicle Emissions Simulator (MOVES), provide settings for the user to run models with any of the three scopes (*6*). There are very few or no models that can accurately predict energy consumption at a large regional level or at the individual vehicle level where only low-resolution data (origin, destination, and routing only) are available.

At the individual vehicle level, progress has been made in quantifying how various road network and driving

[1]National Renewable Energy Laboratory, Golden, CO
[2]Idaho National Laboratory, Idaho Falls, ID

Corresponding Author:
Address correspondence to Jacob Holden: Jacob.Holden@nrel.gov

attributes affect vehicle energy consumption. Several authors have demonstrated energy and emissions sensitivities to attributes such as road grade, vehicle speed, and congestion (*7–11*). Understanding these attributes has informed some previous energy estimation models for various applications (*12–14*). One application is for battery electric vehicles. In battery electric vehicle operation, predicted range and battery state of charge are two of the most important parameters that are presented to drivers for making decisions about the trips they can make before having to recharge (*12*). The estimation methodology typically uses a regression model based on a set of on-road test data or an analytical model that consists of simplified governing equations. These models are often specialized for a particular vehicle type and set of road/driving conditions. There is a need to predict trip energy consumption for different vehicle powertrains and driving behaviors because the vehicle fleet and driver population are so diverse in energy use characteristics.

The capability presented here estimates energy consumption for vehicle trips that are proposed but where the vehicle has not yet driven (and hence detailed driving data over the trip are not available). The method trains on real-world driving data and is relevant for a variety of data sets and applications. A complete model is presented and validated here, but the methodology itself is also an important result of this work.

Data Set

The approach requires a large quantity of real-world driving data to develop a robust model. The source is a collection of vehicle global positioning system (GPS) data from across the United States. The survey data are made available in the Transportation Secure Data Center (TSDC) at the National Renewable Energy Laboratory (NREL) (*15*). The TSDC contains about 150 million vehicle GPS points collected at 1-Hz frequency. The data used for this study are summarized in Table 1.

Methodology

Data at the scale of the TSDC require efficient handling and a suite of tools to pre-process the raw data into useful trips or drive cycles. Once refined drive cycles are obtained, they are provided as input to a powertrain simulation to append simulated energy consumption results to the data. These energy use results are the building blocks of the energy estimation methodology. The energy use data are reduced into different categories of driving by speed, congestion levels, grade, and acceleration trends. The initial data set must be large enough that each category, or bin, contains a representative distribution of driving. Then, the average energy demands for each of these driving bins are determined.

Table 1. Summary of the TSDC Drive Cycles Used to Inform the Energy Estimation Model

Feature	Value
Number of trips	185,588
Number of drivers	10,417
Number of records	133,914,065
Reporting frequency	1 Hz

Data Pre-Processing

A typical GPS point from a household travel survey contains unique identifiers, a timestamp, longitude/latitude coordinates, and instantaneous speed. This information is valuable; however, it is also imperfect. GPS loggers have a certain level of error in their spatial location and speed reporting. The loggers also stop reporting below a certain speed threshold (typically 0.5 mile per hour [mph]). For these reasons, the spatial and speed data are filtered using NREL's drive cycle processing method (*16*). The drive cycle filters identify and correct signal dropout, unrealistic accelerations, time gaps, and other issues in the time series data. The processing also segments raw point data into trips and micro trips. Figure 1*a* illustrates a filtered and an unfiltered speed profile.

Once the GPS data are "clean," elevation and road grade are appended to the drive cycles. Grade has a substantial impact on energy use and is a critical input to the powertrain simulation. Another NREL standard procedure is applied to derive and filter road grade for vehicle GPS data (*17*). There is some necessary filtering and smoothing of the grade profiles as well. An example of the filtered results is shown in Figure 1*b*; potential bridge locations and noise in the raw signals can be seen in the example.

Finally, the GPS data are map matched to a digital road network. To classify driving by various road network-related attributes, the geospatial points must be stitched to the road network. Each point is assigned a road network link that the vehicle is (likely) traveling on. An NREL-developed map-matching method that has been shown to have high accuracy (*18*) is used to map the GPS data to TomTom's Multinet road network (*19*). A trip segment of GPS points that have been map matched to the road network is shown in Figure 1*c*.

Powertrain Simulation

After the drive cycles are pre-processed, they become acceptable inputs to the Future Automotive Systems

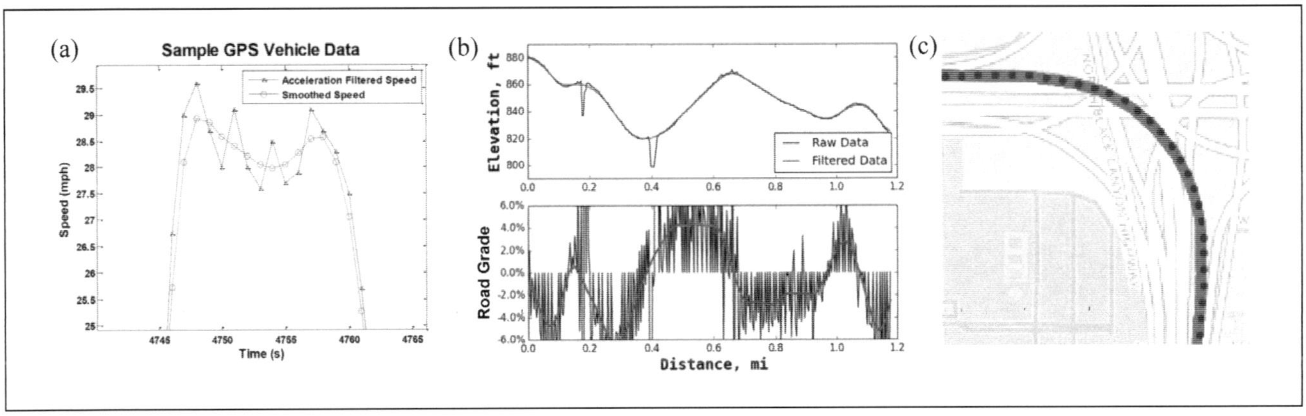

Figure 1. (*a*) Acceleration filter applied to a speed profile, points are smoothed to a more realistic time series (*16*); (*b*) raw and filtered elevation and grade data illustrate how noisy the raw grade signal can be and why it must be filtered (*17*); and (*c*) raw GPS points (blue) plotted on top of the road network link geometries that they are matched to (red).

Technology Simulator (FASTSim) vehicle powertrain simulation model (*20*). For a given vehicle model, FASTSim returns calculated energy consumption at each point in a drive cycle. These 1-Hz results are aggregated to the road network at the link level. Energy consumption values for all points on the same link are summed. These link-level FASTSim results are referred to as "passes." A complete trip drive cycle consists of many passes. The full data set consists of about 13 million passes. Passes have additional attributes joined to them from the road network, such as speed, grade, and link orientation to neighboring links. The pass-level energy results and attributes form the input data for generating the energy estimation model.

Model Generation

The accuracy of the model depends on the road and driving attributes available by which the data are aggregated. One of the strengths of this methodology is the flexibility to add and subtract attributes in the model to accommodate particular data sets or desired levels of accuracy. Additional attributes tend to increase accuracy but reduce the amount of data in each bin (which can create issues with too small sample size). Reducing the number of attributes allows a robust model to be generated from a smaller data set, but also reduces accuracy.

The selection of attributes to include in model generation is critical. Ideal attributes show a distinct correlation to vehicle fuel consumption rate and a low covariance with other attributes. Covariant attributes are to be avoided because they will unnecessarily reduce the amount of driving data in each bin while their effects may already be captured by other attributes. Attribute selection is an iterative process and must be performed by exploring the correlation of each attribute to others and to vehicle fuel economy.

For the model generated in this work, three attributes are selected for binning the pass-level results. The attributes are vehicle average speed, link average road grade, and orientation of the previous link to the current link. The dependence of vehicle energy consumption rate on each of these attributes is shown in Figure 2. A next-link orientation plot is also included to demonstrate why certain attributes are excluded from the model (i.e., fuel consumption did not vary significantly over this attribute). This is not an exhaustive list of attributes that were investigated and excluded.

Vehicle speed is a complex attribute because average speed over a pass is known from the GPS data; however, it cannot be known for each pass before a route is driven. The GPS average speed was used in model generation because it will generate the most accurate model, then on the implementation side, care must be taken to obtain speed predictions for each link pass in a proposed route. Vehicle speed has been identified previously as having a significant impact on fuel economy (*10*) and is again shown for this data set in Figure 2*a*. Bin widths of 10 mph, ranging from 0 to 80 + mph, were selected for the speed binning to strike a balance between resolution and data saturation in each bin.

Road grade has also been shown to be closely correlated to vehicle fuel economy (*17*). Grade contributes to the road load a vehicle experiences and the power that must be supplied to overcome it. Small variations in grade can result in significantly different power demands, which impact the energy consumption rate. This effect is shown in Figure 2*b*. As the grade effect is so pronounced, it is a crucial attribute to consider in the energy estimation model generation.

In addition to speed and grade influences on vehicle road load, acceleration is also an important factor. The geometry of the road network along a particular route

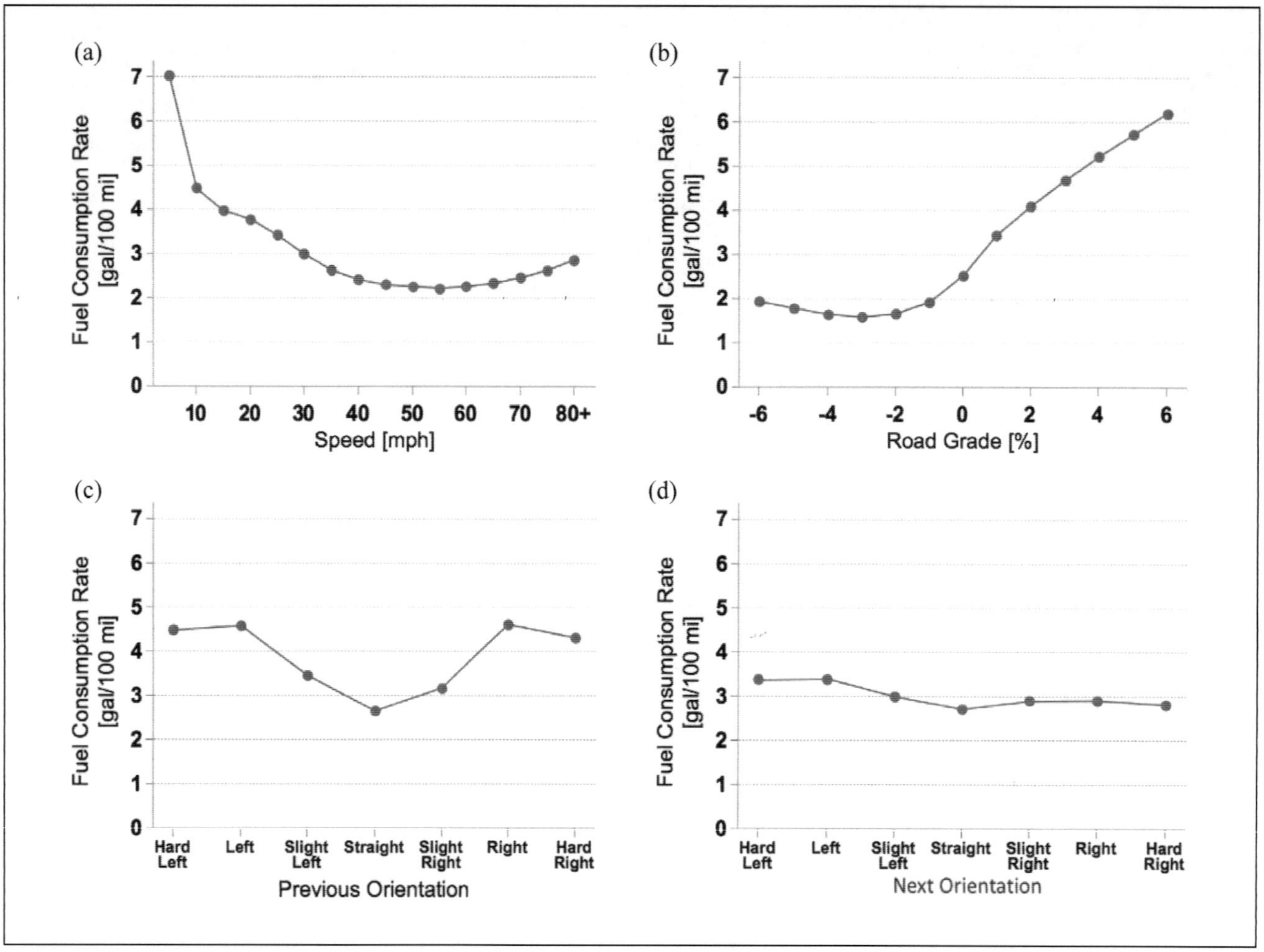

Figure 2. Driving and road network attributes plotted against vehicle fuel use rate for a conventional vehicle to determine the most important attributes to include in the energy estimation model. (*a*) Vehicle average speed over a pass vs. fuel use rate; (*b*) link average road grade vs. fuel use rate; (*c*) severity of turn from the previous link to the current vs. fuel use rate on the current link; and (*d*) severity of turn from the current link to the next link vs. fuel use rate on the current link.

influences a vehicle's acceleration profile. For example, a vehicle coming up on a right-hand turn with no signal and no congestion will decelerate from the free-flow speed, likely to 10 mph or less, execute the turn, then accelerate back up to the free-flow speed. This pattern has significant and predictable energy implications that vary by powertrain but will consistently result in different energy consumption than if a vehicle were cruising on a straight road segment.

The correlations of turn severity onto the current link and onto the proceeding link to vehicle energy consumption rate are shown in Figure 2, *c* and *d*. The previous orientation trends can be explained: if a vehicle has just executed a turn, fuel consumption rate will be higher because the driver is accelerating back up to traffic speed. One hypothesis for the minimal effect seen in the next-orientation plot is that the fuel consumption impacts of slowing while approaching a turn are much less significant than those from accelerating out of a turn on the next road link. The slightly higher fuel consumption shown in Figure 2*d* on links heading into a left turn versus a right turn may be caused by longer idle durations to complete left turns. This subtle effect may be worth including in future iterations of the energy estimation model, but for simplicity in this iteration the next-orientation attribute is excluded given its much smaller impact relative to the other three attributes.

The actual model generation results combine each of the identified attributes to create a multidimensional look-up table of various attribute combinations and the associated energy consumption rate for each combination. The result is a prediction model informed by a large quantity of real-world driving data, which can estimate the energy requirements of a trip with little information.

In summary, the prediction model is built by first categorizing the entire real-world data set and FASTSim

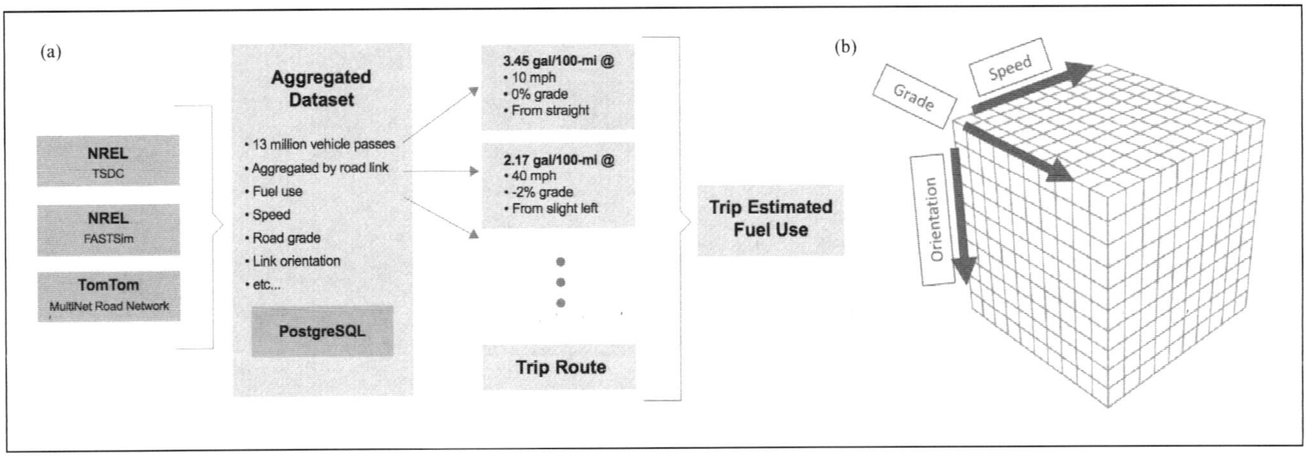

Figure 3. (*a*) From left to right: the data pre-processing and powertrain simulation steps, the results of which are aggregated into a large data set from which an energy estimation look-up for that particular vehicle model can be generated. Trip fuel use is estimated by combining the model results with a proposed trip route; (*b*) Illustration of the multidimensional energy rates look-up table generated from the categorized sub-trips. The examples in the third column of diagram (*a*) would each represent one cell in (*b*).

fuel use by select attributes, then calculating the average fuel use for each category to use as the characteristic fuel consumption rate for that category of driving. A schematic of the entire model generation methodology is shown in Figure 3. The figure also illustrates the resulting multidimensional look-up table. Once this prediction model is built, it can be deployed to estimate fuel use for a particular trip by categorizing each road link in the trip route, looking up the estimated fuel consumption for each road link in the multidimensional table, and summing the results over the entire trip.

Results

In an effort to validate the energy estimation method in a real-world application, an on-road testing program was undertaken in Phoenix, Arizona, in early 2017. For each phase of testing, two identical vehicles simultaneously drove from the same origins to the same destinations at various times of day over different routes. Three different vehicle pairs were tested, including a conventional vehicle (CV), hybrid electric vehicle (HEV), and plug-in hybrid electric vehicle (PHEV). This paper focuses on the development and validation of the energy estimation model for the CV from this testing—a 2014 Mazda 3 with start/stop capability.

The test plan for the two identical Mazda 3 CVs consisted of 430 total miles of driving. The vehicles were instrumented to acquire 10-Hz spatial, telematics, atmospheric, and fuel consumption data. The test matrix included a series of trip pairs both outbound from and returning inbound to downtown Phoenix. Each day, both vehicles ran two inbound-outbound sets. For each trip, one vehicle drove a route over surface streets, and the other drove a primarily highway route. The matrix resulted in 24 total trips for each vehicle type over three days.

A FASTSim powertrain model for the CV was first calibrated to dynamometer test data (*21*). The model was then used to generate an energy estimation model per the method outlined in the previous section by simulating all the driving contained in the TSDC data set. Figure 4 visualizes the results of the energy rates look-up table for the test vehicle. The series of fuel consumption maps show the distinct relationships between road grade, vehicle speed, road link orientation, and fuel consumption rate. The energy estimation model was then used to estimate energy consumption for the on-road test trips where fuel consumption data had been acquired. This allowed for a direct validation of the accuracy of the estimation model to the 24 test trips with vehicle-reported fuel consumption. For comparison, the accuracy of the FASTSim model for making route comparisons was also evaluated (though this relied on the actual vehicle driving profiles over the routes, which would not in practice be available before completing the trip).

As the primary application of the energy estimation methodology is green-routing, the most critical performance metric is the relative error in estimating trip-level energy differences between routes. The structure of the on-road test program allowed for straightforward calculation of relative energy cost/savings between the highway- and surface-dominant routes for each trip. The relative trip energy estimation error is defined below by Equation 1:

$$\mathit{Trip\ Error} = \frac{\left(\left(\frac{A}{B}\right)_{\text{model}} - \left(\frac{A}{B}\right)_{\text{measured}}\right)}{\left(\frac{A}{B}\right)_{\text{measured}}} \tag{1}$$

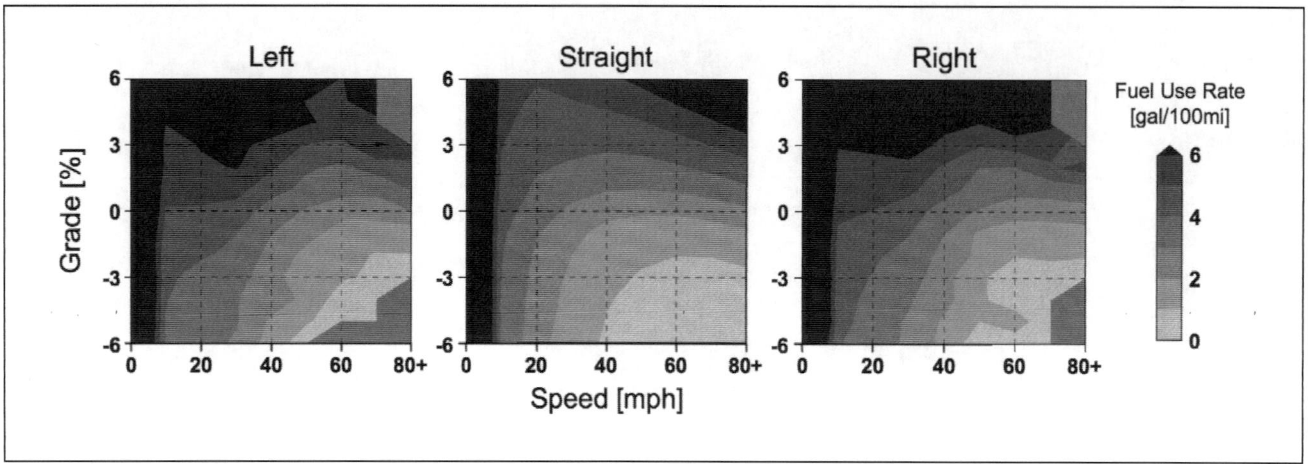

Figure 4. The 2D contour plots show the speed, road grade, and fuel use rate relationships for three previous link orientation bins for the Mazda 3 conventional vehicle model. "Left" indicates a turn in the range of (−100, −80) degrees, "Straight" indicates (−10, 10) degrees, and "Right" indicates (80, 100) degrees.

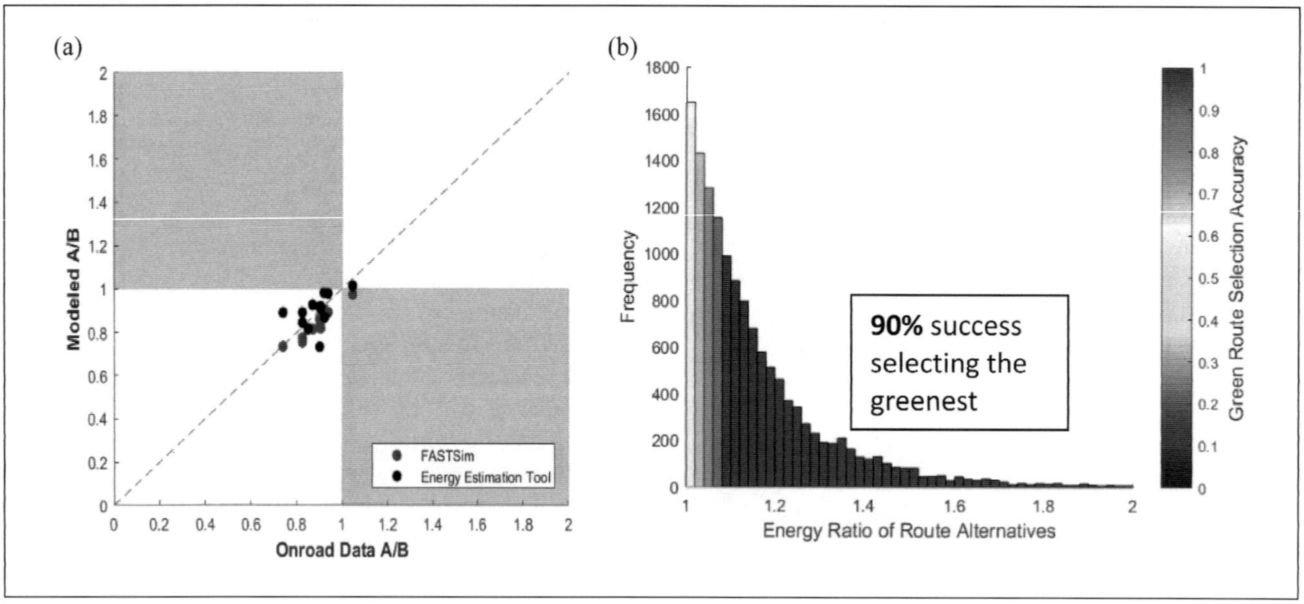

Figure 5. (*a*) The plot highlights regions that indicate correct (white) and incorrect (red) selection of the least energy-consuming route option by the FASTSim (blue) and energy estimation (black) models; (*b*) Estimated success rate identifying the greenest route based on the RMSE from on-road validation of the refined energy estimation model. The distribution demonstrates that the model will accurately select the least energy-consuming route for 90% of real-world trips.

A = surface route;

B = highway route;

FS = FASTSim model (if used directly with actual second-by-second profile);

EE = energy estimation model;

RMSE = root mean squared error over all 12 on-road trip sets;

$\frac{A}{B} > 1$ = highway is "greener" route; and

$\frac{A}{B} < 1$ = surface is "greener" route.

The trip relative energy estimation RMSE was calculated for the FASTSim model and the energy estimation model. The FASTSim results establish a baseline performance for using a validated powertrain model with the actual on-road speed profiles. In contrast, the energy estimation model can be applied without the real on-road driving data. The RMSE value for the FASTSim model is 5.9% and for the energy estimation model is 9.1%. Figure 5 helps to facilitate interpretation of these results and to place them in context. Figure 5*a* illustrates greater scatter in the measured versus estimated route fuel consumption ratios for the energy estimation model than for the FASTSim model, but that the energy estimation model nevertheless accurately selected the least fuel-consuming route option 100% of the time. This plot also shows that the energy consumption differences between the measured route pairs varied from roughly no difference to over 20% difference. Figure 5*b* examines how the model would be expected to perform over a much larger-scale real-world data set. A subset (about 14,000 trips) of real-world data from the TSDC was used in combination with an application programming interface green-routing approach to predict the least energy-consuming route in real-world situations (*3*). Using the RMSE results for the energy estimation model, a probability function is established and demonstrates an expected 90% success of the energy estimation model selection the least energy-consuming route. Note that most trips within this larger real-world sample show energy consumption differences between alternate routes in a similar range as the smaller-scale instrumented vehicle testing conducted for this project, but that there are also route alternatives in the larger set of real-world origin–destination pairs that show a fuel-saving potential greater than 20%.

Conclusion

A robust, data-driven vehicle trip energy estimation technique has been described, demonstrated, and validated. The energy estimation technique is computationally inexpensive to apply, and in addition, the methodology for generating the model is generally agnostic to powertrain type. The model has demonstrated an accurate estimation of trip energy consumption for a CV powertrain, with an RMSE of 9.1% when benchmarked against calibrated on-road fuel consumption data. Additional on-road data for HEV and PHEV powertrains are available for an identical validation exercise, which is planned for completion and publication as part of ongoing work on this topic. Other recommendations for future work include completion of additional on-road testing in a wider variety of driving conditions and climates. Testing for the current effort was conducted in Phoenix, Arizona, for convenience, but testing in additional locations would obviously help explore robustness of the methodology over a wider variety of ambient conditions.

From the literature reviewed, this energy estimation technique is the first of its kind to be generated from such a large quantity of real-world driving data. The evaluation completed thus far indicates that the method is robust (e.g., the method accurately selected the greener of the two route alternatives over all of the CV testing that was completed). The contributions of this energy estimation technique are two-fold. One is a working model that can accurately estimate trip fuel consumption and can be expanded to a broad range of powertrain types. The second is a general methodology to use real-world GPS trajectories to train an energy estimation model for various mobility choices at various project scopes. The methodology is highly extensible—it could generate a model by aggregating simulation data by various explicit attributes, or it could employ more advanced machine learning techniques to infer attributes beyond those that can be explicitly pulled from the road network, driver or vehicle information, to inform vehicle energy consumption. The potential result of using advanced techniques would be to generate a more accurate energy estimation model while maintaining a major benefit of this methodology—namely, that deploying the model requires only a simple, computationally inexpensive look-up process.

Acknowledgments

This work was sponsored by the U.S. Department of Energy's Vehicle Technologies Office under the Systems and Modeling for Accelerated Research in Transportation (SMART) Mobility Laboratory Consortium, an initiative of the Energy Efficient Mobility Systems Program. The authors acknowledge Eric Rask of Argonne National Laboratory for leading the Connected and Automated Vehicles Pillar of the SMART Mobility Laboratory Consortium. Program management provided by David Anderson within the DOE Office of Energy Efficiency and Renewable Energy was important for establishing the project concept, advancing implementation, and providing ongoing guidance. The authors also acknowledge Matt Moniot and other TSDC researchers for their support with data visualization.

Author Contributions

The authors confirm contribution to the paper as follows: study conception and design: Wood, Gonder, Holden, Zhu; data collection: Shirk, Gonder, Holden, Zhu; analysis and interpretation of results: Holden, Van Til, Gonder; draft manuscript preparation: Holden, Van Til. All authors reviewed the results and approved the final version of the manuscript.

References

1. Davis, S. C., S. E. Williams, and R. G. Boundy. *Transportation Energy Data Book*, lst ed. Oak Ridge National Laboratory, Oak Ridge, TN, 2016.
2. Gearhart, C. Implications of Sustainability for the United States Light-Duty Transportation Sector. *MRS Energy & Sustainability: A Review Journal*, Vol. 3, 2016, p. E8. https://doi.org/10.1557/mre.2016.8
3. Zhu, L., J. R. Holden, J. D. Gonder, and E. Wood. Green Routing Fuel Saving Opportunity Assessment: A Case Study on California Large-Scale Real-World Travel Data. Presented at the 28th IEEE Intelligent Vehicles Symposium,

Redondo Beach, CA. National Renewable Energy Laboratory (NREL), Golden, CO, 2017, NREL/CP-5400-68194.
4. Sarkar, R., and J. Ward. DOE SMART Mobility: Systems and Modeling for Accelerated Research in Transportation. In *Road Vehicle Automation 3* (Meyer, G., and S. Beiker, eds.), Springer International Publishing, New York City, NY, 2016, pp. 39–52.
5. D'Orey, P. M., and M. Ferreira. ITS for Sustainable Mobility: A Survey on Applications and Impact Assessment Tools. *IEEE Transactions on Intelligent Transportation Systems*, Vol. 15, No. 2, 2014, pp. 477–493. https://doi.org/10.1109/TITS.2013.2287257
6. U.S. Environmental Protection Agency. *Greenhouse Gas and Energy Consumption Rates for On-Road Vehicles: Updates for MOVES2014* (EPA-420-R-15-003, October 2015). U.S. Environmental Protection Agency, Washington D. C., 2014.
7. Wood, E., E. Burton, A. Duran, and J. Gonder. *Contribution of Road Grade to the Energy Use of Modern Automobiles Across Large Datasets of Real-World Drive Cycles.* SAE Technical Paper 2014-01-1789, January 2014, p. 9. https://doi.org/10.4271/2014-01-1789.Abstract
8. Park, S., and H. Rakha. Energy and Environmental Impacts of Roadway Grades. *Transportation Research Record: Journal of the Transportation Research Board*, 2006. 1987: 148–160.
9. Boriboonsomsin, K., and M. Barth. Impacts of Road Grade on Fuel Consumption and Carbon Dioxide Emissions Evidenced by Use of Advanced Navigation Systems. *Transportation Research Record: Journal of the Transportation Research Board*, 2010. 2139: 21–30.
10. Barth, M., and K. Boriboonsomsin. Real-World Carbon Dioxide Impacts of Traffic Congestion. *Transportation Research Record: Journal of the Transportation Research Board*, 2008. 2058: 163–171.
11. Frey, H. C., K. Zhang, and M. N. Rouphail. Fuel Use and Emissions Comparisons for Alternative Routes, Time of Day, Road Grade, and Vehicles Based on In-Use Measurements. *Environmental Science and Technology*, Vol. 42, No. 7, 2008, pp. 2483–2489. https://doi.org/10.1021/es702493v
12. Wu, X., D. Freese, A. Cabrera, and W. A. Kitch. Electric Vehicles' Energy Consumption Measurement and Estimation. *Transportation Research Part D: Transport and Environment*, Vol. 34, 2015, pp. 52–67. https://doi.org/10.1016/j.trd.2014.10.007
13. Kraschl-Hirschmann, K., and M. Fellendorf. Estimating Energy Consumption for Routing Algorithms. *Proc., 2012 IEEE Intelligent Vehicles Symposium*, Alcala de Henares, Spain, 2012, pp. 258–263. https://doi.org/10.1109/IVS.2012.6232127
14. Lin, Z., and D. Greene. Predicting Individual Fuel Economy. *SAE International Journal of Fuels and Lubricants*, Vol. 4, 2011, pp. 84–95. https://doi.org/10.4271/2011-01-0618
15. National Renewable Energy Laboratory. *Transportation Secure Data Center*. National Renewable Energy Laboratory, Golden, CO, 2017.
16. Duran, A., and M. Earleywine. *GPS Data Filtration Method for Drive Cycle Analysis Applications*. SAE Technical Paper, 2012, pp. 1–743. https://doi.org/10.4271/2012-01-0743
17. Wood, E., E. Burton, A. Duran, and J. Gonder. *Appending High-Resolution Elevation Data to GPS Speed Traces for Vehicle Energy Modeling and Simulation*. Technical Report NREL/TP-5400-61109. National Renewable Energy Laboratory, Golden, CO, 2014.
18. Zhu, L., J. R. Holden, and J. D. Gonder. A Trajectory Segmentation Map-Matching Approach for Large-Scale, High-Resolution GPS Data. *Transportation Research Record: Journal of the Transportation Research Board*, 2016. 2645: 67–75.
19. TomTom. *Delivering a Global Map Database to Power a Wide Range of Location-Based Applications*. TomTom N.V., The Netherlands, 2016.
20. Brooker, A., J. Gonder, L. Wang, E. Wood, S. Lopp, and L. Ramroth. *FASTSim: A Model to Estimate Vehicle Efficiency, Cost and Performance*. SAE Technical Paper 2015-01-0973, 2015. https://doi.org/10.4271/2015-01-0973
21. Advanced Powertrain Research Facility. https://www.anl.gov/energy-systems/group/downloadable-dynamometer-database/conventional-vehicles/2014-mazda-3-ieloop, 2002. Accessed: June, 10, 2017.

The Standing Committee on Transportation Energy (ADC70) peer-reviewed this paper (18-05696).

Article

TRR
JOURNAL OF THE TRANSPORTATION RESEARCH BOARD

Transportation Research Record
2018, Vol. 2672(24) 49–59

DOI: 10.1177/0361198118798461
journals.sagepub.com/home/trr
SAGE

Big Data Fusion to Estimate Urban Fuel Consumption: A Case Study of Riyadh

Adham Kalila[1], Zeyad Awwad[2], Riccardo Di Clemente[1,3], and Marta C. González[1,4,5]

Abstract

Falling oil revenues and rapid urbanization are putting a strain on the budgets of oil-producing nations, which often subsidize domestic fuel consumption. A direct way to decrease the impact of subsidies is to reduce fuel consumption by reducing congestion and car trips. As fuel consumption models have started to incorporate data sources from ubiquitous sensing devices, the opportunity is to develop comprehensive models at urban scale leveraging sources such as Global Positioning System (GPS) data and Call Detail Records. This paper combines these big data sets in a novel method to model fuel consumption within a city and estimate how it may change in different scenarios. To do so a fuel consumption model was calibrated for use on any car fleet fuel economy distribution and applied in Riyadh, Saudi Arabia. The model proposed, based on speed profiles, was then used to test the effects on fuel consumption of reducing flow, both randomly and by targeting the most fuel-inefficient trips in the city. The estimates considerably improve baseline methods based on average speeds, showing the benefits of the information added by the GPS data fusion. The presented method can be adapted to also measure emissions. The results constitute a clear application of data analysis tools to help decision makers compare policies aimed at achieving economic and environmental goals.

In many oil-producing countries with substantial fuel subsidies, a fall in oil revenue and increasing domestic consumption has put increasing strain on government budgets (*1*). Countries in the Gulf Cooperation Council, including Saudi Arabia and the UAE, have launched programs to reduce government expenditure on energy subsidies (*2*). In Saudi Arabia, energy subsidies are estimated at 9.3% of GDP, with 1.4% for petroleum subsidies alone. Decreasing energy subsides can be achieved in a number of ways, but congestion relief offers a simple and direct path to lower fuel consumption. As the burden of fuel subsidies continues to grow, it has become increasingly important for these countries to find simple and accurate methods to quantify the effects of policies on congestion relief and fuel consumption in cities. Recent technological advances in collecting and analyzing big data offer a potential method to measure how policy changes affect fuel consumption. With the advent of ubiquitous sensing devices, transport network companies (TNCs), and new methods of estimating flow, this paper proposes a method to answer such questions that can be further applied anywhere in the world and extended to model emissions and air pollution.

Call detail records (CDRs) produced passively by mobile phones represent a cutting-edge method to estimate travel demand. Most traffic studies currently use local and national household travel surveys to estimate the rate of trip production between different zones of the city, but such surveys are expensive to conduct and only cover a small sample of the population. Leveraging on previous work (*3–5*), CDRs can provide simple and effective methods for estimating origin–destination (OD) flows using location data collected from millions of individual mobile phone users.

In addition to CDRs, the Global Positioning System (GPS) function on smartphones offers a precise method to spatially and temporally track traffic conditions across a city. As smartphone market penetration is almost complete in the TNC industry (*6*), GPS tracking has been successfully used to estimate air pollution (*7*), instantaneous

[1]Department of Civil & Environmental Engineering, MIT, Cambridge, MA
[2]Center for Complex Engineering Systems, King Abdulaziz City for Science and Technology, Riyadh, Saudi Arabia
[3]The Bartlett Centre for Advanced Spatial Analysis (CASA), University College London, London, UK
[4]Department of City and Regional Planning, UC Berkeley, Berkeley, CA
[5]Lawrence Berkeley National Laboratory, Berkeley, CA

Corresponding Author:
Address correspondence to Adham Kalila: kalila@mit.edu

fuel consumption (*8–12*), and traffic conditions (*13–17*). Most studies that use GPS data to estimate fuel consumption have focused on individual user fuel consumption for route optimization (*10*, *12*). This paper employs the method described in Toole et al. (*4*) to model trip demand, then fuses this result with location data from GPS data to calculate fuel consumption during peak and off-peak hours in Riyadh with higher accuracy than models that use average speed.

Fuel consumption and emissions models have been extensively developed in the literature (*18–25*). They are generally split between models that estimate the fuel consumption by balancing the engine's carbon intake and combustion and those that attempt to use mode-specific variables, such as speed and acceleration, to fit a model that estimates fuel consumption (*24*). Of those models that use instantaneous mode-specific variables, some estimate air pollution and emissions (*7*, *19*, *20*, *22*), fuel consumption (*10–12*, *23*), or both (*8*, *9*, *21*). Most previous attempts at estimating instantaneous fuel consumption and emissions do not incorporate GPS data but rely instead on on-board diagnostics devices (OBD-II) that measure fuel consumption and emissions (*8–10*, *21*). The models that have attempted to use GPS data to estimate fuel consumption do so without consideration of the different fuel efficiencies found in today's cars (*11*, *12*, *23*) and do not account for the total demand.

With the aim of filling this gap, the contributions of this work are threefold: First, a previously developed fuel consumption model (StreetSmart) was calibrated and applied to varying fuel efficiencies in car fleets; second, the model is used along with travel demand estimated by CDR-based traffic assignment to approximate fuel consumption rates in Riyadh, Saudi Arabia; third, the paper examines the effects of the proposed method comparing the reduction on fuel consumption in two scenarios: random trip reduction and reduction targeted at the least fuel-efficient trips. The proposed models calibrated with speed and acceleration profiles were found to considerably differ in their fuel consumption estimates at the city scale when verified against estimations from a baseline model with average speeds.

The paper is structured as follows. The methodology used to calibrate the fuel consumption model (StreetSmart) is discussed, then the cleaning of the GPS data and the extraction of speed and acceleration profiles on each street in specific time windows in a typical week to represent different snapshots of traffic throughout the road network of Riyadh is described. Application of the model combined with flow data to visualize the fuel consumption across different time periods is then described, and the results of the targeted and random flow reductions and their effects on fuel consumption via the presented model vs. baseline estimates are presented. Finally the results and the conclusions derived from the study are discussed.

Methodology

StreetSmart Model Calibration

This work uses the StreetSmart model to estimate the fuel consumption of personal vehicles (*26*). It was comprehensively developed by measuring the energy required by vehicles for various movement conditions. It estimates the fuel consumption with data from GPS coordinates from smartphones and ground truth fuel consumption data from OBD-II devices. Using the details of a trip's speed profile, the model successfully predicts fuel consumption with over 96% accuracy (*26*), a substantial improvement over models that only consider constant average speeds. Average speed estimations do not account for the stop-and-go effect of traffic, which is a significant factor leading to an increase in fuel consumption. In other words, average speed simplification results in lower, or more optimistic, fuel consumption estimates because drag is lower at low speeds.

After testing different variables for their use in predicting fuel consumption, the model employs a combination of four variables to predict fuel consumption, as shown in Equation 1. The first term accounts for energy wasted while the car is idling with the engine turned on; the second accounts for energy used with time spent moving; the third accounts for energy used as a result of acceleration and deceleration over a distance; the fourth accounts for energy used with distance traveled. Each term is multiplied by a specific energy index, k_i, which depends on the vehicle efficiency, such that:

$$FC = k_1 \mathrm{T}_{\mathrm{idle}} + k_2 T_{\mathrm{move}} + k_3 \int |a| dx + k_4 L$$

where FC is fuel consumption in US gallon and k_1,k_2,k_3, and k_4 are the energy indices calibrated with data separately for each bin of vehicle efficiency. T_{idle} and T_{move} are time spent idling and moving respectively in seconds, a is acceleration in m/s^2, and L is the distance driven in km.

Following Oehlerking (*26*), this study further tested the StreetSmart model's indices by regression of the idle fuel consumption and moving fuel consumption separately using data from an experiment conducted at the University of Illinois by Wu et al. (*27*). To verify the benefit of using the model, its results were compared with a baseline estimation using average speed and the United States' Department of Energy's (DOE) graph of fuel economy variations by speed, as shown in Figure 1. The StreetSmart model achieved inaccuracies of about 4% whereas the baseline method had inaccuracies of up to 29%. The details of the comparison are summarized in Table 1.

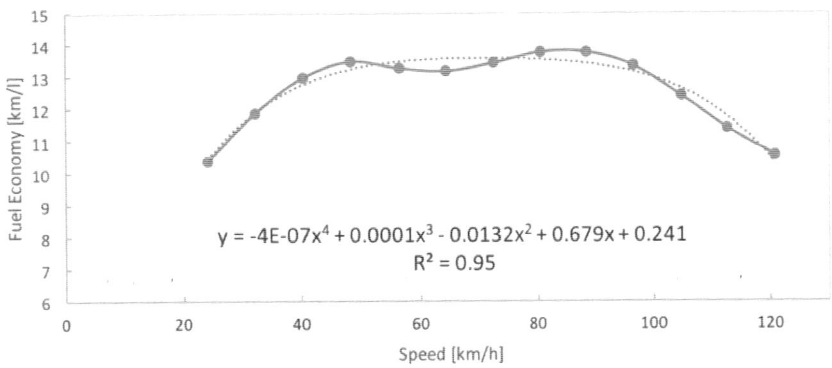

Figure 1. A fitted model of fuel economy vs. speed from DOE as a baseline comparison to the StreetSmart model (*28*).

Table 1. A Comparison of Fuel Consumption Estimates from the StreetSmart Model and the DOE Fuel Economy Fit on Data from the Experiment Conducted by Wu et al. (27)

Car No. (Illinois Test A):	6	7	8	9
OBD-II FC [US Gal]	0.0246	0.0220	0.0356	0.0211
DOE Fitted Curve [US Gal]	0.0247	0.0253	0.0252	0.0245
StreetSmart FC [US Gal]	0.0243	0.0230	0.0371	0.0210
% diff. StreetSmart	−1.2%	4.2%	4.2%	−0.6%
% diff. DOE Fitted Curve	0.4%	14.7%	−29.3%	16.0%

To apply the StreetSmart model on a known distribution of car fuel economies, the first step is to relate the energy indices specific to each car type in order to predict fuel consumption. A sensitivity analysis was conducted to assess the influence of each term on the estimates, and found that the fuel consumption estimates were most affected by the second and fourth terms. This indicates that the variables with the highest influence over fuel consumption were the time moving and the distance traveled. This suggests that the first and third terms, representing the influence of speed profiles, have little impact on the overall estimate of fuel consumption. However, results in the "Fuel Consumption Results" section show that this is not the case. To calibrate the model and obtain energy index values for use on the scale of a city, the fuel economies reported by the Environmental Protection Agency's (EPA) 2016 report were used to arrive at ranges of each index for different cars, categorized by their fuel economy (*29*). The EPA uses a standard speed profile to test for a car's urban fuel economy, the FTP-75, shown in Figure 2*a* (*30*). The study used the mode-specific variables from the FTP-75 speed profile with the reported fuel economy to calibrate the energy index ranges for each bin shown in Table 2. Using these ranges and the distribution of fuel economies found in the EPA report, the distribution of fuel economies was successfully recreated (shown in Figure 2*b*) with the StreetSmart model.

The energy index ranges for each bin are then used to recreate the fuel economy distribution of the fleet of cars in Riyadh using the car makes and models from motor vehicle crash statistics data provided by the city of Riyadh. Data on car crashes from January 2013 until October 2015 in the city of Riyadh were used as a proxy for Riyadh's fleet composition. The different energy indices, which produce different fuel consumption rates, are combined in proportion to Riyadh's fleet bin distribution. For validation, the fuel economy distribution of Riyadh's fleet was compared with two cities in Europe chosen on the basis of similar population size or gross domestic product, and the results show similar trends exist in the relative variety of fuel efficiencies in all cities. The data on fleet compositions of the two European countries were acquired from an in-depth study of the fleets of all European countries (*31*). As can be seen in Figure 2*c*, the comparison in the distributions of fleet fuel economies between the three areas indicates that the car fleet of Riyadh is less energy efficient those of Poland and the UK. The usage of country level fleet composition for Poland and the UK compared with city level for Riyadh represents a limitation in this comparison, but the results adequately verify the credibility of the fleet

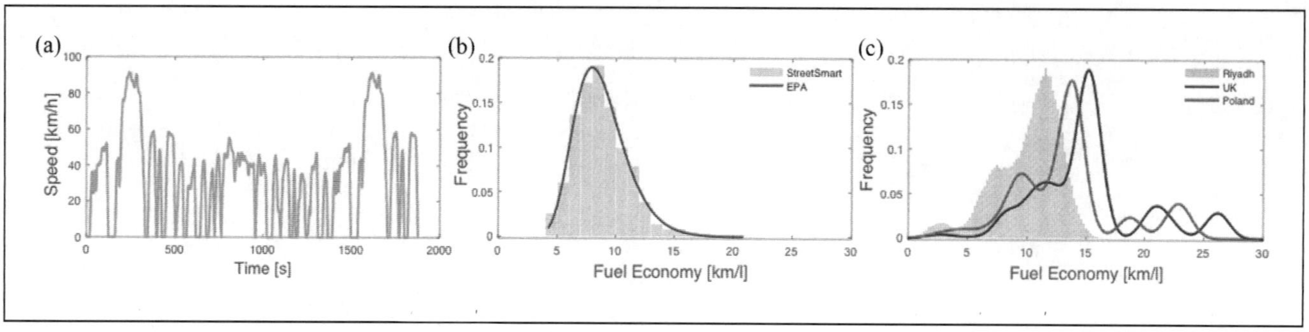

Figure 2. (*a*) FTP-75 EPA's standard speed profile used for calculating the reported inner-city fuel economies of cars. (*b*) The distribution of fuel economies recreated by the StreetSmart model shows the same distribution as that of the reported fuel economies. (*c*) The distribution of fuel economies based on Riyadh's fleet of cars compared with those of Poland and the UK shows that the distributions are similar but shifted from one another. The car fleet of Riyadh is less fuel efficient than that of Poland, which is less than that of the UK.

Table 2. Results of the Calibration of the StreetSmart Model. Ranges of *ki* Parameters for Each Bin of Fuel Economy

Bin	FE Range [MPG]	FE Range [km/l]	k1	k2	k3	k4
1	[10–12)	[4.25–5.10)	37	30–34	1–4.8	2000–2300
2	[12–14)	[5.10–5.95)	34.6	23–32	1–4.8	1300–2300
3	[14–16)	[5.95–6.80)	31.9	21–26	1–4.8	1100–1900
4	[16–18)	[6.80–7.65)	29.5	17.5–24	1–4.8	1000–1600
5	[18–20)	[7.65–8.50)	26.9	15–22	1–4.8	1000–1250
6	[20–22)	[8.50–9.35)	24.3	13–18	1–4.8	980–1250
7	[22–24)	[9.35–10.20)	21.7	12–16	1–4.8	850–1200
8	[24–26)	[10.20–11.05)	19	12–15	1–4.8	750–1050
9	[26–28)	[11.05–11.90)	16.3	11–14	1–4.8	780–900
10	[28–30)	[11.90–12.75)	14	10.5–12.5	1–4.8	710–900
11	>30	>12.75	5	5–14.5	1–4.8	500–1000
12 (Bus)	6.3	2.68	30.0–37.0	30–75	1–4.8	2000–800
13 (Truck)	17.27	7.34	29.0–30.0	12–27	1–4.8	500–2200
14 (Motorcycle)	43.5	18.49	5	6–10	1–4.8	500–600

composition used in this study. With the relative fuel economies of Riyadh's fleet and the calibrated StreetSmart model, this paper next discusses how speed profiles from GPS data were integrated to estimate fuel consumption at the urban scale.

From GPS Data to Speed Profiles

Speed profiles were extracted from a large dataset of GPS tracking points of taxi trips from a local Saudi Arabian TNC company over the period of May 2015 until December 2016. Speed profiles are the result of both driving styles and traffic conditions. Driving styles of taxis may be different from those of personal vehicles, affecting the results to a small extent, but it was assumed that speed profiles are mainly a reflection of traffic conditions, which would affect taxis and personal vehicle trips similarly. The dataset included trip duration and length, pick-up and drop-off times, and a chronologically ordered list of GPS coordinates. To ensure that the traffic is representative of year-round conditions, the rate of trip production during Ramadan of 2015 and 2016 was compared with non-Ramadan trip production rates. It was found that Ramadan trip production rates are much fewer so their impact on the average traffic speed profile for a specific street is negligible. For this reason, the Ramadan trips were retained in the analysis to benefit from the higher amount of data on the street level. A graph of the average number of trips per hour during Ramadan, non-Ramadan, and combined can be seen in Figure 3.

Before use, the data were filtered to remove trips that were outside Riyadh and the GPS routes were cleaned and modified to correct for measurement errors. Errors detected in the GPS points and were fixed using the following algorithm. As the GPS coordinates were given as an ordered list without a timestamp, and are known to be collected at regular intervals, the interval or frequency of recording was calculated as the total duration of the trip divided by the number of points recorded. Errors were detected as spikes in speed that are greater than 160 km/h, because the taxi fleet and the road conditions

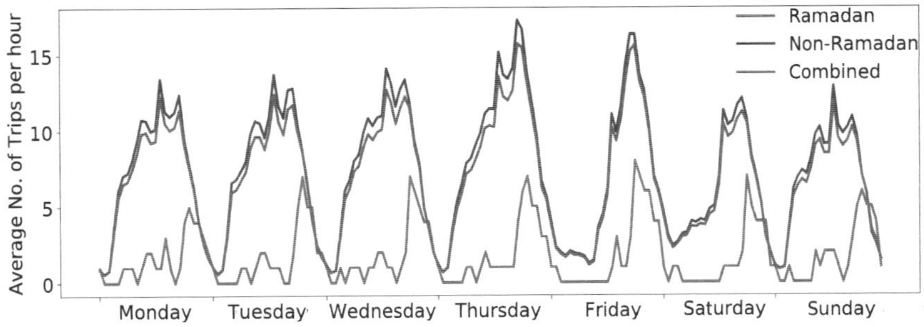

Figure 3. Average hourly taxi trip production rates in Riyadh in Ramadan, non-Ramadan, and combined.

typically preclude speeds above that. Two different causes of error were detected and repaired using different methods. In the first case, errors were caused by missing points resulting from a lack of network signal. This would result in a spike in speed for one segment but not the following segment, which would revert to realistic speeds and GPS points. The number of skipped points was estimated from the average of the speed before and after the single speed spike. Second, for errors caused by a GPS point that is in an obviously anomalous location, the speed spike occurs in two simultaneous segments, one to jump to the wrong location and another to return to the realistic location. This error was fixed by removing the erroneous point. This simple method was able to adequately correct the GPS coordinates.

Taxi data were also cleaned using an algorithm developed by Jiang et al. (*32*) to detect long periods of immobility, or stays, that can be interpreted as parking in the dataset. This may occur if a client asks to keep the meter running while they finish an errand. The reason for the splitting is to not affect the recorded speeds on the streets where the taxi is effectively parked. The minimum thresholds for idling time and distance were chosen by inspecting the distribution of stay durations throughout the week. A minimum value of 2,200s (around 37 min) was chosen to allow for the majority of traffic stays during peak and off-peak hours. A maximum distance for a trip to be split around a stay was 10 m. This number reflects the relatively high precision of the GPS points. In other words, a trip that remained within a 10 m radius for longer than 37 min was split into two trips, removing the 37 min or longer section of immobility.

After filtering and cleaning GPS routes for the city of Riyadh, nearly 43,000 trips were obtained that were analyzed by a custom mapping algorithm to assign full GPS routes to edges in the road network. For more accurate mapping, longer edges in the road network were split into shorter segments to ensure that nodes are no more than 10 m apart. The mapping algorithm was implemented using the following procedure:

1. For each point i in the GPS trajectory, identify the set of nodes (N_i) in the road network that fall within a 25 m radius.
2. Construct a path network G consisting of the nodes N_i.
3. For each point i in the GPS trajectory, the Dijkstra algorithm was used to find the fastest route from every node in N_i to every node in $N_i + 1$. For each route, a representative edge was added to G with the route's total travel time as the weight.
4. For each edge, a time penalty was added based on the distance of the target node to the original GPS coordinate at a rate of 1 second per additional meter past the closest node.
5. Any gaps in G were identified to determine contiguous sequences of paths that represent segments of potential routes.
6. For each contiguous sequence, the fastest path in G was identified as the most likely route taken by the vehicle.

For verification purposes, trip distances, free flow, and observed travel times as well as fuel economy estimates are plotted in Figure 4. Fuel economy is defined as the distance traveled per liter of fuel consumption. In Figure 4*a*, it was verified that the reported distances were generally consistent with the sum of the distances calculated between every two consecutive GPS points in each trip using the Haversine formula. In Figure 4*b* the free-flow times, computed as the sum of the free-flow times of every matched segment in each trip, are compared with the observed flow times as reported in the taxi data. The comparison only considered the observed total times from trips that were successfully matched with streets. The figure shows consistent results, with observed travel times during the morning peak hour being slower than the free-flow times of each trip. As a baseline comparison to using speed profiles from taxi data, fuel economy was calculated with the StreetSmart model assuming a

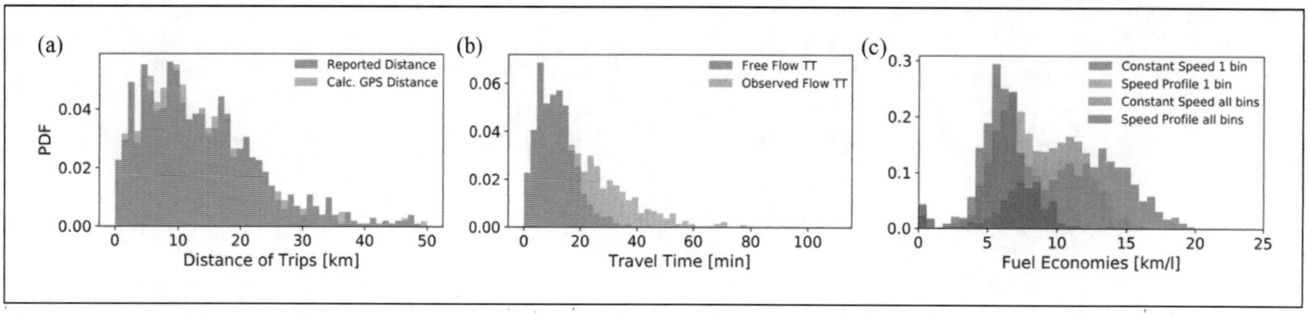

Figure 4. Data verification figures using trips in the morning peak time period of weekdays from 8–9 AM. (*a*) Histogram of reported and calculated trip distances. (*b*) Histogram of free-flow travel time and observed travel time in matched trips. (*c*) Histogram of fuel economies using constant speed, speed profiles, and 1 bin and all bins.

constant speed and one fuel economy bin based on the Hyundai Elantra, the most common car according to the accident data (Figure 4*c*). The results were plotted if speed profiles are used with only one fuel economy bin and if all bins are used in proportion to the fleet of Riyadh from the crash statistics data. This shows a very close distribution to the simpler assumption that all cars are the most common car fuel economy, and in high contrast to the result given by not using the speed profiles. Incorporating speed profiles in the model results in higher fuel consumption, or much lower fuel economy, which is more accurate and important for policy projections. This constitutes the core benefits of integrating the more accurate fuel consumption model.

Fuel Consumption Results

With the StreetSmart model calibrated and speed profiles extracted from GPS data, fuel consumption can now be estimated for three typical time periods representing distinct traffic conditions. The time periods are morning peak (0800–0900 h), midday off-peak (1200–1300 h), and evening peak (1700–1800 h) during weekdays. For every edge in the road network, fuel consumption per car was calculated based on each speed profile matched to that edge. The estimates were computed for each bin of fuel economy in the StreetSmart model. For each edge, the average fuel consumption rate per car was multiplied by the flow of cars per hour as computed by a version of the iterative traffic assignment (ITA) algorithm to obtain a fuel consumption rate per hour, with car volumes included.

The flow was calculated from OD matrices previously derived from CDR data in Riyadh by Toole et al. (*4*), in cars/hour over morning, midday, and evening time periods. Following Toole et al. (*4*), a factor of 1.5 was applied to the average morning flow of the morning and evening time periods to determine peak hour demand. The congested time estimates correlated successfully with the ones acquired from Google Maps (see comparisons in Chodrow et al. [*5*]).

Fuel Consumption Rate. As presented in Chodrow et al. (*5*), the ITA algorithm assigns trips in a series of four increments. It incrementally assigns 40%, 30%, 20%, and 10% of the OD flows to the fastest routes in the road network. After each iteration, the congested time of each edge is updated so that the effects of congestion can be factored into the assignment. The resulting flow volume is multiplied by the fuel consumption rate per car to arrive at hourly fuel consumption as shown in Equation 2, which is normalized by edge length.

$$FC = \frac{flow_e\left[\frac{car}{hr}\right] \times fcr_e\left[\frac{liter}{car}\right]}{L_e[m]}$$

where FC is the rate of fuel consumption in $\left[\frac{liter}{m.hr}\right]$, $flow_e$ is the flow on edge e as estimated by the assignment algorithm, fcr_e is the rate of fuel consumption per car on edge e as estimated by the application of the StreetSmart model, and L_e is the length of the edge in meters.

The results of the model are shown in the choropleth maps in Figure 5 below. For the fuel consumption rate per street, a weighted average of fuel consumption by bin proportional to Riyadh's fleet was used. The maps show that the most fuel-consuming streets are the grid highways of the city. As expected, the flow of cars and total fuel consumption per meter of road is found to be higher in the peak periods of the morning and evening than the midday off-peak. The evening peak shows slightly higher fuel consumption than the morning peak, indicated by more red streets. Quantile breaks on the fuel consumption rate values of all time periods were combined to display the streets that are the most fuel consuming. A high fuel consumption rate per meter of road is a result of high fuel consumption rate per car and high car flow values. King Fahd Road is by far the most fuel-consuming road in the city, especially in the area bounded by the old city

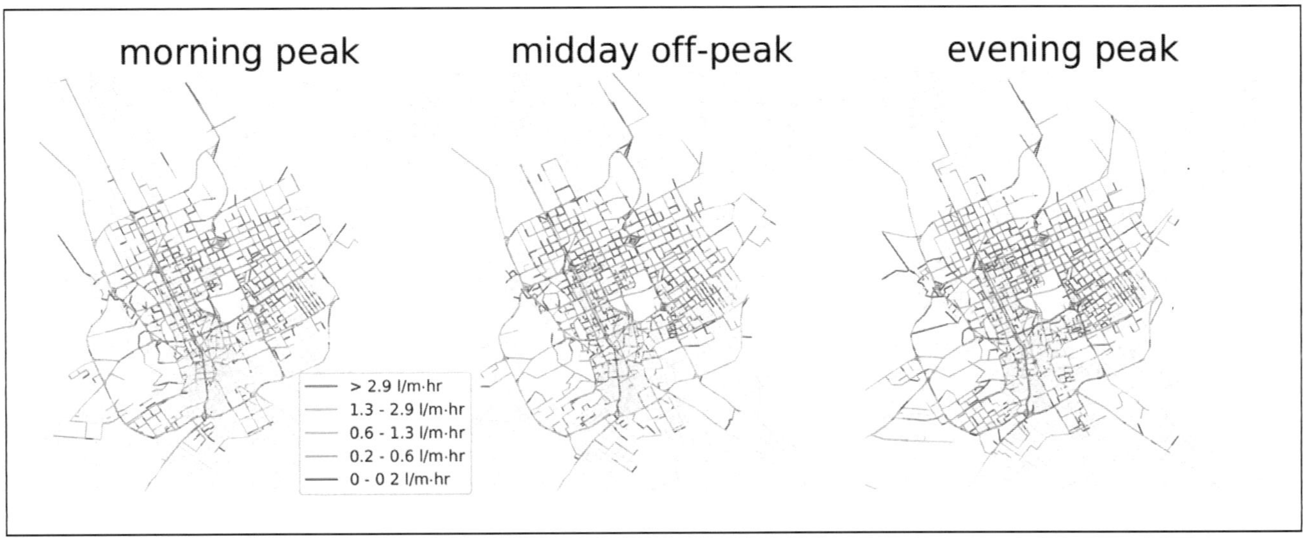

Figure 5. Choropleth Maps of fuel consumption rates [Liter/meter.hour] by the StreetSmart model on streets matched with GPS data for typical time periods morning peak (0800–0900 h) weekdays, midday off-peak (1200–1300 h) weekdays, evening peak (1700–1800 h) weekdays.

center from the south and the Northern Ring road from the north.

The taxi GPS data necessary to calculate a fuel consumption value on each street did not cover the entire network of streets. Similarly, the traffic assignment of ODs did not use all of the city's streets and the majority of roads used were also matched with at least one speed profile from the taxi GPS data. Specifically, 56%, 57%, and 63% of streets were covered for the morning, midday, and evening time periods, respectively.

Relevance of Integrating GPS Data. For the morning time period, a simulation of all OD trips in the city is made and the path, defined as the sequence of edges taken from origin node to destination node, is defined using the shortest path algorithm as in the ITA algorithm. In the simulation, OD pairs with flow values that are less than one car/hour are omitted to ensure each flow represents a discrete trip. A car fuel economy bin was assigned at random to each trip in proportion to the probability of that bin in Riyadh's fleet. Using the StreetSmart model, fuel consumption and trip time were estimated as the sum of the their values on each edge in the trip's path. For verification, total trip times derived via the demand model, which were previously verified against Google Maps estimates (*5*), are plotted against the trip times derived via the GPS data. As shown in Figure 6*a*, travel times via GPS data are generally higher than their counterparts from the simulation but their correlation is satisfactory.

Two sample speed profiles used by the StreetSmart model are plotted with the constant speed assumed by the baseline comparison. As expected, not all trip times are equivalent to their speed profile counterparts, and the constant speed is generally lower than the peaks of the speed profile. Extracted speed profiles do not always end at 0 km/h speed as the originally matched trips on these edges may not be stopping at that edge. This overestimation of speed at the end of a trip represents a negligible increase in fuel consumption estimation. The lower constant speed assumption used in the baseline comparison results in fuel consumption estimates that are consistently lower than those from using speed profiles from GPS data. Thus, the effect of the accelerations and decelerations on the StreetSmart model are observed to be significant and not negligible, which shows the benefit of using GPS data in the fuel consumption model. The distributions of fuel consumption per trip using speed profiles and constant speed are shown in Figure 6*c* along with the fuel economies. It is clear that the acceleration and detailed speed profiles are not negligible in the estimates. These results bring to the urban scale the results of Figure 4*c*.

The effect of reducing flows on overall fuel consumption is shown in Figure 6*d*. Trips were removed from the simulation described above via three rankings and the resulting fuel saving potential is shown, normalized by the total fuel consumption of each method. Random trip reduction results in a perfectly linear fuel saving effect. In contrast, we see the best case scenario, where trip reduction of the worst fuel-consuming trips per meter are ranked. As expected, targeted reduction results in higher fuel savings with the same number of reduced trips. More importantly, when comparing the speed profile estimates (method proposed here) versus constant speed fuel estimates (baseline), differences emerge. Constant

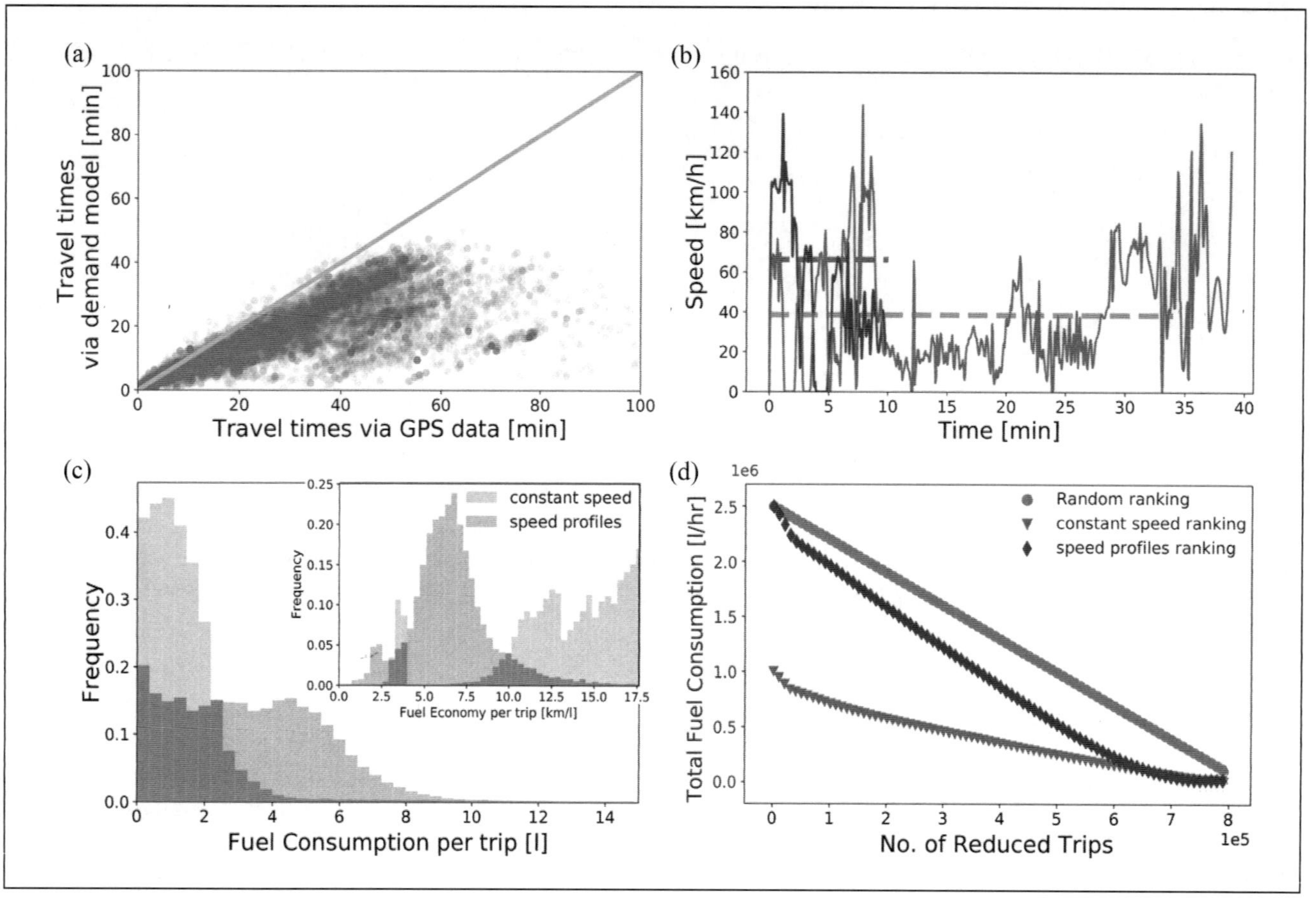

Figure 6. Fuel consumption estimates at urban scale. (*a*) Comparison of the travel time of the routes in the constant speed model via travel demand vs. the input used in this paper's method using GPS data; (*b*) Sample speed profiles in two routes used to estimate fuel consumption overlaid with the constant speed used for comparison; (*c*) Estimates of fuel consumption and fuel economy in the morning peak via this paper's method (speed profiles) and the base line method (constant speed); (*d*) Random and targeted fuel savings vs. number of reduced trips.

speed targeting shows a higher return on fuel savings because the variance of the distribution of fuel consumption estimates using constant speed is less than that of the speed profiles. In other words, the fuel economy distribution of the constant speed estimates shows higher proportions at the high end of the distribution than the high end of the speed profile distribution, shown in Figure 6*c*. This would explain the higher fuel savings observed when the trips are ranked by constant speed fuel consumption per meter. However, because the constant speed estimates are not as accurate as those derived using speed profiles, the fuel savings are spurious.

The real gain in using speed profiles to estimate fuel consumption over the baseline constant speed assumption is in the city-wide total fuel consumption per hour estimate. The constant speed assumption underestimates the city-wide fuel consumption for the morning hour by 60% compared with using the speed profiles.

The relative gain in fuel saving potential of targeting the highest fuel-consuming trips over random trip reduction is approximately 10% if 14.5% of trips are reduced. In other words, if 14.5% of trips are reduced, the policy targeting the highest fuel-consuming trips per meter would save 10% more fuel for every morning peak hour. When 14.5% of targeted trips are reduced, 25% of fuel consumption for the morning peak hour is saved. These results are encouraging, but the ratio is not as high as the effect of similar targeting policies on household energy consumption (*33*), in which the increase over random is 51%. These differences can be explained by the more normal distribution of fuel economy of the trips, which renders the effects of targeting to be much less dramatic than the more broad distribution found in house energy consumption.

Discussion and Conclusion

This paper presents a data fusion method to estimate fuel consumption at the urban scale. A travel demand model is leveraged that uses mobile phone data and integrates speed profiles from taxi GPS data that covered most of the street network of Riyadh. To identify fleet distribution, car crash statistics data were used as a proxy with the assumption that the distribution of car makes and models is representative of Riyadh's car fleet distribution. The method developed here and the calibration of the StreetSmart model for fuel consumption can easily be extended to any other region. It is significantly faster than if speed profiles were simulated by software, which would require a large amount of time and computing resources (*34*).

The fuel consumption model was verified to produce better results than the DOE fuel economy by speed graph, and tested on real OBD-II fuel consumption measurements before being calibrated to be used on any car fuel economy. The resulting calibration results are presented in Table 2. The calibrated StreetSmart model with speed profiles from GPS data were used to produce estimates which were compared with the baseline without GPS data and assuming constant speed. The results show that the differences of using speed profiles is significant, justifying the introduction of the more elaborate model in policy estimates. As a proof of concept, the calibrated StreetSmart model was applied to test a policy of flow reduction, both random and targeted to the least fuel-efficient trips, and simulated the effect on the rate of fuel consumption in the city. It was shown that the difference in fuel consumption reduction between targeted and random schemes was around 10% more fuel savings for 14.5% trip reduction. Interestingly, 25% of city-wide fuel savings potential can be achieved by removing only 14.5% of trips ranked by the worst fuel-consuming trips per meter.

Although this project has demonstrated the potential of data-driven models to estimate the effects of policies on fuel consumption, it can benefit from further study to understand the impacts of several simplifications and assumptions. There are three main areas which require further research:

1. To correctly assess the benefit of this straightforward approximation over the computationally costly alternative, speed profiles of each car across every OD trip should be simulated and the results compared.
2. The GPS data used in this model covered most, but not all, of the street network. A more accurate fuel consumption calculation can be achieved with more data, covering a higher proportion of flow over a longer time period.
3. The trip assignment method used is an efficient and reasonably effective method to estimate overall congestion based on a simplified method of route choice. However, the baseline estimates would benefit from trip assignment derived from a dynamic traffic assignment model.

It is hoped that the method used to obtain the results of the fuel reduction strategies can be implemented for other fuel or emissions-reduction strategies, assuming the conditions remain relatively unchanged, for example, the composition of the Riyadh fleet remains similar and the streets remain unchanged. For future implementations, the composition of the fleet can easily be adjusted as the method uses bins to account for all car types. However, a significant change in the road network layout or the traffic conditions would require new datasets (Road network and GPS routes for speed profile extraction) to recompute modular road-by-road fuel consumption estimates for each bin of car type.

Current fuel consumption models are adapting to the wealth of data available from ubiquitous sensing devices such as smartphones. This project aims to show the relative gain in accuracy of incorporating both fuel economy distribution considerations and speed profiles derived, in this case, from GPS devices used by a local taxi company. The method developed in this paper can be adapted to also measure emissions. It can be further augmented to analyze the economic and environmental impacts of policies targeting specific trips by conducting a network analysis to identify the affected trips. The tools developed in this work have the potential to assess the consequences of a variety of policies under different circumstances and in any region of the world.

Acknowledgments

The research was supported by grants from the MIT Energy Solutions Initiative, and from the Center for Complex Engineering Systems at the King Abdulaziz City for Science and Technology. R. Di Clemente, the Newton International Fellow of the Royal Society, acknowledges support from The Royal Society, The British Academy, and the Academy of Medical Sciences (Newton International Fellowship, NF170505).

Authors' Note

The methods and code used to perform the analysis as well as the figures are available for public access at www.github.com/adhamkalila/riyadhfuelconsump. We encourage the scientific community to reproduce the analysis and notify us of any issues on the GitHub repository directly or by email.

Author Contributions

The authors confirm contribution to the paper as follows: study conception and design: M.C. González, R. Di Clemente; data preparation and cleaning: Z. Awwad, A. Kalila; map Matching:

Z. Awwad; application of methodology: A. Kalila; figure preparation: R. Di Clemente, A. Kalila; analysis and interpretation of results: M.C. González, R. Di Clemente, A. Kalila; draft manuscript preparation: A. Kalila. All authors reviewed the results and approved the final version of the manuscript.

References

1. Kerr, S., Saudi Arabia Looks to Reform Energy Subsidy Programme. *Financial Times*, 2015.
2. Group, O. B. Saudi Arabia Sets Out Strategy to Reform Subsidies and Reduce Domestic Energy Usage. *Oxford Business Group*, 2016.
3. Çolak, S., L. P. Alexander, B. G. Alvim, S. R. Mehndiratta, and M. C. González. Analyzing Cell Phone Location Data for Urban Travel: Current Methods, Limitations, and Opportunities. *Transportation Research Record: Journal of the Transportation Research Board*, 2015. 2526: 126–135.
4. Toole, J. L., S. Çolak, B. Sturt, L. P. Alexander, A. Evsukoff, and M. C. González, The Path Most Traveled: Travel Demand Estimation Using Big Data Resources. *Transportation Research Part C: Emerging Technologies*, Vol. 58, 2015, pp. 162–177.
5. Chodrow, P. S., Z. Al-Awwad, S. Jiang, and M. C. González, Demand and Congestion in Multiplex Transportation Networks. *PLoS One*, Vol. 11, No. 9, 2016, p. e0161738.
6. Do, T. M. T., J. Blom, and D. Gatica-Perez. Smartphone Usage in the Wild: A Large-Scale Analysis of Applications and Context. *Proc., 13th International Conference on Multimodal Interfaces*, ACM, 2011, pp. 353–360.
7. De Nazelle, A., E. Seto, D. Donaire-Gonzalez, M. Mendez, J. Matamala, M. J. Nieuwenhui- jsen, and M. Jerrett. Improving Estimates of Air Pollution Exposure Through Ubiquitous Sensing Technologies. *Environmental Pollution*, Vol. 176, 2013, pp. 92–99.
8. Lei, W., H. Chen, and L. Lu. Microscopic Emission and Fuel Consumption Modeling for Light-duty Vehicles Using Portable Emission Measurement System Data. *World Academy of Science, Engineering and Technology*, Vol. 66, 2010, pp. 918–925.
9. Pelkmans, L., P. Debal, T. Hood, G. Hauser, and M.-R. Delgado. Development of a Simulation Tool to Calculate Fuel Consumption and Emissions of Vehicles Operating in Dynamic Conditions. SAE Technical Paper, 2004.
10. Song, G., L. Yu, and Z. Wang. Aggregate Fuel Consumption Model of Light-Duty Vehicles for Evaluating Effectiveness of Traffic Management Strategies on Fuels. *Journal of Transportation Engineering*, Vol. 135, No. 9, 2009, pp. 611–618.
11. Vilaça, A., A. Aguiar, and C. Soares. *Estimating Fuel Consumption from GPS Data*, Springer International Publishing, Cham, 2015, pp. 672–682.
12. Ericsson, E., H. Larsson, and K. Brundell-Freij. Optimizing Route Choice for Lowest Fuel Consumption-Potential Effects of a New Driver Support Tool. *Transportation Research Part C: Emerging Technologies*, Vol. 14, No. 6, 2006, pp. 369–383.
13. Work, D. B., and A. M. Bayen. *Impacts of the Mobile Internet on Transportation Cyberphysical Systems: Traffic Monitoring Using Smartphones*. National Workshop for Research on High-Confidence Transportation Cyber-Physical Systems: Automotive, Aviation, & Rail, 2008, pp 18–20.
14. Work, D. B., O.-P. Tossavainen, Q. Jacobson, and A. M. Bayen. Lagrangian Sensing: Traffic Estimation with Mobile Devices. *Proc., American Control Conference, 2009, ACC'09*, IEEE, 2009, pp. 1536–1543.
15. Donovan, B., and D. B. Work. Using Coarse GPS Data to Quantify City-Scale Transportation System Resilience to Extreme Events. *arXiv preprint arXiv:1507.06011*, 2015.
16. Herrera, J. C., D. B. Work, R. Herring, X. J. Ban, Q. Jacobson, and A. M. Bayen. Evaluation of Traffic Data Obtained via GPS-enabled Mobile Phones: The Mobile Century Field Experiment. *Transportation Research Part C: Emerging Technologies*, Vol. 18, No. 4, 2010, pp. 568–583.
17. Mazaré, P.-E., O.-P. Tossavainen, A. Bayen, and D. Work. Trade-offs Between Inductive Loops and GPS Probe Vehicles for Travel Time Estimation: A Mobile Century Case Study. *Transportation Research Board 91st Annual Meeting (TRB'12)*. Washington, D.C., Vol. 349, 2012.
18. Bowyer, D. P., R. Akçelik, and D. Biggs. *Guide to Fuel Consumption Analyses for Urban Traffic Management*, Vol. 32, 1985.
19. Joumard, R., P. Jost, J. Hickman, and D. Hassel. Hot Passenger Car Emissions Modelling as a Function of Instantaneous Speed and Acceleration. *Science of the Total Environment*, Vol. 169, No. 1–3, 1995, pp. 167–174.
20. Jimenez-Palacios, J. L. *Understanding and Quantifying Motor Vehicle Emissions with Vehicle Specific Power and TILDAS Remote Sensing*. Massachusetts Institute of Technology, Cam- bridge, 1998.
21. Cappiello, A., I. Chabini, E. K. Nam, A. Lue, and M. A. Zeid. A Statistical Model of Vehicle Emissions and Fuel Consumption. *Proc., IEEE 5th International Conference on Intelligent Transportation Systems*, IEEE, 2002, pp. 801–809.
22. Rakha, H., K. Ahn, and A. Trani. Development of VT-Micro Model for Estimating Hot Stabilized Light Duty Vehicle and Truck Emissions. *Transportation Research Part D: Transport and Environment*, Vol. 9, No. 1, 2004, pp. 49–74.
23. Ribeiro, V., J. Rodrigues, and A. Aguiar. Mining Geographic Data for Fuel Consumption Estimation. *Proc., 16th International IEEE Conference on Intelligent Transportation Systems (ITSC 2013)*, 2013, pp. 124–129.
24. Zhou, M., H. Jin, and W. Wang. A Review of Vehicle Fuel Consumption Models to Evaluate Eco-driving and Eco-routing. *Transportation Research Part D: Transport and Environment*, Vol. 49, 2016, pp. 203–218.
25. Louhghalam, A., M. Akbarian, and F.-J. Ulm. Carbon Management of Infrastructure Performance: Integrated Big Data Analytics and Pavement-vehicle-interactions. *Journal of Cleaner Production*, Vol. 142, 2017, pp. 956–964.
26. Oehlerking, A. L. *StreetSmart: Modeling Vehicle Fuel Consumption with Mobile Phone Sensor Data Through a Participatory Sensing Framework*. PhD thesis. Massachusetts Institute of Technology, 2011.

27. Wu, F., R. Stern, M. Churchill, M. L. Delle Monache, K. Han, B. Piccoli, and D. Work. Measuring Trajectories and Fuel Consumption in Oscillatory Traffic: Experimental Results. *Presented at 96th Annual Meeting of the Transportation Research Board*, Washington, D.C., 2017.
28. Davis, S., S. William, and R. Boundy. *Light Vehicles and Characteristics*. U.S. Department of Energy, 2016, p. 4.30.
29. *Fuel Economy Guide: Model Year 2016*. U.S. Department of Energy, 2016.
30. *EPA Federal Test Procedure*, 2017.
31. Papadimitriou, G., L. Ntziachristos, P. Wutrich, B. Notter, M. Keller, E. Fridell, H. Winnes, L. Styhre, and A. Sjodin. *Transport Data Collection Supporting the Quantitative Analysis of Measures Relating to Transport and Climate Change (TRACCS)*, 2013.
32. Jiang, S., G. A. Fiore, Y. Yang, J. Ferreira, Jr, E. Frazzoli, and M. C. González. A Review of Urban Computing for Mobile Phone Traces: Current Methods, Challenges and Opportunities. *Proceedings of the 2nd ACM SIGKDD International Workshop on Urban Computing*, ACM, 2013, p. 2.
33. Qomi, M. J. A., A. Noshadravan, J. M. Sobstyl, J. Toole, J. Ferreira, R. J.-M. Pellenq, F.- J. Ulm, and M. C. Gonzalez. Data Analytics for Simplifying Thermal Efficiency Planning in Cities. *Journal of the Royal Society Interface*, Vol. 13, No. 117, 2016, p. 20150971.
34. Cetin, N., A. Burri, and K. Nagel. A Large-Scale Agent-Based Traffic Microsimulation Based on Queue Model. *Proc., Swiss Transport Research Conference (Strc)*, Monte Verita, Ch, Citeseer, 2003.

The Standing Committee on Transportation Energy (ADC70) peer-reviewed this paper (18-05948).

TRR
JOURNAL OF THE TRANSPORTATION RESEARCH BOARD

Article

Transportation Research Record
2018, Vol. 2672(24) 60–74
© National Academy of Sciences: Transportation Research Board 2018
Article reuse guidelines: sagepub.com/journals-permissions
DOI: 10.1177/0361198118796401
journals.sagepub.com/home/trr
SAGE

Low-Level Automated Light-Duty Vehicle Technologies Provide Opportunities to Reduce Fuel Consumption

Saeed Vasebi[1], Yeganeh M. Hayeri[1], Constantine Samaras[2], and Chris Hendrickson[2]

Abstract

Gasoline is the main source of energy used for surface transportation in the United States. Reducing fuel consumption in light-duty vehicles can significantly reduce the transportation sector's impact on the environment. Implementation of emerging automated technologies in vehicles could result in fuel savings. This study examines the effect of automated vehicle systems on fuel consumption using stochastic modeling. Automated vehicle systems examined in this study include warning systems such as blind spot warning, control systems such as lane keeping assistance, and information systems such as dynamic route guidance. We have estimated fuel savings associated with reduction of accident and non-accident-related congestion, aerodynamic force reduction, operation load, and traffic rebound. Results of this study show that automated technologies could reduce light-duty vehicle fuel consumption in the U.S. by 6% to 23%. This reduction could save $60 to $266 annually for the owners of vehicles equipped with automated technologies. Also, adoption of automated vehicles could benefit all road users (i.e., conventional vehicle drivers) up to $35 per vehicle annually (up to $6.2 billion per year).

Gasoline is the main source of energy used for surface transportation in the United States. According to the U.S. Energy Information Administration, Americans use more than a gallon of gasoline per person each day, adding up to 392 million gallons of gasoline per day in 2016, of which 90% is solely used for light-duty vehicles (*1*). Reducing gasoline consumption could significantly improve air quality, health, energy security, and greenhouse gas (GHG) emissions. Emerging automated vehicle technologies can accelerate the pace of realizing some of these benefits. Since driving patterns, commute times, and the design of metropolitan regions could all be affected by automated vehicles, there is a great potential for saving energy as these emerging technologies gain market share.

In this study, we examine the impacts of various low-level automated vehicle technologies on fuel consumption. In addition, we examine the time that would take to realize such impacts based on market penetration rate of the technologies.

The scope of this study comprises three main elements:

1. Different groups of people that could benefit from automated technologies (e.g., owners of the vehicles equipped with automated technologies),
2. Benefit and cost categories (e.g., fuel saving benefits due to accident reduction),
3. Types of automated technologies (e.g., adaptive cruise control).

The scope of this study is illustrated in Figure 1, which considers fuel saving possibilities from two perspectives defined as affected groups: (a) those driving vehicles equipped with automated technologies and (b) all other road users including those driving conventional vehicles. The first group, automated vehicle owners (AVO), benefits directly from fuel saving potentials of the technologies. This is also the group that incurs cost of purchasing the automated packages. The second group, all road users (ARU), potentially benefit from the efficient flow of traffic (e.g., fewer accidents caused by human error, more efficient route guidance). The first group is an enabler for the second group's benefits.

[1]School of Systems and Enterprises, Stevens Institute of Technology, Hoboken, NJ
[2]Department of Civil and Environmental Engineering, Carnegie Mellon University, Pittsburgh, PA

Corresponding Author:
Address correspondence to Yeganeh M. Hayeri: yhayeri@stevens.edu

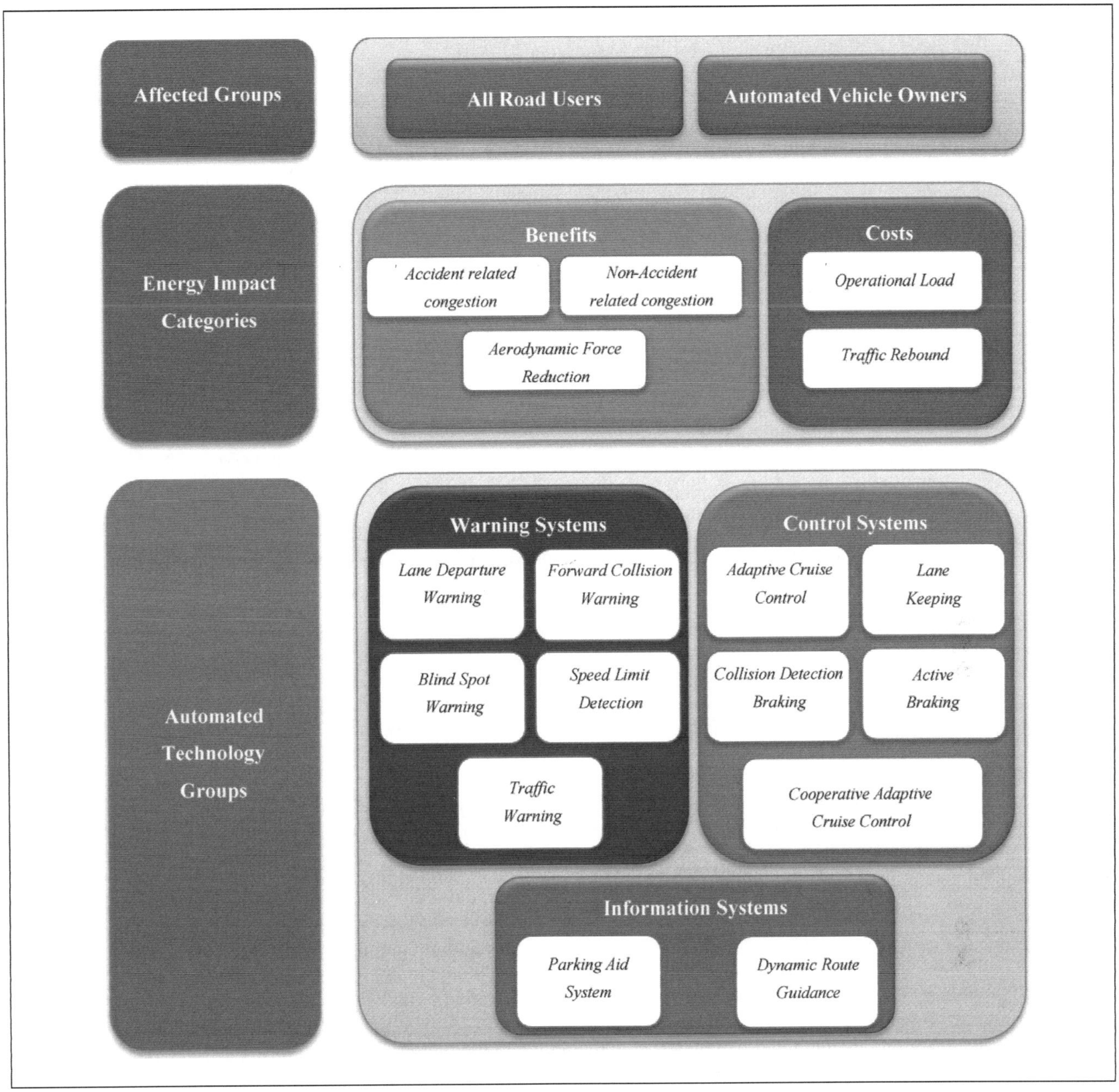

Figure 1. Scope and boundaries of this study.

In order to estimate the net benefit of automated technologies, we analyze the benefits and costs associated with each technology specified in Figure 1. For each automated technology, we estimate benefits resulting from congestion reduction (accident related and non-accident related), and aerodynamic force reduction. To estimate costs, we develop a model to include: (a) operational load, which is the energy that a vehicle consumes for the operation of computers, sensors, and other automated equipment installed on the vehicle, and (b) rebound effects, which in this paper are considered to be induced demand and fuel efficiency. Owners of automated vehicles are affected by all these benefit and cost categories. Other vehicles are affected only by the reduction of accident and non-accident congestion categories and rebound effects.

We categorize automation technologies into three groups: (a) warning systems, (b) control systems, and (c) information systems (Figure 1). Automated warning systems are those generating warnings (e.g., sounds, lights) inside the vehicle to alert the driver of a potential hazard. Warning systems considered in this study include lane departure warning (LDW), forward collision warning (FCW), blind spot warning (BSW), speed limit detection, and traffic warning. Control systems not only warn drivers but also may take partial control of vehicle, for example, to pass a threat or to improve driving experience. Control systems included in this study are: adaptive cruise control (ACC), cooperative adaptive cruise control (CACC), lane keeping assistance (LKA), forward collision avoidance (FCA), and active braking. Information systems provide useful driving information to reduce travel time and delay. Information systems considered in this study include parking aid system and dynamic route guidance. The complete list of automated technologies and their detailed definition is available from the authors.

In the next section, we provide a thorough review of the existing impact analyses of automated vehicle systems conducted to date. The following section explains our model for fuel saving analysis of automated vehicle technologies. We continue by providing estimates of net energy benefit of automated light-duty vehicle technologies as well as uncertainties involved with our estimates and discussion of the results. In the last section, we provide steps for future enhancement of our model. This paper's contribution is to realize the near- to mid-term energy impact of automated vehicle technologies.

Existing Impact Assessments

A few recent studies have shed lights on the potential benefits and costs of automated vehicle technologies (*2–5*) (Table 1). Harper et al. determined that if all light-duty vehicles in the U.S. were equipped with LDW, FCW, and BSM systems, it would result in $4 billion to $202 billion annual net social/safety benefit cumulatively (*4*). Jermakian reviewed accident records from the National Automotive Sampling System and the General Estimates System and Fatality (*2*). He estimated LDW individually could prevent 3% of crashes and save up to 7,529 lives and BSW could reduce accidents by 6.8% and save 393 lives annually. Findings of other general and safety impact studies can be found in Table 1.

Most fuel and GHG emission studies have concentrated on fully autonomous vehicles (AV) (*9*, *11*, *12*) or SAV (*7*, *8*) (Table 1). Brown et al. examined the potential energy impact of AV (*9*). They concluded that AV may increase fuel demand by 173% or decrease it by 95%. MacKenzie and Wadud also analyzed the energy impact of light-duty and heavy-duty AV in the U.S. (*11*, *12*). Due to uncertainty in future prediction of AV, they developed four scenarios. The most optimistic scenario estimates 40% reduction in total fuel demand of the transportation sector and the most pessimistic scenario predicts significant growth (i.e., more than 100%) in fuel demand. Stephens et al. studied the potential impacts of CAV and vehicle ridesharing on fuel consumption (*8*). They estimated 10.68% decrease to 13.59% increase in fuel demand for partial AV (i.e., driver assistant feature with human involvement) and 58.25% decrease to 204.85% increase and 64.08% decrease to 194.17% increase in fuel consumption for fully CAV and fully rideshared CAV respectively.

Although the safety benefits of automated vehicle technologies have been investigated in previous studies (*2–5*), the energy impacts of low-level automated features remain uncertain. Most fuel impact studies have concentrated on AV (*9*, *11*, *12*), which may not be commercialized with a strong enough market share until the 2070s (*13*). Our study is the first energy impact investigation focusing on lower levels of automation and individual technologies that are either already introduced or will be in the next few years.

Method for Estimating Energy Impacts of Automated Technologies

The paper considers five main energy impact categories resulting from the use of automated vehicle technologies:

1. Accident-related congestion reduction impact,
2. Non-accident-related congestion reduction impact,
3. Aerodynamic force reduction impact,
4. Equipment operation load impact,
5. And traffic rebound impact.

We estimate savings and costs associated with each technology as a percentage of average gasoline consumption, which is about 508 gallons per vehicle annually in the U.S. (*14*).

This paper analyzes only the energy impact of automated vehicles. Since the energy used in vehicle manufacturing is small relative to driving energy (*15*), it will not be considered in this paper. We evaluate direct vehicle fuel savings, as these are the most tangible to consumers and policymakers. While automation may affect different types of drivers and vehicles differently, our analysis pertains to average implications in the U.S. We focus on the cumulative effect of each technology listed on Figure 1. We estimate potential fuel saving benefits for automated vehicle owners (AVO) and all road users (ARU).

Fuel Saving for Automated Vehicle Owners

We estimate net fuel consumption change of the automated vehicles as:

Table 1. Existing Assessments of Safety and Energy Impacts of Automated and Autonomous Vehicles'

Study	Scope	Model	Results
Harper et al. (*4*)	Safety/social benefits of automated vehicle features	Studied safety/social impacts of LDW, FCW, and BSW based on crash and insurance data.	If all light-duty vehicles in the U.S. were equipped with these warning systems, it would result in $4 billion to $202 billion annual net benefit cumulatively per year.
Bayly et al. (*3*)	Safety/social benefits of automated vehicle features	Reviewed crash avoidance impact of 138 vehicle safety features in European Union; including passive, active, and automated safety features.	The study claims that automation safety features could prevent most human errors and reduce accidents, injuries, and fatalities.
Jermakian (*2*)	Safety/social benefits of automated vehicle features	Reviewed accident recodes of National Automotive Sampling System and the General Estimates System and Fatality.	LDW individually could prevent 3% of crashes and save up to 7,529 lives and BSW could reduce accidents by 6.8% and save 393 lives per year.
Anderson et al. (*6*)	Safety/social benefits of automated vehicle features	Analyzed safety benefit/cost of FCA, alcohol interlocks, and fatigue management systems.	These technologies could avoid 31% of passenger car fatalities in Australia. This study estimates the technologies provide benefit-cost ratios of 0.5 to 1.3.
Carsten et al. (*5*)	Safety/social benefits of automated vehicle features	Studied installation of speed limit detection technology at advisory, voluntary, and mandatory level.	Speed limit detection technology at advisory, voluntary, and mandatory level could reduce fatal accidents by 18% to 24%, 19% to 32%, and 37% to 59% respectively. Their safety cost and benefit ratios are between 5.0 and 16.7.
Greenblatt et al. (*7*)	Energy impact of SAV	Studied SAV influence on fuel demand and GHG emissions in the U.S.	If SAV is substituted for 10% of VMT in the U.S., it will result in reduction of 65 to 75 metric tons of CO_2 and 25.2 million gallons of gasoline annually. This reduction equals a saving of $56.7 million in fuel or $ 0.21 per vehicle per year.
Stephens et al. (*8*)	Energy impact of light-duty partial CAV (i.e., driver assistant feature with human involvement) and shared CAV	Studied potential impacts of CAV and vehicle ridesharing on VMT, fuel consumption, and costs of the users.	They estimated −10.68% to 13.59% increase in fuel demand for partial AV and also −58.25% to 204.85% and −64.08% to 194.17% increase in fuel consumption for fully CAV and fully rideshared CAV respectively.
Brown et al. (*9*)	Energy impact of light-duty AV	Considered several potential impacts of AVs, including air drag reduction (i.e., platooning), efficient routing, efficient driving, more travel, vehicle size optimization, and fast parking.	They concluded that fully autonomous vehicles (NHTSA levels 3 and 4 (*10*)) may increase fuel demand by 173% or decrease it by 95%.
MacKenzie et al. & Wadud et al. (*11*, *12*)	Energy impact of heavy-duty and light-duty AV	Analyzed impacts of AV on VMT and energy demand. They considered several parameters (e.g., eco-driving, higher speed travel, vehicle right size, congestion reduction, and crash avoidance). Due to uncertainty in future prediction of AV (e.g., acceptance of AV technology, traffic elasticity, traffic policies), they developed four scenarios.	The most optimistic scenario estimates 40% reduction in total fuel demand of transportation sector and the most pessimistic scenario predicts significant growth (i.e., more than 100%) in fuel demand when AV reaches levels 3 and 4 of automation. However, if AV is stopped at NHTSA's level 2 automation (*10*), it will reduce total fuel consumption on road transportation by 9%.
Our study	Energy impact of light-duty AVs	Analyzed impacts of each automated technology in detail for the vehicles' owners and all road users.	Detailed results and comparison with existing literature in Table 4.

Note: CAV = connected autonomous vehicles; SAV = shared autonomous vehicles; VMT = vehicle miles traveled.

$$\Delta\hat{\gamma}^{TOT} = \left(1 - \left(1 - \Delta\hat{\gamma}^{AC}/\gamma^{AGC}\right) \times \left(1 - \Delta\hat{\gamma}^{NAC}/\gamma^{AGC}\right) \times \left(1 - \Delta\hat{\gamma}^{AFFS}/\gamma^{AGC}\right) \times \left(1 - \Delta\hat{\gamma}^{EQP}/\gamma^{AGC}\right) \times \left(1 - \Delta\hat{\gamma}^{TR}/\gamma^{AGC}\right)\right) \times \gamma^{AGC} \quad (1)$$

where $\Delta\hat{\gamma}^{TOT}$ stands for total net fuel saving for an AVO, $\Delta\hat{\gamma}^{AC}$ is fuel saving from reduction in accident-related congestion, $\Delta\hat{\gamma}^{NAC}$ is fuel saving from reduction in non-accident-related congestion, $\Delta\hat{\gamma}^{AFFS}$ is fuel saving from aerodynamic force reduction, $\Delta\hat{\gamma}^{EQP}$ is equipment's operation load fuel consumption, $\Delta\hat{\gamma}^{TR}$ is traffic rebound fuel consumption, and γ^{AGC} indicates average annual fuel consumption of each vehicle in the U.S. In general, γ stands for the categories' fuel consumption, ρ is the technologies' fuel saving impacts, ϕ is the models' parameters, M is the market penetration rate, and random variables (stochastic parameters) are identified using the ∩ symbol. The superscripts used in this paper are: AC (accident-related congestion reduction), NAC (non-accident-related congestion), NACA (non-accident congestion of ARU), NACAV (non-accident congestion of AVO), AFFS (aerodynamic force reduction), EQP (equipment's operation load), TR (traffic rebound), IDR (induced rebound), and FER (fuel efficiency rebound). Each of these items will be described and calculated in the following subsections.

Accident-Related Congestion Reduction. It is estimated that 21% of the U.S. traffic congestion delay is the result of accidents, and an average vehicle consumes about 3.8 gallons of fuel annually due to accident-related congestion (*16–18*). Most of the automated technologies have the potential to reduce the number and severity of traffic accidents, which would reduce traffic congestion and consequently energy consumption. The reduction of accident-related congestion benefits all road users—whether they drive conventional cars or automated cars. Using estimates from the literature, the impacts of each of the specified technologies on safety and energy savings were estimated for the technologies. The change in fuel consumption from reduction of accident-related congestion of technology i is estimated as:

$$\Delta\hat{\gamma}_i^{AC} = -\gamma^{CGN}\phi^{AC}\hat{\rho}_i^{AC}M_i \quad (2)$$

where $\Delta\hat{\gamma}_i^{AC}$ is the change in fuel consumption due to accident-related congestion reduction capability of automated technology i, $\gamma^{CGN} = 19\text{gal/veh/yr}$ is fuel consumed as a result of congestion in U.S. (*16*), $\phi^{AC} = 21\%$ is the portion of congestion delays due to accidents (*16–18*), $\hat{\rho}_i^{AC}$ is the portion of accident-related congestion reduced by technology i, and M_i is the market penetration rate of technology i. We assume that when technologies are combined the portion of accidents avoided (from the accidents remaining after other technologies have been implemented) is independent of the presence of other technologies, so total fuel savings ($\Delta\hat{\gamma}^{AC}$) from implementation of a set of technologies $\mathcal{T}$ is determined as:

$$\Delta\hat{\gamma}^{AC} = -\gamma^{CGN}\phi^{AC}\left(1 - \prod_{i\in\mathcal{T}}\left(1 - \hat{\rho}_i^{AC}M_i\right)\right) \quad (3)$$

Non-Accident-Related Congestion Reduction. All vehicles on the road benefit from the reduction of non-accident-related congestion brought about by automated vehicles. Reduction in non-accident congestion is achieved by reducing the traffic load due to better navigation, less travel time, or more roadway capacity. Both parking aid systems and dynamic route guidance systems can reduce non-accident congestion by improving the efficiency of the national roadway system. In addition, CACC increases road capacity by reducing the headway of vehicles in a platoon (*19*). Non-accident-related congestion of ARUs is handled similarly:

$$\Delta\hat{\gamma}_i^{NACA} = -\gamma^{CGN}\left(1 - \phi^{AC}\right)\hat{\rho}_i^{NACA}M_i \quad (4)$$

where $\hat{\rho}_i^{NACA}$ stands for non-accident congestion reduction of technology i.

In addition to the non-accident congestion benefits for ARUs, AVOs also benefit directly from the energy saving capabilities of their vehicles when using a parking aid system or dynamic routing guidance. Parking aid technology assists drivers to find parking lots faster than conventional vehicles. Drivers may also utilize dynamic routing guidance to find faster routes with less congestion. The fuel saving potential for AVOs from non-accident-related congestion is estimated as:

$$\Delta\hat{\gamma}_i^{NACAV} = -\gamma^{AGC}\left(\hat{\mu}_i^{NACAV}\right) \quad (5)$$

where $\hat{\rho}_i^{NACAV}$ is the fuel saving capability of technology i for the vehicle driver due to less driving. Although the non-accident congestion reductions for ARUs and AVOs are independent, arithmetic summation of the values is used. Since the values are small, their correlation is assumed to be negligible. So, the non-accident congestion reduction fuel saving for technology i is calculated as:

$$\Delta\hat{\gamma}_i^{NAC} = \Delta\hat{\gamma}_i^{NACA} + \Delta\hat{\gamma}_i^{NACAV} \quad (6)$$

The total fuel savings of non-accident congestion reduction ($\Delta\hat{\gamma}^{\mathrm{NAC}}$) for a set of technologies $\mathcal{T}$ is estimated as:

$$\Delta\hat{\gamma}^{\mathrm{NAC}} = \left(1 - \prod_{i\in\mathcal{T}}\left(1 - \frac{\Delta\hat{\gamma}_i^{\mathrm{NAC}}}{\gamma^{\mathrm{AGC}}}\right)\right) \times \gamma^{\mathrm{AGC}} \quad (7)$$

It is important to note that in estimating total fuel savings of accident and non-accident congestion reduction all individual technology effects are essentially aggregated.

Aerodynamic Force Related Fuel Saving. Aerodynamic force reduction has been estimated to cut fuel consumption significantly (*20*, *21*). Technologies such as ACC and CACC have the capability to capture a portion of these savings by improvement of efficient driving. The aerodynamic related fuel saving capability of ACC and CACC is calculated as:

$$\Delta\hat{\gamma}_i^{\mathrm{AFFS}} = -(\gamma^{\mathrm{AGC}} - \gamma^{\mathrm{CGN}})\phi^{\mathrm{FWY}}\hat{\mu}_i^{\mathrm{AFFS}}M_i \quad (8)$$

where $\phi^{\mathrm{FWY}} = 55\%$ is the portion of VMT traveled on the highway (*22*), and $\hat{\rho}_i^{\mathrm{AFFS}}$ is the fuel efficiency parameter for the vehicle driver with technology i.

Operation Load. Automated features aid drivers by warning, controlling, or providing information; however, these features consume energy to provide these functions. Operation load is defined as the energy used by sensors, radars, or steering or braking systems. Operation of automation features requires use of energy-consuming equipment (e.g., sensors, processors, and actuators). Based on the Bosch automotive handbook, higher comfort and safety features have significantly increased the electricity consumption of vehicles (*23*). Hence, theses operation loads should be considered as a negative energy impact for automated features. The following formula is utilized to determine operation load.

$$\Delta\hat{\gamma}_{ij}^{\mathrm{EQP}} = \frac{P_{ij}^{\mathrm{EQP}} \times \hat{\phi}_{ij}^{\mathrm{ACT}}}{u^{\mathrm{GAS}} \times \alpha} \times t^{\mathrm{DRV}} \times 3600 \quad (9)$$

where $\Delta\hat{\gamma}_i^{\mathrm{EQP}}$ is the change in gasoline consumption due to operation of equipment type j in technology i, P_{ij}^{EQP} is the average power consumption of equipment type j in technology i while active, $t^{\mathrm{DRV}} = 227\mathrm{hr/yr/veh}$ is the amount of time spent driving (*22*, *24*), $u^{\mathrm{GAS}} = 118\mathrm{MJ/gal}$ is the energy density of gasoline (*21*), $\hat{\phi}_{ij}^{\mathrm{ACT}}$ is the portion of driving time when equipment type j in technology i is active, and $\alpha = 22\%$ stands for fuel efficiency of light-duty internal combustion engines (*25*).

Traffic Rebound. Reduction of traffic congestion and travel cost can potentially induce new travel demands (*26*). Rebound effect is defined as new VMT generated due to improved mobility or reduction in travel costs (*27*). As an illustration, studies of travel patterns show that when gasoline prices decrease, people tend to drive more and VMT increases (*26*, *28*). Similarly, when mobility improves, which could result in a more pleasant driving experience due to less congestion and consequently less fuel consumption, people tend to drive more (induced demand) (*27*, *29*, *30*). Automated vehicle features have the potential to result in traffic rebound due to congestion reduction and enhanced fuel efficiency. Safety and information systems could reduce traffic congestion. Hence, ARUs spend less time in traffic congestion and may consequently drive more. Moreover, when automated features (e.g., CACC and ACC) increase fuel efficiency, it reduces driving cost per mile and AVOs are likely to drive more.

Improvement of traffic flow induces more people to drive their own cars. We estimate fuel consumption from induced demand as:

$$\Delta\hat{\gamma}_i^{\mathrm{IDR}} = \frac{-\left(\Delta\hat{\gamma}_i^{\mathrm{AC}} + \Delta\hat{\gamma}_i^{\mathrm{NACA}}\right)}{\gamma^{\mathrm{AGC}}} d\left(\hat{\phi}^{\mathrm{IDR}}\right)\eta^{-1} \quad (10)$$

where $\Delta\hat{\gamma}^{\mathrm{IDR}}$ is the change of fuel consumption due to induced traffic demand, $\hat{\phi}^{\mathrm{IDR}}$ is elasticity of induced demand for mobility improvement, d is the average annual VMT, and η is the average vehicle efficiency (miles per gallon). Because $d\eta^{-1} = \gamma^{\mathrm{AGC}}$, the formula simplifies to

$$\Delta\hat{\gamma}_i^{\mathrm{IDR}} = -\left(\Delta\hat{\gamma}^{\mathrm{AC}} + \Delta\hat{\gamma}^{\mathrm{NACA}}\right) \times \hat{\phi}^{\mathrm{IDR}} \quad (11)$$

On the other hand, the automated features increase fuel efficiency. These features reduce the cost of driving per mile and could generate extra travel. Since the fuel efficiency impact of low-level automation technology is estimated to be relatively small, we used a linear approximation; however, large fuel efficiency changes could have non-linear impact on the demand. We estimate the effect of fuel efficiency rebound on fuel consumption as:

$$\Delta\hat{\gamma}_i^{\mathrm{FER}} = -\frac{\left(\Delta\hat{\gamma}_i^{\mathrm{AC}} + \Delta\hat{\gamma}_i^{\mathrm{NACA}} + \Delta\hat{\gamma}_i^{\mathrm{NACAV}} + \Delta\hat{\gamma}_i^{\mathrm{AFFS}} + \Delta\hat{\gamma}_i^{\mathrm{EQP}} + \Delta\hat{\gamma}_i^{\mathrm{IDR}}\right)p}{\gamma^{\mathrm{AGC}}p}\left(\hat{\phi}^{\mathrm{FER}}\right)d\eta^{-1} \quad (12)$$

where $\Delta\hat{\gamma}^{FER}$ is the change in fuel consumption due to the fuel efficiency rebound, $\hat{\phi}^{FER}$ is the elasticity fuel efficiency, and p is the price of gasoline. Since $d\eta^{-1} = \gamma^{AGC}$, the expression simplifies to

$$\Delta\hat{\gamma}_i^{FER} = -\left(\Delta\hat{\gamma}_i^{AC} + \Delta\hat{\gamma}_i^{NACA} + \Delta\hat{\gamma}_i^{NACAV} + \Delta\hat{\gamma}_i^{AFFS} + \Delta\hat{\gamma}_i^{EQP} + \Delta\hat{\gamma}_i^{IDR}\right)\hat{\mu}^{FER} \tag{13}$$

In conclusion, net traffic rebound of automated vehicles is estimated in our model as:

$$\Delta\hat{\gamma}^{TR} = \Delta\hat{\gamma}^{IDR} + \Delta\hat{\gamma}^{FER} \tag{14}$$

Details of the model, including all the above-mentioned parameters as well as the underlying assumptions and sources used in this analysis, are available as part of the supporting documents upon request. Table 2 provides a summary of the key parameter values by category and technology. Due to limitation of space, documentation on how these parameters were driven is omitted from the main text, but Table 2 provides a general idea of the magnitude of these values and the importance of each technology and category.

Fuel Saving for All Road Users

Automated vehicle technologies not only save fuel for their owners but they could also save fuel for ARU, even conventional vehicle drivers. Automated vehicles could influence the fuel consumption of ARU by reduction in accident-related congestion and non-accident-related congestion, but also by induced traffic demand. Total change in fuel consumption (saving) of automated vehicles for ARU is determined as:

$$\Delta\hat{\gamma}^{TOTA} = \left(1 - \left(1 - \Delta\hat{\gamma}^{AC}/\gamma^{AGC}\right) \times \left(1 - \Delta\hat{\gamma}^{NACA}/\gamma^{AGC}\right) \times \left(1 - \Delta\hat{\gamma}^{IDR}/\gamma^{AGC}\right)\right) \times \gamma^{AGC} \tag{15}$$

where $\Delta\hat{\gamma}^{TOTA}$ is total fuel consumption change of automated vehicles for ARU.

Results

We used stochastic simulation to trace the changes with respect to uncertain variables within our model. Triangular probability distributions were used based on minimum, maximum, and mode values. We estimated these values from a thorough review of the existing literature. We also considered market penetration rate in our model and analysis, as it can significantly change the magnitude of the impact of automated features. To simplify market penetration analyses, the penetration rate of all technologies is assumed to be equal (e.g., 30% penetration rate for all automated technologies). We simulated our model several times using different penetration rates.

Fuel Saving For Automated Vehicle Owners

Figure 2 shows the fuel saving impacts of automated technologies for AVOs based on 50% penetration rate (i.e., when half of all vehicles on the road are equipped with all the automated technologies considered in this study). The impacts of warning systems, lane keeping, collision detection breaking, and active braking systems are negligible (i.e., less than ±0.12%). While these features save a small amount of fuel by reducing accident-related congestion, their operation load offsets most of the gained benefit. Parking aid systems and dynamic route guidance (i.e., information systems) could save 3.5% and 4.5% of total fuel consumption, respectively. These savings result from less non-accident-related congestion. The original savings in this category (4.1% and 5.2% for parking aid systems and route guidance respectively) are reduced by the rebound effect by 13.5%. ACC and CACC have the highest energy saving potentials for AVOs; 6.8% and 4.6% respectively. The original fuel savings are estimated to be 7.1% and 4.9% for ACC and CACC respectively. However, ACC and CACC savings are reduced by traffic rebound and operation load.

We conducted trend analysis to realize the effect of the market penetration rate of automated technologies on energy savings. Figure 3 illustrates the energy saving trends for the automated technology groups (i.e., warning systems, information systems, and control systems). The fuel consumption of warning systems is higher than their fuel saving. Their net energy saving varies from −0.51% to −0.38% for 1% to 100% penetration rate of warning technologies. Information systems could save 6% to 9.9% for AVOs based on different penetration rates of information systems. Due to overlap of ACC and CACC technologies in the control system category (i.e., CACC and ACC cannot operate simultaneously), which might cause double counting, we distinguished between the two control systems:

1. Control systems with ACC (excluding CACC)
2. Control systems with ACC in a connected environment (CACC)

Fuel saving from control systems with ACC is not associated with the number of vehicles equipped with

able 2. Key Parameters Used for the Analysis of Low-Level Automation Technologies

T	% of accident-related congestion reduced by technology $i\,\hat{\rho}_i^{AC}$				% of non-accident-related congestion reduced by technology $i\,\hat{\rho}_i^{NAC}$				% of fuel saving capability of technology $i\,\hat{\rho}_i^{NACAV}$				% of fuel saving capability of technology $i\,\hat{\rho}_i^{AFFS}$				Average energy consumption rate (watts) of automated technology equipment, and portion of driving time when equipment is active (P_{ij}^{EQP}, $\hat{\phi}_{ij}^{ACT}$)			
	Low	Mode	High	Rationale and source	Low	Mode	High	Rationale and source	Low	Mode	High	Rationale and source	Low	Mode	High	Rationale and source	Camera	Radar	Steering	Brake
W LD	0%	1.5%	3%	1.21% (4); 7% of fatal crashes (6) equal to 0.035% of all crashes (*18*); 3% (2)													15, 1			
Collision warning	2%	3%	4%	7% of rear-end (equal to 1.9% of all accidents) (*31*); 3.97%(4)														20, 1		
BSW	0.5%	3.65%	6.8%	0.53% (4); 6.8% (2)														20, 1		
Speed limit detection	2.8%	3.2%	3.6%	10–13% of injury crashes and 18–24% of fatal crashes (*5*); equal to 2.8–3.6% of all crashes (*18*)													15, 1			
raffic warning	0%	0.5%	1%	0.5% (*32*)																
TCC A	1.7%	4.5%	7.25 %	4–7% (*6*); 5.9% of rear-end equal to 1.71% of all (3); 25% of rear-end equal to 7.25% of all (33)									7%	13.5%	20%	7.59–19.69% (*34*) 13.38 % (*35*)		20, 0.55		60, 0.55
CC CA	1.7%	4.75 %	8.75 %	A conservative range between ACC and platooning accident reduction capabilities (*3, 6, 33, 36*)	0%	50%	100%	It is varying from 0% to 100% for different market penetration rates (*19*)					0%	18.5%	37%	Up to 37% fuel saving (*35*)		20, 0.55×M_i		60, 0.55×M_i
LKA	2.5	3.2	4.2	8.9 – 15.2% fatality and injury crashes; equal to 2.5–4.2% of all crashes (*18, 37*) 2.9% all crash (*38*)													15, 1		60, negligible	
Collision detection	3	3.5	4	7% fatality and injury of rear-end and fixed obstacle crashes equal to braking 3.5% of all crashes (*37*)														20, 1		60, negligible
Active braking	5%	15%	25%	Range aggregated 5–25% (*3*)														20, 1		60, negligible
Parking aid system					1%	6%	11%	15% of all VMT on CBD (*22, 39*).	2%	4%	6%	4% fuel saving (*9*).								
Dynamic oute guidance					10%	15%	20%	(*3*)	3%	5%	7%	5% fuel saving (*9, 20*).								

Note: Blank areas assumed negligible impact by the technology on fuel consumption.
ϕ_{IDR} = induced traffic elasticity of automated vehicles assumed values of $\hat{\phi}^{IDR}$ = (0, 0.11, 0.26) (*27, 29, 30, 40, 41*).
ϕ_{FER} = (.05, .1, 0.25) (*18*) Automated features increase fuel efficiency therefore reduce cost of driving. As a result, more trips can be generated. In regulatory impact assessment of new corporate average fuel economy regulations, NHTSA (*18*) reviewed recent rebound research showing a declining rebound effect over time and used a baseline rebound value of 10%, with a sensitivity analysis using 5% and 25% bounds.

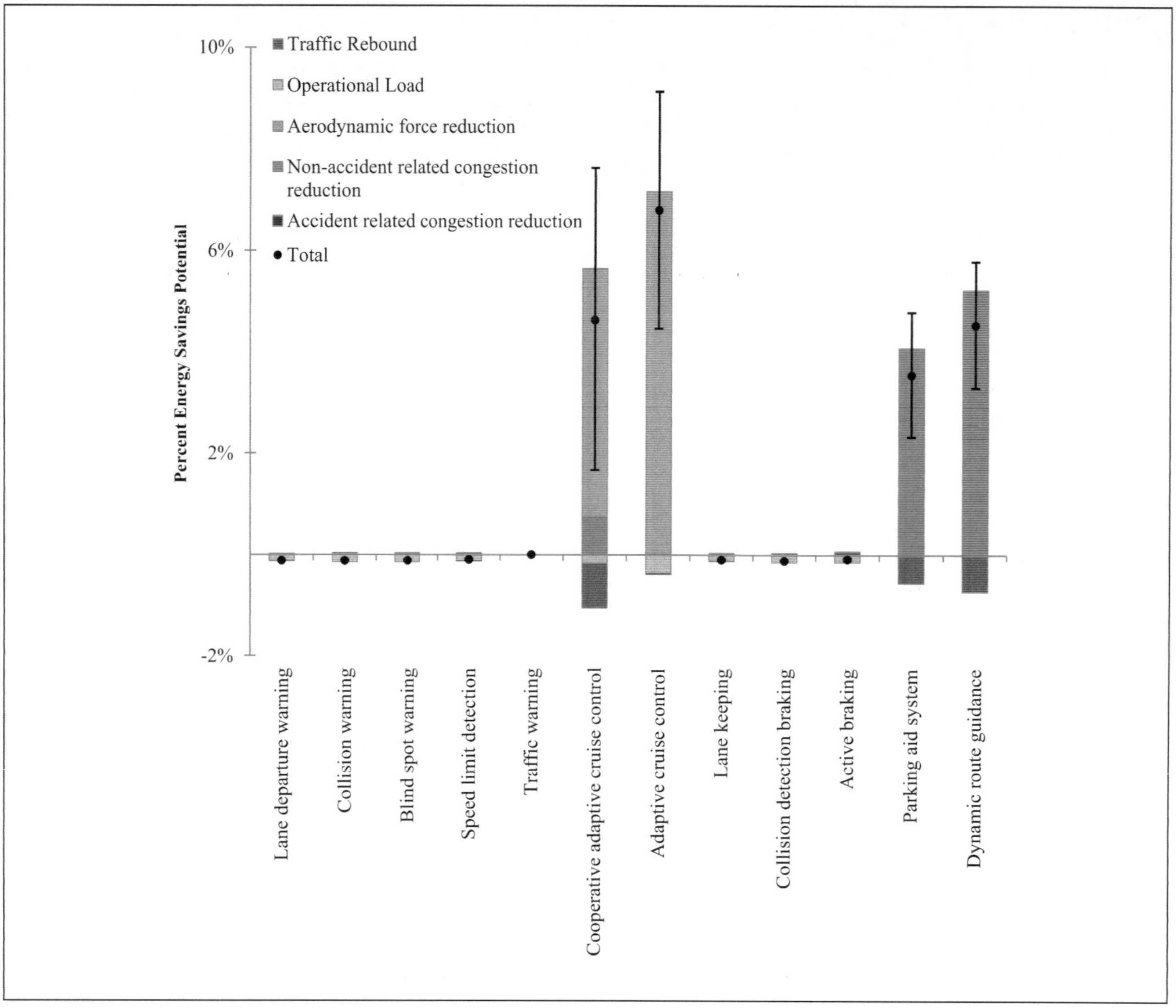

Figure 2. Fuel saving potential of automated technologies at 50% market penetration rate with 90% confidence interval.

these features. ACC provides 6.5% fuel saving at any penetration rate. In contrast, fuel saving of ACC in a connected environment is strongly correlated with adoption of the technologies. Whereas its energy saving is −0.3% at 1% penetration, it could reduce fuel consumption by up to 15.2% when connected control systems are installed on all vehicles.

Figure 4 presents fuel savings for both ACC and CACC. All automated features with ACC show a smooth trend (i.e., varies a small amount, from 13.3% to 13.8%, for 1% to 100% penetration rates respectively) (Figure 4*a*). In contrast, total energy saving of automated vehicles in a connected environment that would include CACC is strongly correlated with penetration rate (i.e., varies from 7% to 16.2% saving for 1% to 100% penetration rates respectively) (Figure 4*b*). Hence, vehicles equipped with all the automated features considered in this study (excluding CACC) should expect 13.3% to 13.8% of net fuel saving, regardless of the number of automated vehicles in the market. However, vehicles equipped with all the automated features considered in this study (excluding ACC) could expect higher potential fuel savings as the number of automated vehicles in the market increases (i.e., from 7% to 16.2% net saving).

Fuel Saving for All Road Users

The fuel saving impact of automated vehicles on ARU is shown in Figure 4*c–d* for different market penetration rates. Fuel saving of ARUs is highly influenced by the market penetration rate of the automated vehicle technologies. Fuel saving of automated features with ACC (i.e., excluding CACC) is less than 0.2% at a low penetration rate (i.e., less than 25%); however, vehicles

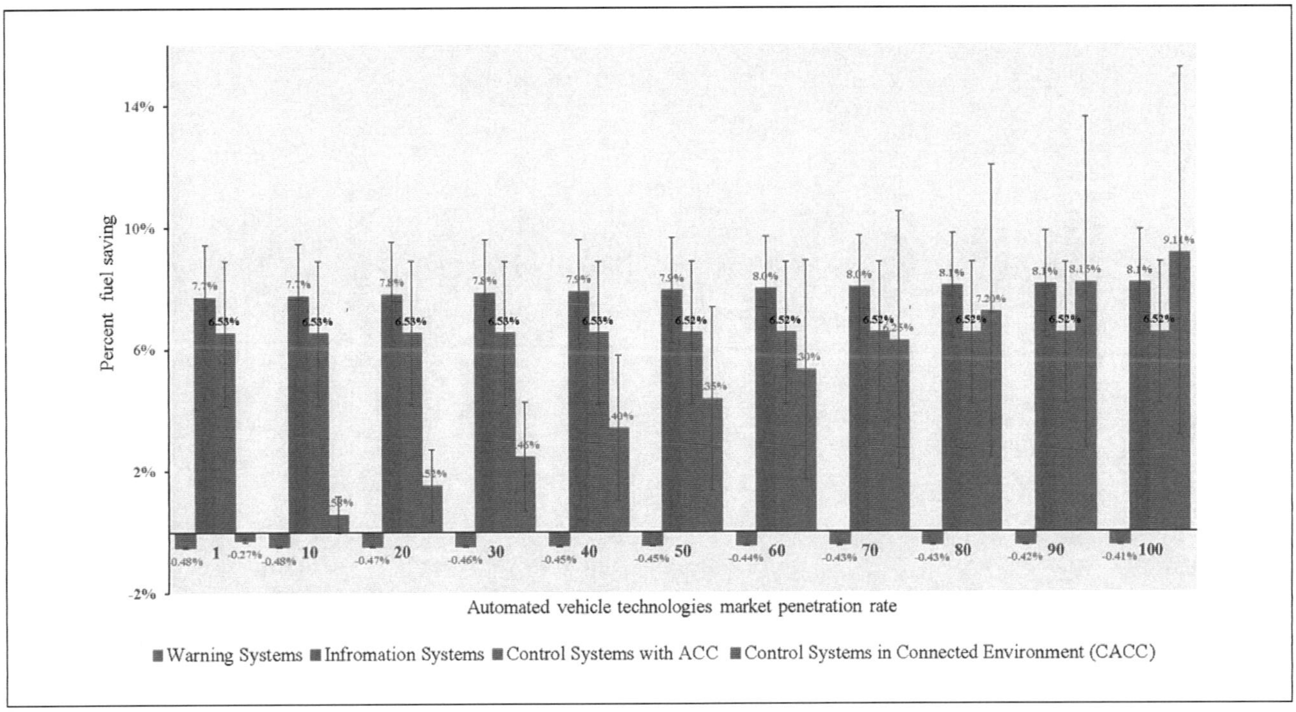

Figure 3. Fuel saving (%) for AVOs based on automated technology groups (with 90% confidence interval).

equipped with automated features could save up to 0.76% for ARU at higher penetration rates (Figure 4*c*). In addition, automated features in a connected vehicle-to-vehicle (V2V) environment (i.e., excluding ACC) could save up to 1.9% of total fuel for all vehicles on road (Figure 4*d*).

Discussion

Combined low-level automated vehicle technologies (those considered in this study) could reduce fuel consumption of light-duty vehicles by 5% to 23%. This saving is a result of reduction of accident-related and non-accident-related congestion, reduction of aerodynamic force, traffic rebound, and operation load. Table 3 presents the ranges of fuel savings for these benefit-cost categories for AVO. Moreover, low-level automated vehicles could reduce the fuel consumption of all vehicles on the road. Our results demonstrate that low-level automated vehicles could save up to 9.8 gallon/vehicle/year for ARU. Table 3 shows the potential fuel saving for ARU for each benefit-cost category and compares the saving ranges for AVO and ARU.

Table 4 summarizes these studies while comparing their results with the result of our study. The table describes what type of benefit-cost analysis was conducted for each study. It shows which study considered fully AV versus individual automated vehicle systems. We also identified what types of impacts were included in each study, followed by the results. In our study, owners of low-level automated vehicles (those equipped with all technologies considered in this study) could reduce fuel consumption by 27 to 119 gallon/year/vehicle. This saving equals from 6% to 23% of average fuel consumption in the U.S. and it could save \$60 to \$266 (in 2016 values) each year for vehicle owners (*42*).

Table 3. Comparison of Fuel Savings for ARU and AVO

Benefit/cost category	ARU (percentage saving)	AVO (percentage saving)
Reduced accident-related congestion	0 to 0.3	0 to 0.3
Reduced non-accident-related congestion	0 to 2.1	8.8 to 10.7
Aerodynamic force reduction	0	7.2 to 16.3
Operational load	0	−1.3 to −1.7
Traffic rebound	−0.5 to 0	−1 to −3.2

In comparison with other studies, our result demonstrates that, regardless of the penetration rate, automated vehicles equipped with a combination of warning, control, and information systems always result in fuel savings. The results of other studies include both potential negative and positive impacts (*8*, *9*, *11*, *12*). Their analyses predict that AV might increase annual fuel

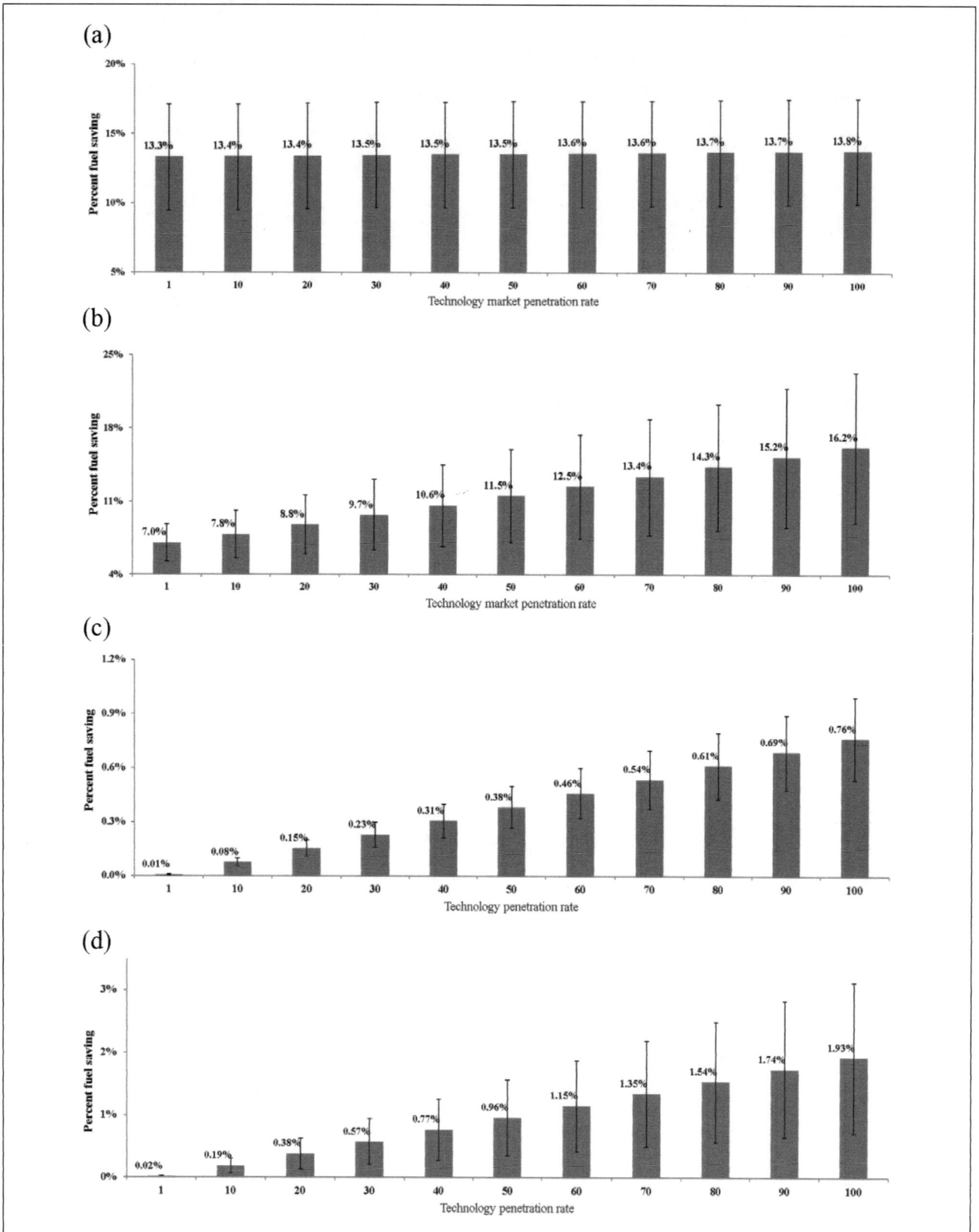

Figure 4. Fuel savings (%) of AVO and ARU for different penetration rates of the technologies: (*a*) fuel saving of all automated features with ACC for AVO; (*b*) fuel saving of all automated features in a connected V2V (vehicle-to-vehicle) environment for AVO; (*c*) fuel saving of all automated features with ACC for ARU; and (*d*) fuel saving of all automated features in a connected V2V environment for ARU.

Table 4. Comparison of Energy Impact Analysis of Automated and AV Systems

Study	Type of impact analysis	Scope	Analysis factors	Cost savings in $\$_{2016}$/VMT
This Study	Energy saving	Individual low-level automated light-duty vehicle	Accident congestion reduction Non- accident congestion reduction Air drag reduction Operation load Travel rebound	\$60 to \$266 saving per year per vehicle
Brown et al. (2014) (*9*)	Energy saving	Autonomous light-duty Vehicles	Efficient driving Air drag reduction Travel rebound Efficient routing Travel by underserved population Vehicle size optimization Fast parking Higher occupancy Vehicle electrification	−\$2063 to \$1132 saving per year per vehicle
MacKenzie et al. (2014) & Wadud et al. (2016) (*11*, *12*)	Energy saving	Autonomous heavy-duty and light-duty vehicles	Improved crash avoidance Congestion mitigation Eco-driving Air drag reduction De-emphasized performance, Vehicle right-sizing Higher highway speed Reduction in generalized costs New user groups Car-sharing Increased features Potential for low carbon transition	−\$975 to \$542 saving per year per vehicle
Stephens et al. (2016) (*8*)	Energy saving	Partial and fully connected shared/non-shared autonomous light-duty vehicles	Fast travel Air drag reduction V2I communication Collision avoidance Smooth traffic flow Vehicle resize Empty vehicles Ridesharing Search for parking spot Easier travel Travel mode change New users	−\$764 to \$2442 saving per year per vehicle
Greenblatt et al. (2015) (*7*)	Energy saving	SAV	VMT covered by shared AV	\$ 0.21 per road user
Harper (2016) (*4*)	Safety	Individual low-level automated light-duty vehicle	Crash reduction benefits Equipment installation costs	\$20 to 877 saving per year per vehicle

consumption due to increase in average VMT. AV also might induce new user groups to travel on the roads (e.g., elderly people, or teenagers). Our study demonstrates that although low-level automated vehicles increase traffic rebound and annual VMT, this increase is less than the fuel saving potentials associated with low-level automated vehicles. It is important to note that automated vehicles, which have limited control abilities, are not able to attract new user groups as AV might do.

In this study, we conducted a comprehensive review of the literature on the energy and safety impacts of automated features. This thorough assessment provided precise data for our analyses. As a result, the uncertainty of calculations is significantly reduced in comparison with similar studies. Our estimation (i.e., annual \$60 to \$266 saving) narrows down the uncertainty of predicting energy impact in comparison with similar studies (Table 4).

Conclusion

In this paper, we assessed the energy impact of low-level automated vehicle technologies. We conducted a comprehensive literature review of the safety, congestion, fuel efficiency, operation load, and traffic rebound impacts of automated features, including research papers, real experiments, and road testing data associated with individual automated technologies (*2, 4, 6, 9, 17, 19, 20, 31, 36, 40, 43*). The thorough literature review made detailed modeling of the impacts possible. We developed a stochastic method to consider both induced demand and fuel efficiency rebound effects, which is one of the main contributions of the paper. In addition, we modeled the operation load of automated features. These types of potential negative impacts of automated vehicles have not been thoroughly considered in previous studies. We also used market penetration rate of automated technologies to predict the current, and near- to mid-term, impacts of these technologies. Analysis of trends in automated features based on penetration rate can assist decision makers to estimate the fuel savings to be gained from the technologies at any given time.

We estimate that low-level automated vehicles could reduce fuel consumption from 27 to 119 gallon/year/vehicle. These technologies could save \$60 to \$266 (2016 values) annually for AVO. Automated vehicles could also reduce fuel consumption of other vehicles. It could save up to about 10 gallon/vehicle/year which has a total value of up to \$6 billion annually. Our study demonstrates the positive impact of automated features on energy savings and this impact becomes more significant as penetration increases. Low-level automated vehicles provide environmental benefits for society (i.e., reduce GHG emission). Hence, authorities should promote development and deployment of low-level automated vehicle technologies.

This study has a few limitations and future work opportunities. First, we do not consider interactions of technologies and their dependencies. Automated features might have interactions with each other that could affect the overall saving potential (perhaps lowering it). For instance, LDW and LKA might have some interactions in their impact levels. Interactions and dependencies should be considered in the next versions of the model. The exact timeline of market penetration of automated technologies and their penetration trends are not clear and in a few studies they were just estimated (*43, 44*). We considered equal penetration rates for all technologies. However, different levels of penetration of automated technologies could cause varying results (e.g., negative total fuel impact). These scenarios should be investigated in future studies. Moreover, we only analyzed the energy impacts of automated vehicles. Ideally, the model should be expanded to include a holistic analysis of automated vehicles; one that could capture not only potential energy savings but also social, economic, mobility, and safety impacts.

Acknowledgments

The authors sincerely thank Professor Jeremy Michalek from Carnegie Mellon University, who co-developed some of the early models adopted for this study.

Author Contributions

The authors confirm contribution to the paper as follows: study conception and design: Yeganeh M. Hayeri, Chris H. Hendrickson; data collection: Saeed Vasebi, Yeganeh M. Hayeri; analysis and interpretation of results: Yeganeh M. Hayeri, Saeed Vasebi, Constantine Samaras; draft manuscript preparation: Yeganeh M. Hayeri, Saeed Vasebi. All authors reviewed the results and approved the final version of the manuscript.

References

1. *Use of Gasoline*. U.S. Energy Information Administration, 2017. https://www.eia.gov/energyexplained/index.cfm?page=gasoline_use. Accessed June, 2017.
2. Jermakian, J. S. Crash Avoidance Potential of Four Passenger Vehicle Technologies. *Accident Analysis & Prevention*, Vol. 43, No. 3, 2011, pp. 732–740.
3. Bayly, M., B. Fildes, M. Regan, and K. Young, Review of Crash Effectiveness of Intelligent Transport Systems. *Emergency*, Vol. 3, No. 14, 2006.
4. Harper, C. D., C. T. Hendrickson, and C. Samaras. Cost and Benefit Estimates of Partially-Automated Vehicle Collision Avoidance Technologies. *Accident Analysis & Prevention*, Vol. 95, 2016, pp. 104–115.

5. Carsten, O. M., and F. N. Tate. Intelligent Speed Adaptation: Accident Savings and Cost–Benefit Analysis. *Accident Analysis & Prevention*, Vol. 37, No. 3, 2005, pp. 407–416.
6. Anderson, R., T. P. Hutchinson, B. Linke, and G. Ponte. *Analysis of Crash Data to Estimate the Benefits of Emerging Vehicle Technology*. Centre for Automotive Safety Research, The University of Adelaide, Australia, 2010.
7. Greenblatt, J. B., and S. Saxena. Autonomous Taxis Could Greatly Reduce Greenhouse-Gas Emissions of US Light-Duty Vehicles. *Nature Climate Change*, Vol. 5, No. 9, 2015, pp. 860–863.
8. Stephens, S., J. Gonder, Y. Chen, Z. Lin, C. Liu, and D. Gohlke. *Estimated Bounds and Important Factors for Fuel Use and Consumer Costs of Connected and Automated Vehicles*. National Renewable Energy Laboratory, report number NREL/TP-5400-67216, Golden, CO, 2016.
9. Brown, A., Gonder, J. and Repac, B., An analysis of Possible Energy Impacts of Automated Vehicle. Road Vehicle Automation, Lecture Notes in Mobility, pp. 137-153, Springer, Switzerland, 2014, DOI: 10.1007/978-3-319-05990-7_13.
10. National Highway Traffic Safety Administration. *Preliminary Statement of Policy Concerning Autonomous Vehicles*. U.S. Department of Transportation, 2013. http://www.nhtsa.gov/About+NHTSA/Press+Releases/U.S.+Department+of+Transportation+Releases+Policy+on+Automated+Vehicle+Development
11. MacKenzie, D., Z. Wadud, and P. Leiby. A First Order Estimate of Energy Impacts of Automated Vehicles in the United States. Presented at Annual Meeting of the Transportation Research Board, Washington, D.C., 2014.
12. Wadud, Z., D. MacKenzie, and P. Leiby. Help or Hindrance? The Travel, Energy and Carbon Impacts of Highly Automated Vehicles. *Transportation Research Part A: Policy and Practice*, Vol. 86, 2016, pp. 1–18.
13. Shladover, S. Research and Policy Actions Needed To Support Implementation. Presented at Annual Meeting of the Transportation Research Board, Washington, D.C., 2017.
14. *How Much Does the United States Consume?* U.S. Energy Information Administration, 2016. http://www.eia.gov/tools/faqs/faq.cfm?id=23&t=10 Accessed July, 2017.
15. MacLean, H. L., and L. B. Lave. Peer Reviewed: A Life-Cycle Model of an Automobile. *Environmental Science & Technology*, Vol. 32, No. 13, 1998, pp. 322A–330A.
16. Schrank, D., B. Eisele, T. Lomax, and J. Bak. *2015 Urban Mobility Scorecard*. Texas A&M Transportation Institute, College Station, TX, 2015.
17. Blincoe, L. J., T. R. Miller, E. Zaloshnja, and B. A. Lawrence. *The Economic and Societal Impact of Motor Vehicle Crashes, 2010. (Revised)*. DOT HS 812 013. Department of Transportation, National Highway Traffic Safety Administration, 2015. https://crashstats.nhtsa.dot.gov/Api/Public/ViewPublication/812013
18. National Highway Traffic Safety Administration. *Traffic Safety Facts 2015*. FARS GES Annual Report. Department of Transportation, 2016. https://crashstats.nhtsa.dot.gov/#/DocumentTypeList/12
19. Shladover, S., D. Su, and X.-Y. Lu. Impacts of Cooperative Adaptive Cruise Control on Freeway Traffic Flow. *Transportation Research Record: Journal of the Transportation Research Board*, 2012. 2324: 63–70.
20. Gonder, J., M. Earleywine, and W. Sparks. Analyzing Vehicle Fuel Saving Opportunities through Intelligent Driver Feedback. *SAE International Journal of Passenger Cars: Electronic and Electrical Systems*, Vol. 5, No. 2012-01-0494, 2012, pp. 450–461.
21. Department of Energy. *Where the Energy Goes: Gasoline Vehicles*. U.S. Department of Energy. http://www.fueleconomy.gov/feg/atv.shtml Accessed October, 2017.
22. Federal Highway Administration. *Highway Statistics 2015: Vehicle-Miles of Travel, by Functional System, 1980-2014*. U.S. Department of Transportation, 2015. https://www.fhwa.dot.gov/policyinformation/statistics/2015/.
23. Robert Bosch GmbH. *Bosch Automotive Handbook*, 9th ed. Bentley Publishers, Cambridge, MA, 2014.
24. Triplett, T., R. Santos, S. Rosenbloom, and B. Tefft. *American Driving Survey: 2014–2015*. 2016. https://www.aaafoundation.org/sites/default/files/AmericanDrivingSurvey2015.pdf
25. Department of Energy. *Quadrennial Technology Review 2015: Internal Combustion Engines*. U.S. Department of Energy, 2015. https://energy.gov/sites/prod/files/2015/11/f27/QTR2015-8C-Internal-Combustion-Engines.pdf
26. National Highway Traffic Safety Administration. *Final Regulatory Impact Analysis: Corporate Average Fuel Economy for MY 2017-MY 2025 Passenger Cars and Light Trucks*. U.S. Department of Transportation, 2012. http://www.nhtsa.gov/staticfiles/rulemaking/pdf/cafe/FRIA_2017-2025.pdf
27. Hymel, K. M., K. A. Small, and K. V. Dender. Induced Demand and Rebound Effects in Road Transport. *Transportation Research Part B: Methodological*, Vol. 44, 2010, pp. 1220–1241.
28. Goodwin, P., J. Dargay, and M. Hanly. Elasticities of Road Traffic and Fuel Consumption with Respect to Price and Income: A Review. *Transport Reviews*, Vol. 24, No. 3, 2004, pp. 275–292.
29. Barr, L. C. Testing for the Significance of Induced Highway Travel Demand in Metropolitan Areas. *Transportation Research Record: Journal of the Transportation Research Board*, 2000. 1706: 1–8.
30. Haughton, J., and S. Sarkar. Gasoline Tax as a Corrective Tax: Estimates for the United States 1970–1991. *The Energy Journal*, Vol. 2, 1996, pp. 103–126.
31. Regan, M. A., E. Mitsopoulos, N. Haworth, and K. Young. *Acceptability of In-Vehicle Intelligent Transport Systems to Victorian Car Drivers*. Monash University Accident Research Centre, Melbourne, Australia, 2002.
32. Lind, H. *An Efficient Visual Forward Collision Warning Display for Vehicles*. SAE Technical Paper No. 2007- 01-1105. Society of Automotive Engineers, Detroit, 2007.
33. Abele, J., C. Kerlen, S. Krueger, and H. Baum, T. Geißler, S. Grawenhoff, J. Schneider, and W.H. Schulz. *Exploratory Study on the Potential Socio-Economic Impact of the Introduction of Intelligent Safety Systems in Road Vehicles*. Final Report. SEISS, Teltow, Germany, 2005.
34. Luo, L. H., H. Liu, P. Li, and H. Wang. Model Predictive Control for Adaptive Cruise Control with Multi-

Objectives: Comfort, Fuel-Economy, Safety and Car-Following. *Journal of Zhejiang University SCIENCE A*, Vol. 11, No. 3, 2010, pp. 191–201. https://doi.org/10.1631/jzus.A0900374.

35. Park, B., K. Malakorn, and J. Lee. *Quantifying Benefits of Cooperative Adaptive Cruise Control toward Sustainable Transportation System*. Center Transportation Studies, University of Virginia, Charlottesville, VA, 2011.
36. Kunze, R., M. Haberstroh, R. Ramakers, K. Henning, and S. Jeschke. Automated Truck Platoons on Motorways – A Contribution to the Safety on Roads. *Automation, Communication and Cybernetics in Science and Engineering 2009/2010*, 2011, Springer, Berlin, Heidelberg, pp. 415–426.
37. Wilmink, I., W. Janssen, E. Jonkers, K. Malone, M. van Noort, G. Klunder, P. Rämä, N. Sihvola, R. Kulmala, A. Schirokoff, G. Lind, T. Benz, H. Peters, and S. Schönebeck. *Impact Assessment of Intelligent Vehicle Safety Systems*. eIMPACT Deliverable D4. Version 1.0, 2008. https://trimis.ec.europa.eu/sites/default/files/project/documents/20130401_140253_77423_eIMPACT_D9_D10_v2.0.pdf
38. Kulmala, R., P. Rämä, and N. Sihvola. Safety Impacts of Cooperative Systems. *Proc., 21st International Co-operation on Theories and Concepts in Traffic Safety Workshop*, Vienna, Austria 2008, Vienna, Austria, pp. 288–293. http://citeseerx.ist.psu.edu/viewdoc/download?doi=10.1.1.533.5412&rep=rep1 & type=pdf
39. Shoup, D. C. Cruising for Parking. *Transport Policy*, Vol. 13, No. 6, 2006, pp. 479–486.
40. Fulton, L. M., R. B. Noland, D. J. Meszler, and J. V. Thomas. A Statistical Analysis of Induced Travel Effects in the US Mid-Atlantic Region. *Journal of Transportation and Statistics*, Vol. 3, No. 1, 2000, pp. 1–14.
41. Cervero, R., and M. Hansen. Induced Travel Demand and Induced Road Investment: A Simultaneous Equation Analysis. *Journal of Transport Economics and Policy*, Vol. 36, No. 3, 2002, pp. 469–490.
42. Gasprices. *Gas Prices*. 2017. http://gasprices.aaa.com/ Accessed January 10, 2017.
43. Kyriakidis, M., C. Van de Weijer, B. Van Arem, and R. Happee. The Deployment of Advanced Driver Assistance Systems in Europe. Presented at 22nd ITS World Congress, Bordeaux, France, 2015. http://dx.doi.org/10.2139/ssrn.2559034
44. Navigant Research. *Autonomous Vehicles, Self-Driving Vehicles, Autonomous Parking, and Other Advanced Driver Assistance Systems, Global Market Analysis and Forecasts*, 2013. http://www.navigantresearch.com/wp-content/uploads/2013/08/ASDV-13-Executive-Summary.pdf

The Standing Committee on Transportation Energy (ADC70) peer-reviewed this paper (18-06702).

TRR
JOURNAL OF THE TRANSPORTATION RESEARCH BOARD

Article

Transportation Research Record
2018, Vol. 2672(24) 75–86

DOI: 10.1177/0361198118787087
journals.sagepub.com/home/trr
SAGE

The Joint Effects of Income, Vehicle Technology, and Rail Transit Access on Greenhouse Gas Emissions

Marlon G. Boarnet[1], Raphael W. Bostic[2], Andrew Eisenlohr[1], Seva Rodnyansky[1], Raúl Santiago-Bartolomei[1], and Huê-Tâm Webb Jamme[1]

Abstract

This paper examines the relationship between income, vehicle miles traveled (VMT), and greenhouse gas (GHG) emissions for households with varying access to rail transit in four metropolitan areas—Los Angeles, the San Francisco Bay Area, San Diego, and Sacramento—using data from the 2010–2012 California Household Travel Survey. Daily vehicle GHG emissions are calculated using the California Air Resources Board's 2014 EMFAC (emission factors) model. Two Tobit regression models are used to predict daily VMT and GHG by income, rail transit access (within or outside 0.5 miles of a rail transit station in Los Angeles and the Bay Area, and linear distance to rail in San Diego and Sacramento), and metropolitan area. Comparing predicted VMT and GHG emissions levels, this paper concludes that predicted VMT and GHG emission patterns for rail access vary across metropolitan areas in ways that may be related to the age and connectivity of the areas' rail systems. The results also show that differences in household VMT due to rail access do not scale proportionally to differences in GHG emissions. Regardless, the fact that GHG emissions are lower near rail transit for virtually all income levels in this study implies environmental benefits from expanding rail transit systems, as defined in this paper.

Many metropolitan areas see rail transit expansion as an opportunity to pursue multiple policy goals simultaneously. Such policy goals could be: (1) building more affordable housing, taking advantage of infill opportunities near rail transit stations, and (2) reducing vehicle miles traveled (VMT) and greenhouse gas (GHG) emissions. The latter is achieved, in theory, by providing near-transit residences for lower income persons who are more likely to depend on rail transit and drive less, whereas higher income persons drive more and take transit less (*1*). Recent evidence suggests a potential challenge to this approach as higher income households might reduce their driving more if they live near rail precisely because they drive more when living far from rail, and therefore can make larger reductions when near it. In fact, Chatman et al. show in cross-sectional comparisons that the VMT difference among households within and beyond a half-mile area from rail stations is 9.86 miles per day for households below $50,000 in income, and 20.47 miles per day for households above $100,000 in income (*2*).

We expand on Chatman et al. by considering GHG emissions directly instead of using VMT as a proxy. Older vehicles are generally more polluting, so assuming that vehicle vintage and technology are closely related to income (*3*), greatest VMT reduction may not necessarily parallel GHG reduction. This study's objective is twofold: (1) To begin to bridge the gap between VMT and GHG emissions in travel behavior studies, and (2) To examine the relationship between income, VMT, vehicle GHG emissions, and rail transit access. This study observes four California metropolitan areas; considers how the relationships between income, VMT, vehicle GHG emissions, and rail transit access vary by metropolitan area when accounting for vehicle technology; and posits mechanisms suggested by the varying relationships.

Using data from the 2010–2012 California Household Travel Survey (CHTS), two Tobit regression models are used to predict the effects of household characteristics (income and location, i.e., rail transit access) first on household VMT and then on household vehicle GHG emissions. The primary finding is that the pattern of GHG emissions across households and incomes, living near versus far from rail transit, is different from the pattern of VMT. This indicates that VMT may be an incomplete policy proxy when analyzing environmental impacts of travel behavior for those living near rail transit. That pattern also varies across regions in ways that

[1]Department of Urban Planning and Spatial Analysis, Sol Price School of Public Policy, University of Southern California, Los Angeles, CA
[2]Federal Reserve Bank of Atlanta, Atlanta, GA

Corresponding Author:
Address correspondence to Andrew Eisenlohr: eisenloh@usc.edu

suggest that differences in metropolitan land use and rail transit networks are important for the income–VMT–GHG relationship. It would be valuable for the literature to return to analyses that include vehicle GHG emissions and vehicle stock more directly.

Literature Review

VMT and Income. VMT tends to increase with income (*4*, *5*). Zegras argues that this is because higher income households tend to own more vehicles, not because higher income households have higher levels of VMT per vehicle (*4*). Holtzclaw et al. suggest a more direct relationship: both the number of household vehicles and the level of household VMT are directly related to per capita income (*5*).

VMT and Urban Form. Kain and Fauth showed that differences in urban form help explain household vehicle ownership, which is correlated with VMT (*6*, *7*). Cervero and Kockelman found that neighborhoods with higher density, mixed-use structures, and pedestrian-friendly areas all are statistically significantly associated with lower levels of VMT (*8*). The reviews by Ewing and Cervero and Zhang et al. produced similar results (*9–11*). Baum-Snow and Kahn concluded that households who live closer to public transit tend to use public transit more (and holding other factors constant, have higher property values and pay higher rents) (*12*). These conclusions regarding the built environment support the hypothesis that VMT should be lower near rail transit stations.

Income and Rail Transit Access. Boarnet et al. show that in Los Angeles, higher income households (annual incomes above $100,000) living within 0.5 miles of rail transit have less VMT than their income peers living outside 0.5 miles, and the near-versus-far VMT gap is highest for high income households (*3*). This highlights a challenge for research and practice given the current efforts in many metropolitan areas to increase the stock of affordable housing and reduce GHG emission by locating such units near transit. This strategy is based on the now contested assumption that lower income households will reduce VMT the most when living near rail transit (*13*). This approach has used VMT as a proxy for GHG emissions—a common practice in land use and transportation research.

VMT and GHG. Previous studies indicate that a substantial portion of urban GHG stems from personal vehicle use (*14*, *15*). However, recent literature rarely considers the type of cars driven by households, which may also depend on income. An exception on this topic is Brownstone and Golob, who found that residents of denser neighborhoods drive fewer miles, drive more fuel-efficient cars, and thereby consume less fuel than residents of less dense neighborhoods (*16*).

Research Questions and Hypotheses

This paper examines two research questions. First, does the relationship between VMT and income of households inside and outside 0.5 miles of rail transit (Half-Mile Areas) vary across four California metropolitan areas? Using CHTS data from the Los Angeles, San Francisco Bay Area, Sacramento, and San Diego metropolitan areas, differences across metropolitan areas are expected, mostly because of the significant variety of land use and rail transit development.

Second, does the trend in VMT reduction by income level align with that of GHG emission reductions? The core hypothesis of this paper is that the marginal effect of VMT reductions on GHG emissions may be larger for lower income households, who: (1) tend to drive older, and therefore more polluting vehicles; and (2) are less likely to drive non-gasoline-powered vehicles.

Methods

The research questions are addressed by combining three data sources and measuring the effects of income, location relative to rail transit, and household characteristics on daily VMT and daily vehicle GHG emissions. These measurements are conducted within a 22-county study area of four metropolitan areas in California.

Study Area

California's four largest metropolitan areas have developed or expanded their rail transit systems during the past three decades. Sacramento and San Diego are both growing, monocentric cities. Los Angeles and the San Francisco Bay Area (Bay Area) are large, polycentric regions. As a result, these four metropolitan areas provide an opportunity to observe VMT and GHG emission changes near rail transit across different urban forms.

The universe of households comes from survey respondents to the 2010–2012 CHTS in the 22 counties within the four metropolitan areas in the study: Los Angeles, Orange, Ventura, Riverside, San Bernardino, Imperial, San Diego, Sacramento, El Dorado, Placer, Sutter, Yolo, Yuba, Alameda, Contra Costa, Marin, Napa, San Francisco, San Mateo, Santa Clara, Solano, and Sonoma.

Data

Rail Transit Access Data. The research compares treatment households in Half-Mile Areas (within a half-mile of rail transit) with control households outside Half-Mile Areas. This definition derives from analysis of transit-oriented development (TOD) catchment areas, which concluded that households outside of 0.5 miles utilize transit offered by TOD at a significantly lower rate than households within

0.5 miles (*17*). Access to rail transit was defined specifically as access to a well-served intra-city rail-transit station, that is, subways and light rail.

Both light rail and heavy rail transit modes are ideal candidates for considering changes in travel behavior due to their high volume, frequent service, and intra-city orientation. Of the 702 rail stations of all types in California operating in or prior to 2013, 310 fit this definition of rail transit access. These include the Los Angeles County Metropolitan Transportation Authority subway and light rail, San Diego Trolley light rail, Sacramento Light Rail, Bay Area Rapid Transit subway, and the Santa Clara Valley Transportation Authority light rail. Location data for stations in these transit systems were obtained from Southern California Association of Governments, San Diego Association of Governments, and Metropolitan Transportation Commission directly and using a generalized transit feed specification file for Sacramento Light Rail.

The 392 remaining stations are excluded due to specifics of their mode type and incompatibility with the TOD definition based on passenger volume, service frequency, number of stops, or destination type. For example, cable cars, Muni stops (typically a series of streetcar stops), bus rapid transit, and other bus stations are excluded from the rail transit treatment category due to their low passenger volume, high number of stops, or lack of fixed-route infrastructure.

Household Demographics, VMT, and Vehicle Data. Cross-sectional data on household demographics, travel diary data, and vehicle characteristics were obtained from the 2010–2012 CHTS in the 22-county study area (*18*). CHTS data were combined with rail station location data in a geographic information system for the 310 treatment stations. Household address distance to the nearest rail station was used to determine a household's treatment status, that is, inside or outside of Half-Mile Area, and also served as an explanatory factor in some regressions. Income categories were based on the income distribution for sampled households; for example, insufficient data were available to develop distinct income categories above $100,000.

The 2010–2012 CHTS dataset contains information for 42,426 households in California. Ultimately, 18,892 households were excluded: 5,718 households had empty travel diary data; 3,010 lacked income information; and 10,164 households fell outside the study areas or were missing geographic location information. The final observation counts by metropolitan area were: 12,362 households in Los Angeles; 7,923 in the Bay Area; 1,899 in Sacramento; and 1,350 in San Diego, or 23,534 total observations.

The household travel diary's trip length variable was used to obtain VMT estimates. When estimating VMT for each household, lengths from trips made across three modes were considered for each household member: driving a personal vehicle; riding as a passenger in a personal vehicle; and riding (as driver or passenger) on a motorcycle, scooter, or moped (*18*). VMT for each household was estimated by aggregating trip lengths for all trips across these three modes; trips by a second, third, or more members of a household in the same vehicle were excluded to avoid double counting. Households with travel diary data but no reported trips across the three aforementioned modes were assigned a VMT of zero.

Emission Data. This paper estimates emissions using carbon dioxide (CO_2) emission rates from the California Air Resource Board's (ARB) EMFAC 2014 model for gasoline and diesel vehicles and the EPA's online Fueleconomy.gov database (*19*). Carbon dioxide (CO_2) comprises 82% of U.S. GHG emissions (*20*). EMFAC 2014 inputs—county of residence, model year, fuel type, and vehicle category—were obtained from the CHTS for 33,929 vehicles driven by households living in the 22-county study area. Sedans, SUVs, pickup trucks, coupes, convertibles, hatchbacks, wagons, minivans, and vans were included in the analysis and were matched to the vehicle categories available in the EMFAC 2014 framework based on vehicle type and weight (*19, 21*).

In contrast, 4,210 vehicles were removed from the analysis because motorcycles (only 0.6% of total sample household VMT), recreational vehicles, mopeds, scooters, uncategorizable, and unknown vehicle types did not have EMFAC emissions data available. Emission rates are only available for gasoline, diesel, hybrid, and electric vehicles. We thus excluded natural gas, biofuel, and other, unknown and uncategorized fuel types, which eliminated 100 more vehicles. The 687 vehicles for which the model year was unknown, unreported, or earlier than 1972 were also excluded. Afterwards, 28,932 vehicles remained in the sample, driven by 19,616 households.

Each vehicle's emissions estimates were obtained using trip and VMT data from the CHTS multiplied by emissions rates from the EMFAC 2014 model. A unique emissions rate based on the vehicle body type, county of residence, fuel type, and model year was calculated for each vehicle in the sample. Other EMFAC 2014 parameters were kept constant across vehicles including season and speed. Emission rates vary by speed of driving, which is not in CHTS data, so the "aggregated speed" option was selected to account for an aggregate distribution of speeds. Emissions also vary by season, but season was not recorded in CHTS so the "annual" option was selected as a season-averaged rate.

The EMFAC model yielded emission rates for two processes: STREX, the emissions associated with starting a car, and RUNEX, the emissions associated with driving a car. To obtain total daily CO_2 emissions for vehicle j, the number of trips was multiplied by STREX and the VMT was multiplied by RUNEX; for household emissions, all of the household vehicles' emissions were summed:

Equation 1: Gasoline and Diesel Daily Emissions Calculation

$$Daily\ household\ emissions\left(grams\ CO_2\right)_i = \sum_j [daily\ trips_j * CO_2STREX_j\left(\frac{grams}{trip}\right) + daily\ VMT_j * CO_2RUNEX_j] \quad (1)$$

For hybrid and electric vehicles, it was not possible to use the EMFAC 2014 model so EPA's fueleconomy.gov was used. Hybrid and electric emissions were calculated using fueleconomy.gov's online interface for each vehicle *j* based on model year, make, body type, and fuel type. Per-mile tailpipe emissions rates were recorded for hybrids, and both tailpipe and upstream emissions rates were noted for electric vehicles. Daily average emissions were obtained for these vehicles using VMT from the CHTS travel diaries, demonstrated in Equation 2 (*18*). Household emissions were the summed household vehicles' daily emissions:

Equation 2: Hybrid and Electric Daily Emissions Calculation

$$Daily\ household\ emissions\left(grams\ CO_2\right)_i = \sum_j [daily\ VMT_j * CO_2\ daily\ emissions\ rate_j\left(\frac{grams}{mile}\right)] \quad (2)$$

where *i* represents individual household and *j* represents an individual household vehicle.

Model Specifications

VMT Estimations. A regression model for each metropolitan area was used to predict the effect of household characteristics on VMT. A Tobit regression specification was appropriate as the VMT data from CHTS are censored—7.5% of households in the entire restricted dataset had a VMT of zero on the survey day (6.8% for Los Angeles, 9.1% for the Bay Area, 5.8% for Sacramento, and 8.0% for San Diego). Explanatory factors are those used in present literature and comprise relevant household characteristics, including household income stratum, number of household vehicles, household size, number of household members who are employed, and the distance of a household to the nearest rail transit station. Housing price is not available for use as a control from the CHTS data. Equation 3 describes the model:

Equation 3: Daily VMT Tobit prediction model

$$Y_i \begin{cases} Y_i^* \text{ if } Y_i^* > 0 \\ 0 \text{ if } Y_i^* \leq 0 \end{cases}$$

$$\text{where } Y_i^* = \beta_0 + \sum_{j=1}^{m}\beta_j W_{ij} + \sum_{k=1}^{n}\beta_k X_{ik} + \sum_{j=1}^{m}\sum_{k=1}^{n}\beta_{jk} W_{ij} X_{ik} + \sum_{l=1}^{p}\beta_q Z_i + \varepsilon_i \quad (3)$$

Equation 3 above shows a two-step process for each of the metropolitan area-level models. Y_i* is a latent variable related to Y_i, which is the observed diary-day daily VMT for household *i*. Y_i* is regressed on independent variables:

- W_{ij} = income band dummy variables for income groups *j* ($0–$25,000; $25,001–$50,000; $50,001–$100,000; and >$100,000) for household *i*
- X_{ik} = Half-mile dummy variable for Los Angeles and Bay Area, and continuous distance to nearest rail station *k* for Sacramento and San Diego, for household *i*
- $W_{ij}X_{ik}$ = income band dummy and half-mile dummy/distance interaction terms
- Z_i = set of characteristics for household *i*: household size, number of household vehicles, and number of household members employed

The interaction variables between income strata and living within Half-Mile Areas (or linear distance to rail transit) enable comparisons of VMT estimates inside and outside of Half-Mile Areas by income group. Due to outliers in sampled VMT, the four metropolitan area Tobit models exclude households in the top 1% of VMT. Analyzing the inter-quartile range of the VMT distribution by metropolitan area confirmed this approach for removing outliers.

The regression results for the four metropolitan areas were used to predict VMT levels for households. These predicted VMT levels were based on the regression models' estimated parameters and the households' characteristics observed in the 2010–2012 CHTS. Depending on the nature of the dependent variable, there are several ways to obtain predicted values from Tobit models (*22*). For this analysis, a household's VMT is always greater than or equal to zero. As a result, expected VMT values for households were generated by fitting them to a distribution with minimum value zero. As with the Tobit regressions, households in the top 1% of VMT were excluded when generating predicted VMT values.

GHG Estimations. GHG emissions were estimated directly using household and vehicle characteristics. The dependent variable is average daily household emissions. As with VMT, the Tobit specification was used with the same form as Equation 3. The dependent variable in the emissions model is household vehicle CO_2 emissions on the diary day in grams. Carless (zero-vehicle) households were excluded from this regression because they have zero emissions. As with the VMT regressions, households in the top 1% of VMT were excluded from the regression analysis.

Results

Descriptive Statistics

VMT by Income Level Within and Outside Half-Mile Areas. Across the four metropolitan areas, Table 1 shows

Table 1. Actual Daily VMT for All Households in Regression Models

	Average actual VMT		Sample size		Difference in daily VMT
	Outside 0.5 miles	Within 0.5 miles	Outside 0.5 miles	Within 0.5 miles	(within minus outside)
Bay Area					
$0–$25,000	26.07	13.60	560	88	−12.47
$25,001–$50,000	34.89	24.83	1,043	101	−10.07
$50,001–$100,000	44.87	21.78	2,238	191	−23.09
>$100,000	56.54	37.49	3,379	244	−19.05
All income levels	47.43	27.26	7,220	624	−20.17
Los Angeles					
$0–$25,000	30.05	14.32	1,809	148	−15.73
$25,001–$50,000	40.28	25.76	2,261	121	−14.52
$50,001–$100,000	51.76	36.39	3,920	128	−15.37
>$100,000	61.09	55.31	3,790	62	−5.78
All income levels	49.22	29.03	11,780	459	−20.20
Sacramento					
$0–$25,000	31.21	18.68	201	13	−12.53
$25,001–$50,000	41.26	25.98	356	17	−15.28
$50,001–$100,000	53.38	70.58	727	21	17.19
>$100,000	65.38	66.39	529	17	1.02
All income levels	52.04	48.46	1,813	68	−3.58
San Diego					
$0–$25,000	21.47	15.21	166	29	−6.25
$25,001–$50,000	34.60	32.51	231	14	−2.08
$50,001–$100,000	51.24	50.98	425	19	−0.26
>$100,000	67.70	41.93	436	17	−25.77
All income levels	49.96	32.63	1,258	79	−17.33

Note: Table includes all households in VMT Models, for households within the 99th percentile of VMT.

that, on average, household VMT increases as household income increases, both for households living within and outside Half-Mile Areas. Average unadjusted VMT calculations show that households living within Half-Mile Areas drive from 20 fewer miles per day in Los Angeles to 4 miles fewer in Sacramento compared to non-TOD households (Table 1, rightmost column). Both of the above phenomena are consistent with the existing literature on the relationship between household VMT, household income, and household proximity to rail transit (*4–8*).

Each metropolitan area displays its own nuanced VMT–income–rail access relationship. In San Diego, the households in the highest income stratum have a much larger "within minus outside Half-Mile Area" difference (26 miles) than households with lower incomes (Table 1, panel 4). In the Bay Area, households with $50,001–$100,000 in annual income show the highest "within" to "outside" VMT difference. In Los Angeles and Sacramento, households with incomes at or below $100,000 have the greatest net VMT differences. As a note of caution, the sample sizes for households within Half-Mile Areas by individual income stratum are small in Sacramento and San Diego. This is why linear distance to nearest rail transit station served as an explanatory factor within Tobit regressions for these two metropolitan areas.

Vehicle Characteristics Within and Outside Half-Mile Areas. Across metropolitan areas, nearly 95% of the vehicles in the sample are gasoline-powered. For the four metropolitan areas combined, among households earning below $25,000 per year, 98.8% of their vehicles were gasoline-powered, while among households earning above $100,000 per year, 93.5% of their vehicles were gasoline-powered. This pattern likely reflects the relative recency of alternative-fuel vehicles on the market and their relatively high cost compared with conventionally powered vehicles.

In each region, higher income households drove newer cars, consistent with expectations. Households living within Half-Mile Areas tended to drive older cars than households outside, except in San Diego.

GHG Emissions Within and Outside Half-Mile Areas. Daily household carbon dioxide emission average levels vary by region. Sacramento's are the highest and the Bay Area's are the lowest (Table 2). Higher income households have higher average daily emissions in every region. In all regions, households within Half-Mile Areas demonstrate substantially lower emissions than households outside but the magnitude of difference is greatest in the Bay Area. The pattern of emissions differences by income and proximity to rail transit varies across the metropolitan areas.

Table 2. Average Model Year of All Sampled Vehicles and Calculated Average Daily Household CO_2 Emissions (grams)

	Average model year		Difference in average age (within minus outside)	Average daily emissions		Difference in daily emissions (Within minus outside)
	Outside 0.5 miles	Within 0.5 miles		Outside 0.5 miles	Within 0.5 miles	
Bay Area						
$0–$25,000	2000	1999	1.3	13,471	13,254	−217
$25,001–$50,000	2001	2000	1.7	14,743	13,492	−1,250
$50,001–$100,000	2003	2003	0.4	18,412	11,059	−7,352
>$100,000	2004	2003	0.9	20,322	15,710	−4,611
All income levels	2003	2002	1.1	18,570	13,716	−4,854
Los Angeles						
$0–$25,000	2001	1998	2.2	13,620	10,802	−2,818
$25,001–$50,000	2002	2002	0.2	15,122	14,146	−976
$50,001–$100,000	2004	2003	1.0	18,728	15,782	−2,945
>$100,000	2005	2005	0.2	21,254	24,423	3,169
All income levels	2004	2002	1.4	18,337	15,965	−2,372
Sacramento						
$0–$25,000	1999	1998	1.6	12,525	8,523	−4,002
$25,001–$50,000	2002	2004	−1.7	15,625	9,730	−5,894
$50,001–$100,000	2003	2002	1.4	19,915	23,094	3,179
>$100,000	2005	2004	1.3	22,095	29,656	7,561
All income levels	2003	2002	1.0	19,107	19,017	−91
San Diego						
$0–$25,000	2000	2002	−2.3	12,678	8,627	−4,051
$25,001–$50,000	2002	2001	1.1	15,413	16,867	1,455
$50,001–$100,000	2003	2006	−3.3	21,473	19,438	−2,035
>$100,000	2005	2005	0.4	24,105	19,356	−4,749
All income levels	2004	2004	−0.5	20,442	16,506	−3,936

Note: Table includes all vehicles in Emissions Models, for households within the 99th percentile of VMT.

Regression Results

VMT Regression Results. Regression results show a statistically significant relationship for incomes at or below $50,000 and reduced household VMT in all four metropolitan areas. Residence within a Half-Mile Area is negatively associated with VMT in the Bay Area; similarly, for San Diego, living closer to rail stations (the linear distance variable) is associated with lower VMT. Both relationships are statistically significant. Interaction terms are only statistically significant in one case: lowest income (below $25,000 per year) households in Los Angeles living within Half-Mile Areas are associated with lower VMT. A test of joint significance indicates that each metropolitan area's variables for income, rail transit access (half-mile dummy or linear distance), and their interaction are jointly significant in their effect on VMT (Table 3).

Emissions Regression Results. Regression results show a negative and statistically significant emissions effect for households with income at or below $50,000 in Los Angeles and Sacramento as well as a positive and significant emissions effect on $50,000–$100,000 income households in the Bay Area. Interaction effects are only individually statistically significant in Los Angeles; they indicate that households with incomes below $25,000 and $50,000–$100,000 have a negative emissions effect from living within Half-Mile Areas relative to beyond. A test of joint significance indicates that each metropolitan area's income and distance to rail transit variables are jointly significant in their effect on GHG emissions (Table 4).

Predicted Values from Regression Results. Predicted values by income band for both daily VMT and GHG emissions highlight several patterns (Table 5). As Table 5 demonstrates, the Tobit regression models' average predicted values of daily VMT are very similar to all four metropolitan areas' average actual values of VMT (see Table 1).

Predicted emission results show that those living within Half-Mile Areas emit fewer daily grams of CO_2 than their peers outside Half-Mile Areas in all regions (Table 5). Differences in predicted emissions magnitudes between households within and outside Half-Mile Areas by income group are generally similar to actual emissions, but the patterns between income groups are clearer. In the Bay Area and San Diego, the magnitude of emissions differences increases with income (to $100,000 income in the Bay Area). Los Angeles and Sacramento's residents within Half-Mile Areas

Table 3. Daily VMT Tobit Regression Results

	Bay Area	Los Angeles	Sacramento	San Diego
Half-mile indicator	−13.99***	−2.844		
	(−4.40)	(−0.42)		
Distance			−0.183	0.715***
			(−1.46)	(3.21)
Income <$25,000	−13.05***	−17.93***	−25.47***	−28.78***
	(−5.64)	(−10.65)	(−3.67)	(−4.97)
Income $25,000–$50,000	−6.851***	−10.17***	−16.61***	−15.39***
	(−3.94)	(−7.00)	(−3.02)	(−2.97)
Income $50,000–$100,000	−3.982***	−3.388***	−8.008*	−6.004
	(−3.06)	(−2.82)	(−1.86)	(−1.32)
Half-mile * income <$25,000	−5.942	−17.91**		
	(−0.86)	(−2.16)		
Half-mile * income $25,000–$50,000	−1.096	−13.48		
	(−0.18)	(−1.60)		
Half-mile * income $50,000–$100,000	−3.795	−8.957		
	(−0.78)	(−1.09)		
Distance * income <$25,000			0.283	−0.294
			(1.20)	(−0.80)
Distance * income $25,000–$50,000			0.204	−0.464
			(1.10)	(−1.32)
Distance * income $50,000–$100,000			0.157	−0.292
			(0.98)	(−0.99)
Household size	3.857***	3.133***	2.539**	4.810***
	(7.88)	(8.21)	(2.19)	(4.69)
# vehicles in household	15.90***	11.93***	11.16***	10.77***
	(23.49)	(19.34)	(6.34)	(6.62)
# of employed in household	6.340***	5.922***	6.291***	5.612***
	(8.27)	(8.94)	(3.39)	(3.16)
Constant	0.108	13.21***	23.35***	7.473
	(1.770)	(8.17)	(4.37)	(1.46)
Sigma	46.26***	51.76***	57.18***	46.40***
	(0.391)	(0.345)	(0.969)	(0.944)
N	7,844	12,239	1,881	1,337
Log-likelihood	−3.8E+04	−6.2E+04	−9.8E+03	−6.5E+03
F-statistic	191.73***	202.98***	20.41***	39.42***
Prob>F	0.0000	0.0000	0.0000	0.0000

Note: Dependent variable: VMT. Omitted category: income >$100,000. *t*-statistics in parentheses, other than for Sigma, where standard error is in parentheses. Distance used instead of half-mile indicator where too few households observed within 0.5 miles of rail transit (for Sacramento and San Diego). ***$p < 0.001$. **$p < 0.01$. *$p < 0.1$.

also display lower emissions, especially those with incomes below $50,000, but income-emission patterns are less distinct than for the other two regions.

Furthermore, the predicted differences in average daily VMT and GHG emissions for households living within versus outside Half-Mile Areas follow dissimilar patterns across income bands (Figure 1). And these disparate trends vary across metropolitan areas (Figure 1).

The amount by which households in various income categories are predicted to reduce their VMT when living closer to rail transit varies by metropolitan area (Tables 3 and 5). In the Bay Area, households with incomes between $50,001 and $100,000 have the largest VMT difference; in the Los Angeles and Sacramento metropolitan areas, the lowest income households have the largest VMT difference; and in San Diego, it is the highest income households with the largest difference. Emission levels predicted from the CHTS and households' car characteristics (Tables 2, 4, and 5) exhibit distinct patterns as well.

Regardless of metropolitan area, higher income households tend to drive newer cars (Table 2) and are more likely to drive non-gasoline-fueled vehicles. Second and similar to VMT, the pattern of "within" minus "outside" GHG emission difference in income categories varies across metropolitan areas. Predicted differences in VMT and GHG emissions for the same income band in a given metropolitan area are frequently dissimilar. Consider, for example, the lowest two income bands for the Bay Area (Table 5): the VMT difference

Table 4. Daily Emissions Tobit Regression Results

	Bay Area	Los Angeles	Sacramento	San Diego
Half-mile indicator	−2,250.0^	4,496.8*		
	(−1.79)	(2.06)		
Distance (miles)			5.802	200.8*
			(0.14)	(2.43)
Income <$25,000	107.1	−2,333.8***	−4,717.0^	−2,320.2
	(0.11)	(−4.10)	(−1.94)	(−0.99)
Income $25,000–$50,000	−20.57	−1,905.3***	−3,287.2^	−1,254.0
	(−0.03)	(−4.11)	(−1.85)	(−0.64)
Income $50,000–$100,000	1,086.2*	−372.7	−712.1	−193.0
	(2.25)	(−0.98)	(−0.51)	(−0.11)
Half-mile * income <$25,000	1,695.5	−6,236.1*		
	(0.49)	(−2.02)		
Half-mile * income $25,000–$50,000	317.9	−4,658.7		
	(0.13)	(−1.64)		
Half-mile * income $50,000–$100,000	−1,810.5	−6,715.1*		
	(−0.92)	(−2.50)		
Distance * income <$25,000			50.86	−140.0
			(0.62)	(−0.97)
Distance * income $25,000–$50,000			50.88	−54.84
			(0.83)	(−0.42)
Distance * income $50,000–$100,000			39.35	114.8
			(0.76)	(1.05)
Household size	72.73	188.8	−381.1	620.2
	(0.39)	(1.51)	(−0.99)	(1.53)
# vehicles in household	11,907.4***	10,289.4***	12,008.6***	10,980.2***
	(31.80)	(36.92)	(14.04)	(12.49)
# of employed in household	1,694.8***	1,049.3***	1,250.0*	1,151.4^
	(5.86)	(4.86)	(2.02)	(1.70)
Constant	−2,084.1**	1,942.2***	2,069.7	−1,742.7
	(−2.98)	(3.65)	(1.18)	(−0.89)
Sigma	16,243.9***	15,632.0***	17,404.4***	16,244.8***
	(114.08)	(142.29)	(55.89)	(46.99)
N	6,548	10,183	1,576	1,113
Pseudo R^2	0.010	0.009	0.008	0.012
F-statistic	161.62***	232.38***	31.58***	35.14***
Prob>*F*	0.0000	0.0000	0.0000	0.0000

Note: Dependent variable: daily CO_2 emissions (grams). Omitted category: income >$100,000. Excluding households with VMT greater than 99th percentile. *t*-statistics in parentheses, other than for Sigma, where standard error is in parentheses. Distance used instead of half-mile indicator where too few households observed within 0.5 miles of rail transit (for Sacramento and San Diego).
$^{***}p < 0.001$. $^{**}p < 0.01$. $^{*}p < 0.1$.

is larger for households earning less than $25,000, but the GHG difference is substantially larger for households earning $25,001–$50,000. VMT may be an imperfect proxy for GHG emissions in terms of environmental impact (Figure 1).

It should be noted that Figure 1 does not show whether the differences across metropolitan areas are statistically significant—nor is this easy to assess from predicted Tobit values. As an alternative, significance tests were run for each metropolitan area on the linear combinations of the Half-Mile Area or distance variable and their interaction-with-income variables. These tests provided estimates of the joint income and proximity to rail effect on household VMT and GHG emissions.

At each income stratum, the authors assessed whether the 95% confidence interval for a metropolitan area's joint effect overlapped with the 95% confidence interval of another metropolitan area's joint effect. Doing so provided insight into whether the VMT and GHG differences across metropolitan areas (as shown in predicted values in Figure 1) reflect statistically significant differences between the metropolitan areas. Comparisons were made in pairs, between Los Angeles and the Bay Area and between Sacramento and San Diego, because each pair's metropolitan areas share the same rail access transit variables.

Briefly, Los Angeles and the Bay Area's joint effects for VMT always overlap, while VMT joint effects for Sacramento

Table 5. Predicted Average Daily Household VMT (miles) and Average Daily CO_2 Emissions (grams) from Tobit Regressions

	Average VMT		Difference in VMT (within minus outside)	Average emissions		Difference in emissions (within minus outside)
	Outside 0.5 miles	Within 0.5 miles		Outside 0.5 miles	Within 0.5 miles	
Bay Area						
$0–$25,000	29.80	13.62	−16.18	15,470	15,338	−133
$25,001–$50,000	39.46	26.34	−13.12	16,567	15,648	−919
$50,001–$100,000	48.39	25.62	−22.77	19,753	13,614	−6,140
>$100,000	58.68	39.10	−19.58	21,453	17,440	−4,012
All income levels	50.47	29.32	−21.16	19,911	15,799	−4,112
Los Angeles						
$0–$25,000	34.69	18.67	−16.02	15,416	13,185	−2,231
$25,001–$50,000	45.66	30.70	−14.96	16,707	15,876	−831
$50,001–$100,000	55.89	42.09	−13.80	19,837	17,369	−2,468
>$100,000	64.03	56.37	−7.66	22,104	24,966	2,862
All income levels	53.29	33.47	−19.83	19,525	17,552	−1,974
Sacramento						
$0–$25,000	37.14	31.28	−5.86	14,912	13,630	−1,282
$25,001–$50,000	47.59	42.74	−4.85	17,503	13,877	−3,626
$50,001–$100,000	58.61	57.81	−0.80	21,271	21,626	355
>$100,000	68.33	69.56	1.23	23,446	21,906	−1,540
All income levels	56.90	51.91	−4.99	20,651	18,247	−2,404
San Diego						
$0–$25,000	25.83	21.44	−4.39	14,643	13,490	−1,153
$25,001–$50,000	39.32	39.03	−0.30	17,092	18,674	1,582
$50,001–$100,000	54.60	44.88	−9.72	22,558	18,067	−4,491
>$100,000	68.70	51.96	−16.73	24,953	18,838	−6,114
All income levels	52.88	36.76	−16.12	21,637	17,360	−4,277

Note: This table excludes households with VMT greater than 99th percentile.

and San Diego overlap except when incomes exceed $100,000. For GHG, joint effects for Los Angeles and the Bay Area overlap except when incomes exceed $100,000; Sacramento and San Diego's joint effects overlap except for incomes between $50,000 and $100,000. These trends mirror patterns shown in Figure 1. More details on this joint effect comparison are available from the authors on request.

Discussion

We suggest some hypotheses and directions for future research based on the results. First, note that, from Figure 1, the "within minus outside" half-mile VMT gap is smaller for higher income households in Los Angeles and Sacramento, while the pattern is reversed in the Bay Area and San Diego. It is likely that the Bay Area and San Diego rail transit systems (the oldest systems studied) provide greater accessibility to employment centers and/or local amenities than the Los Angeles and Sacramento systems. Higher income households also make more trips (*1*), tend to consume more goods, and are more likely to have jobs—regardless of their proximity to transit. But there may be level differences across metropolitan areas. Thus, one hypothesis is that the varying pattern of VMT differences by income reflects the more mature networks in the Bay Area and San Diego, which may allow high income households to drive less when living near rail transit in those places.

Second, a vehicle technology effect may explain why VMT and GHG emission reduction patterns diverge in each metropolitan area. Within and outside Half-Mile Areas, higher income households tend to drive newer vehicles (Table 2) and are more likely to drive non-gasoline-powered vehicles. A comparison of Figure 1's GHG and VMT graphs suggests that interrelationships between vehicle technology, age, and household characteristics may imply that translations of VMT reduction into GHG reduction are not directly proportional.

Note also that the increase in GHG for Los Angeles households earning at least $100,000 and living within Half-Mile Areas appears to stem from the data and not a causal mechanism. Los Angeles households earning at least $100,000 own a much larger share (62%) of Los Angeles' hybrid vehicles than any other income stratum. At incomes in excess of $100,000, drivers of hybrid vehicles in Los Angeles: (1) have

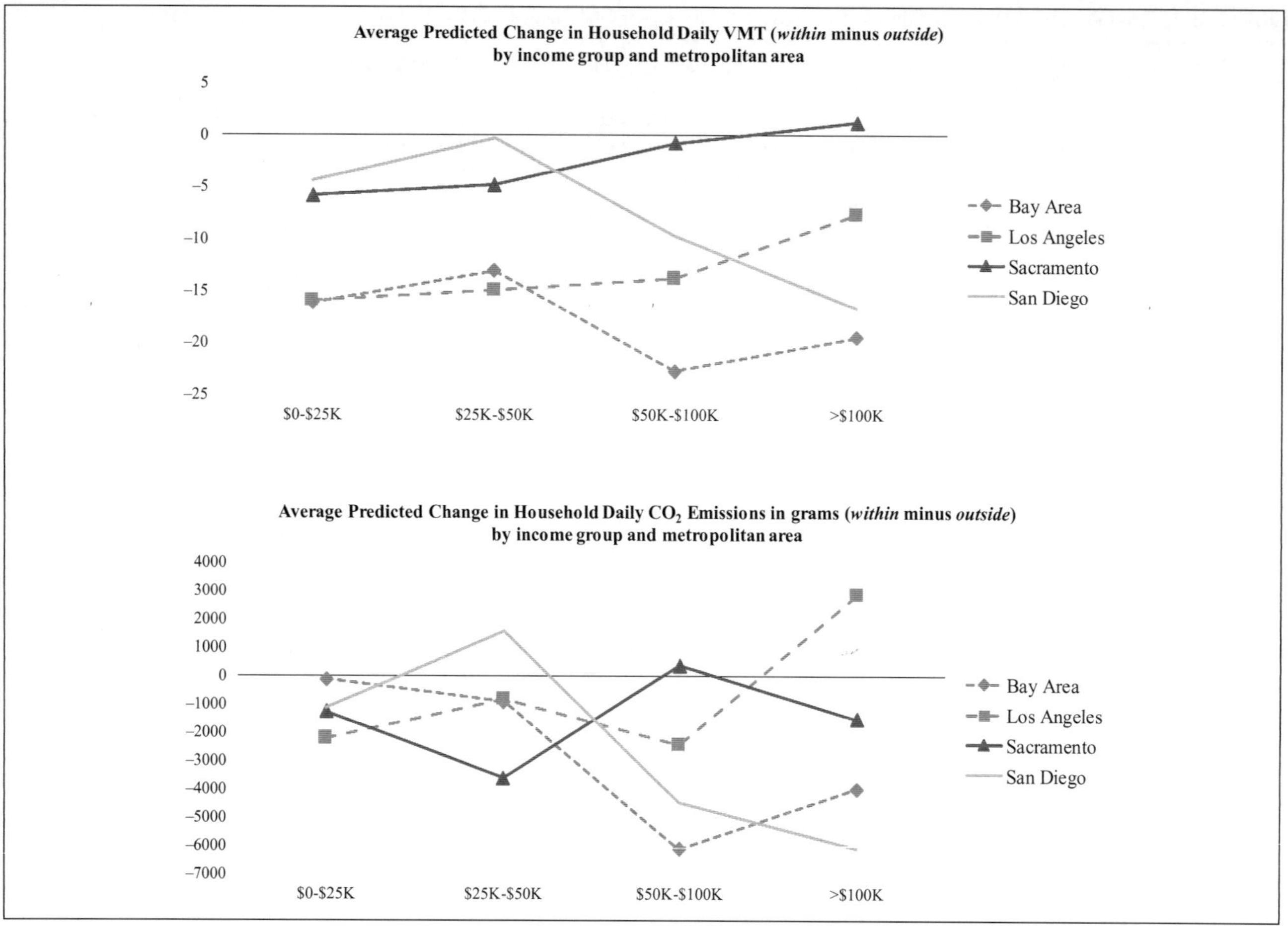

Figure 1. Predicted differences in average daily VMT and average daily CO_2 emissions across income bands, by metropolitan area.

a significantly higher average daily VMT than drivers of gasoline-fuel vehicles (a gap of 5.3 miles more for hybrid than gasoline drivers, two times the size of the 2.6 miles gap in the Bay Area), and (2) almost entirely reside outside of Half-Mile Areas (98%). Thus, this compositional effect of hybrid vehicle ownership may be responsible for a particularly low emissions estimate for Los Angeles households earning at least $100,000 and living outside Half-Mile Areas. In turn, this is likely responsible for the demographic group's positive predicted GHG difference in Figure 1.

Third, a land use effect is possible as well. This study does not incorporate measurements of land use in areas surrounding surveyed households, which the literature (*6–11*) indicates may play a significant role in travel behavior (e.g., housing density, pedestrian friendliness).

Conclusion

The cross-sectional analysis of households presented in Figure 1 suggests two causal mechanisms of interest. First, a network accessibility effect may explain why household reductions in VMT due to proximity to rail transit vary by income strata and metropolitan area. Specifically, higher income households may have higher demand for local amenities and employment centers; and households in areas with more comprehensive rail transit systems may reduce their VMT to a greater degree when traveling for amenity or employment purposes. Second, a vehicle technology and age mechanism, which correlates with household characteristics like income, appears to explain the imperfect translation of VMT reductions by income strata into GHG reductions by income strata.

Admittedly, moving from cross-sectional results to questions of causality is difficult, and the lack of longitudinal travel behavior data has required that most studies of this sort use cross-sectional travel surveys. Nonetheless, recent research suggests that household location choice may not be nearly as endogenous to travel behavior decisions as previously suspected. For evidence that cross-sectional estimates are likely a good approximation and that residential selection bias is likely small, see, for example, Brownstone (*23*) and Cao et al. (*24*). Duranton and Turner, using national land use and travel survey data, estimate that the bias on density in a regression of household travel of the sort used here ranges from zero to one-sixth of the land use effect, and the "no residential selection bias" result is within the confidence interval of their estimates (*25*). This suggests that cross-sectional regressions of this sort are useful, although future research should continue to consider how these findings relate to residential moves.

Three additional limitations are of note. First, and specifically for Sacramento and San Diego, the number of observed households falling within 0.5 miles of rail transit (as defined in this study) was too low for a rail transit access dummy variable to be used in Tobit regressions. Future studies should attempt to replicate the dummy approach for other metropolitan areas. Second, these future studies should develop more sophisticated estimates for GHG by analyzing emissions from vehicle "cold starts" and emissions of driving separately. Third, this model assumed the attributes of a household's vehicle are exogenous; future approaches could first estimate a vehicle choice model based on household characteristics, and then predict VMT and emissions effects from household and land use characteristics.

Acknowledgments

We gratefully acknowledge funding from the METRANS transportation research center for this project. We thank Dr. Nesamani Kalandiyur at the Air Resources Board and Dr. Matthew Barth at University of California-Riverside for helpful comments and suggestions. We thank Anna Sporrong for exemplary research assistance.

Author Contributions

The authors confirm contributing to the paper as follows: study conception and design: Marlon Boarnet, Raphael Bostic; data collection: Huê-Tâm Webb Jamme, Raúl Santiago-Bartolomei, Andrew Eisenlohr, Seva Rodnyansky; analysis and interpretation of results: Marlon Boarnet, Raphael Bostic, Andrew Eisenlohr, Seva Rodnyansky, Raúl Santiago-Bartolomei, Huê-Tâm Webb Jamme; draft manuscript preparation: Marlon Boarnet, Raphael Bostic, Andrew Eisenlohr, Seva Rodnyansky, Raúl Santiago-Bartolomei, Huê-Tâm Webb Jamme. All authors reviewed the results and approved the final version of the manuscript.

References

1. Santos, A., N. McGuckin, H. Y. Nakamoto, D. Gray, and S. Liss. *Summary of Travel Trends: 2009 National Household Travel Survey*. Publication FHWA-PL-II-022. Federal Highway Administration, U.S. Department of Transportation, Washington, D.C., 2011.
2. Chatman, D. G., R. Xu, J. Park, and A. Spevack. The Effects on Auto Use of Household Displacement from Rail Station Areas. In *Developing a New Methodology for Analyzing Potential Displacement (Chapter 4)* (K. Chapple, A. Loukaitou-Sideris, P. Waddell, D. Chatman, and P. Ong, eds.), Publication ARB No. 13-310. University of California, Berkeley, University of California, Los Angeles, and California Air Resources Board, 2017.
3. Boarnet, M. G., R. W. Bostic, S. Rodnyansky, R. Santiago-Bartolomei, and C. Leslie. *Environmentally Sustainable and Affordable Housing Near Transit in Los Angeles*. California Department of Transportation, 2017a. https://www.metrans.org/sites/default/files/research-project/15-13%20Final%20Report_Bostic.pdf
4. Zegras, C. The Built Environment and Motor Vehicle Ownership and Use: Evidence from Santiago de Chile. *Urban Studies*, Vol. 47, No. 8, 2010, pp. 1793–1817.
5. Holtzclaw, J., R. Clear, H. Dittmar, D. Goldstein, and P. Haas. Location Efficiency: Neighborhood and Socio-economic Characteristics Determine Auto Ownership and Use-studies in Chicago, Los Angeles and San Francisco. *Transportation Planning and Technology*, Vol. 25, No. 1, 2002, pp. 1–27.
6. Kain, J. F., G. R. Fauth, and G. K. Ingram. *The Impacts of Urban Land Use and Transportation Investments on Household Auto Ownership Decisions and Journey to Work Mode Choices*. Department of City and Regional Planning, Harvard University, 1976.
7. Kain, J. F., and G. R. Fauth. The Effects of Urban Structure on Automobile Ownership and Journey to Work Mode Choices. *Transportation Research Record: Journal of the Transportation Research Board*, 1977. 658: 9–17.
8. Cervero, R., and K. Kockelman. Travel Demand and the 3Ds: Density, Diversity, and Design. *Transportation Research Part D: Transport and Environment*, Vol. 2, No. 3, 1997, pp. 199–219.
9. Ewing, R., and R. Cervero. Travel and the Built Environment: A Synthesis. *Transportation Research Record: Journal of the Transportation Research Board*, 2001. 1780: 87–114.
10. Ewing, R., and R. Cervero. Travel and the Built Environment: A Meta-Analysis. *Journal of the American Planning Association*, Vol. 76, No. 3, 2010, pp. 265–294. http://dx.doi.org.libproxy1.usc.edu/10.1080/01944361003766766.
11. Zhang, L., J. Hong, A. Nasri, and Q. Shen. How Built Environment Affects Travel Behavior: A Comparative Analysis of the Connections Between Land Use and Vehicle Miles Traveled in US Cities. *Journal of Transport and Land Use*, Vol. 5, No. 3, 2012, pp. 40–52.
12. Baum-Snow, N., and M. Kahn. The Effects of New Public Projects to Expand Urban Rail Transit. *Journal of Public Economics*, Vol. 77, No. 2, 2000, pp. 241–263.
13. Boarnet, M. G., R. W. Bostic, D. Williams, R. Santiago-Bartolomei, S. Rodnyansky, and A. Eisenlohr. *Building Affordable Housing in Transit-Oriented Developments: Impacts on Driving and Policy Approaches*. National Center for Sustainable Transportation, 2017. https://ncst.ucdavis.edu/wp-content/uploads/2015/10/NCST-TO-027-Boarnet-Bostic-Affordable-TOD-White-Paper_FINAL.pdf
14. Brown, M., F. Southworth, and A. Sarzynski. *Shrinking the Carbon Footprint of Metropolitan America*. Metropolitan Policy Program, Brookings Institution, Washington, D.C., 2008.
15. Ewing, R., K. Bartholomew, S. Winkelman, J. Walters, D. Chen, B. Mccann, and D. Goldberg. *Growing Cooler: The Evidence on Urban Development and Climate Change. Specifically Section 1.7—The Impact of Compact Development on VMT and CO_2 Emissions*, Urban Land Institute, 2008.
16. Brownstone, D., and T. F. Golob. The Impact of Residential Density on Vehicle Usage and Energy Consumption. *Journal of Urban Economics*, Vol. 65, No. 1, 2009, pp. 91–98.
17. Guerra, E., R. Cervero, and D. Tischler. Half-mile Circle: Does it Best Represent Transit Station Catchments? *Transportation Research Record: Journal of the Transportation Research Board*, 2012. 2276: 101–109.

18. Kunzmann, M. *2010-2012 California Household Travel Survey Final Report Appendix, Table 8.3.1*. California Department of Transportation, 2013. http://www.dot.ca.gov/hq/tpp/offices/omsp/statewide_travel_analysis/files/CHTS_Final_Report_June_2013.pdf
19. California Air Resources Board (ARB). *Vehicle Categories*, 2015. https://www.arb.ca.gov/msei/vehicle-categories.xlsx
20. United States Environmental Protection Agency (EPA). *Inventory of U.S. Greenhouse Gas Emissions and Sinks: 1990-2015, Figure ES-4*. Publication EPA 430-P-17-001. United States Environmental Protection Agency (EPA), 2017. https://www.epa.gov/ghgemissions/inventory-us-greenhouse-gas-emissions-and-sinks-1990-2015
21. Boarnet, M. G., X. Wang, and D. Houston. Can New Light Rail Reduce Personal Vehicle Carbon Emissions? A Before-After, Experimental-Control Evaluation in Los Angeles. *Journal of Regional Science*, Vol. 57, No. 3, 2017b, pp. 523–539.
22. Amemiya, T. Tobit Models: A Survey. *Journal of Econometrics*, Vol. 24, No. 1–2, 1984, pp. 3–61.
23. Brownstone, D. *Special Report 298: Driving and the Built Environment: The Effects of Compact Development on Motorized Travel, Energy Use, and CO_2 Emissions*. Transportation Research Board, Washington, D.C., 2008.
24. Cao, X., P. L. Mokhtarian, and S. L. Handy. The Relationship Between the Built Environment and Nonwork Travel: A Case Study of Northern California. *Transportation Research A: Policy and Practice*, Vol. 43, No. 5, 2009, pp. 548–559. https://doi.org/10.1016/j.tra.2009.02.001.
25. Duranton, G., and M. Turner. Urban Form and Driving: Evidence from US Cities. Presented at Multiple Conferences and Seminars, 2016-17.

The Standing Committee on Transportation Energy (ADC70) peer-reviewed this paper (18-03004).

The views expressed are those of the authors and do not necessarily reflect the views of the Federal Open Market Committee, Federal Reserve Board of Governors, or others in the Federal Reserve System.

Article

TRR
JOURNAL OF THE TRANSPORTATION RESEARCH BOARD

Transportation Research Record
2018, Vol. 2672(24) 87–97

DOI: 10.1177/0361198118798238
journals.sagepub.com/home/trr
SAGE

Pathways Toward Zero-Carbon Campus Commuting: Innovative Approaches in Measuring, Understanding, and Reducing Greenhouse Gas Emissions

Bruce Appleyard[1], Alexander R. Frost[2], Eduardo Cordova[1], and Jeremy McKinstry[1]

Abstract

While universities are significant generators of transport-related greenhouse gas (GHG) emissions, they are uniquely qualified to encourage more sustainable travel behaviors toward achieving rapid decarbonization to meet our global climate action goals. Building on previous university climate action plans, this paper contributes to the literature by describing new and innovative methods in surveying and geospatial analytics. The new geospatial methods involve mapping individuals' routes to campus, and the creation of Commute/Policy Zones—walk, bike, transit, and motorized—proving useful in (1) quantifying emissions per mode and weighting the survey sample; (2) exploring and testing various policy scenarios; and (3) guiding policy discussions with key stakeholders throughout the campus community. This study also tests four synergistic policy options that, if implemented, could realize a pathway to zero-carbon campus commuting for San Diego State University (SDSU) by 2030, including: (1) more on-campus student housing; (2) expanding public transit use; (3) electrifying vehicles; and (4) electrifying public transit. Finally, the economic and health costs of auto-commuting for SDSU are considerable: Annually, personal auto-commuting is estimated to cost \$46.7 million, healthcare costs from non-GHG air pollutants are \$1.67 million, and \$81 million worth of time (5.4 million hours) is lost in car commutes to campus. SDSU is also estimated to subsidize each parking space at over \$700 per year. By reducing the demand for internal combustion engine vehicles, increasing housing near campus, improving access to walking, biking, public transit, and shared mobility technology, and repowering everything through renewable energy, a zero-carbon transportation network is within reach.

The State of California is a global leader in climate action and aims to reduce its greenhouse gas (GHG) emissions 40% below the 1990 level by 2030, while the City of San Diego has a goal to reduce locally generated emissions by 50% below the 1990 level by 2035 (*1*, *2*). The primary source of GHG emissions is burning of fossil fuels. Transportation-related GHGs are now the largest source of emissions in the United States (32.1%), the State of California (39%), and the City of San Diego (55%) (*2–4*). A rapid transition away from dirty fuel sources is required to maintain a livable climate and protect Americans' current and future quality of life.

San Diego State University (SDSU) is the largest and oldest educational institution in the San Diego Region in California, with an enrollment of over 35,000 students during the academic year 2016–2017. SDSU has set a target to reach "net-zero" operational GHG emissions by 2045, completing its first Climate Action Plan (CAP) in spring 2017. The transportation sector is the second largest source of SDSU's campus carbon footprint (*5*).

This paper describes the development of a methodology to calculate the annual transportation-related GHG emissions and economic and health costs to SDSU commuters, and establishes a pathway toward zero-carbon campus commuting. The findings outlined in this paper improve the understanding of various aspects of university transportation systems to inform researchers, planners, decision-makers, students, faculty, and staff on how to improve and decarbonize campus mobility options and services.

[1]San Diego State University, San Diego, CA
[2]City of San Diego, CA

Corresponding Author:
Address correspondence to Bruce Appleyard: bappleyard@mail.sdsu.edu

Background

Previous literature has shown transportation choice to be a combination of both social and market factors (*6–9*), with public policies and incentives (such as raising the cost of parking) being shown to effectively influence driving behavior and auto ownership, as well as walking and bicycling rates (*10–14*). Low-cost parking often encourages people to drive alone, which increases auto-related GHG emissions (*15*). For university campuses this is often accomplished through parking passes, which are often fully paid upfront at the beginning of the semester. Because of this, many see driving as a "free" behavior, that it is therefore selected over walking, biking, or riding transit, which are oftentimes cheaper, healthier, and more environmentally friendly (*16*).

Of the various documented approaches to calculating university commuting emissions, one of the most favored is through the development and deployment of a comprehensive travel survey to all university affiliates (*17*, *18*). These surveys are generally used to help measure mode split and frequency of commutes to the university and are useful for understanding travel behavior, which then allow universities to develop strategies and policies to help reduce total transportation emissions.

Several resources are available online for calculating university transportation emissions, including the University of New Hampshire's Campus Carbon Calculator, Torrie Smith Associates' Greenhouse Gas Strategy Software, and the California Climate Action Registry (*19–21*). These tools allow universities to input variables obtained from their travel surveys to calculate baseline GHG emissions (*22*, *23*). The Campus Carbon Calculator was chosen for this study because of its popularity—over 90% of universities who have publicly reported their GHG emissions did so through its use (*20*).

A similar study was performed by McGill University (in Montreal, Canada). Transportation GHG emissions were estimated through a travel survey and GIS analysis by using zip codes to geographically weigh the survey sample by season, gender, mode of travel, and affiliation, while presenting several emission-reduction policy options (*17*).

This paper contributes to the literature by using a synthesis approach to calculate campus GHG emissions. The addresses of university affiliates were obtained from the travel survey, which were then geolocated and contextualized to a series of commute/policy zones (Figure 1) emanating from the SDSU campus, using ArcGIS's standard deviation ellipse function (described in more detail below). These catchment areas (also known as "Commute/Policy Zones" or CPZs) allow researchers to (1) quantify emissions per mode for each zone and help weight the survey samples; (2) explore and test various policy scenarios; and (3) guide policy discussions when working with stakeholders throughout the campus community.

Methodology

To calculate emissions for all SDSU campus commuters, a multifaceted approach was undertaken. First, a travel survey was designed and deployed to the entire university population in spring 2017, using Survey Gizmo. Second, the survey results were exported to the SPSS statistical analysis tool for coding and comparison. Outliers and inappropriate responses were removed to clean up the dataset. Third, GIS spatial analyses were conducted for all university affiliates and survey respondents to better understand the spatial distribution and mode of commute (these are the CPZs, mentioned above). Fourth, results from geospatial and statistical analyses were combined to calculate the sampling rate and weights by affiliation and geographic areas for five CPZs using Microsoft Excel. Determining the population-weighted mode share by student, staff, and faculty for the five CPZs provides better accuracy to calculate the GHG emissions. For weighting, we used student registration and faculty/staff human resources data for the entire population in each CPZ, which was then calibrated by the stated addresses obtained from the travel survey. Fifth, a comprehensive set of policy scenarios were developed and tested to explore pathways toward a zero-carbon transport system for SDSU.

Survey Design

A comprehensive online travel survey developed during the 2015–2016 school year was refined and modified for distribution in spring 2017. The purpose of this survey was to establish the frequency of travel and mode of commute to SDSU, and to gather information on various aspects of current university transportation behavior.

The survey was deployed via email to all faculty, staff, and auxiliaries in April 2017. An original invitation was sent and then followed up with a second and a third email seeking additional responses. The survey period concluded on May 15, 2017. The survey contained 44 questions with the first 10 questions designed to determine general information and commuting patterns. Several questions were included to gain a better understanding of parking behavior, including willingness-to-pay and parking and transit pass pricing tradeoff questions. The final portion of the survey attempted to understand travel behavior and incentives necessary for creating a significant mode shift.

Survey Response

A total of 2,230 survey responses were gathered with an overall 5.92% response rate from the total population, and a survey completion rate of 79.1%. There were responses from 1,456 students, 177 faculty, and 495 staff, while 102 people did not state their affiliation. The staff had the highest sampling rate of 23.1%, while faculty had a sampling rate of 10.2%. Students had the lowest sampling rate of 4.3%. Survey responses were separated by affiliation and descriptive statistics were gathered for mode split and frequency of commute. Overall, the staff and faculty were over-represented in the survey, which is common for a university survey (*17*), but the students, staff, and faculty were well represented all around.

Geospatial Analysis and the Creation of Campus CPZs

Mode of commute, distance by mode, and frequency of commute are required to calculate transportation-related GHG emissions. Through an iterative geostatistical analysis using ArcGIS's standard deviation ellipse function, the commute sheds and GHG emissions of survey respondents were calculated through five CPZs—walking, biking, transit one, transit two, and motorized—which have proven useful in (1) quantifying emissions per mode for each zone; (2) exploring and testing various policy scenarios; and (3) guiding policy discussions with key stakeholders (Figure 1).

As these CPZs are geostatistically defined by the explicitly stated and geolocated travel behavior of bicyclists, pedestrians, or transit riders, they are useful in providing an underlying sense of the environmental quality (including, at least by proxy, the quality of the facilities, such as bike lanes, context, and terrain) that either encourages or discourages the use of these modes. By helping identify commuter sheds with similar travel behaviors (mode splits), these zones also helped weight the survey sample.

Caveats. Whenever researchers deal with student addresses, they should be aware that students often provide the address of their parents. If there is any bias, it is

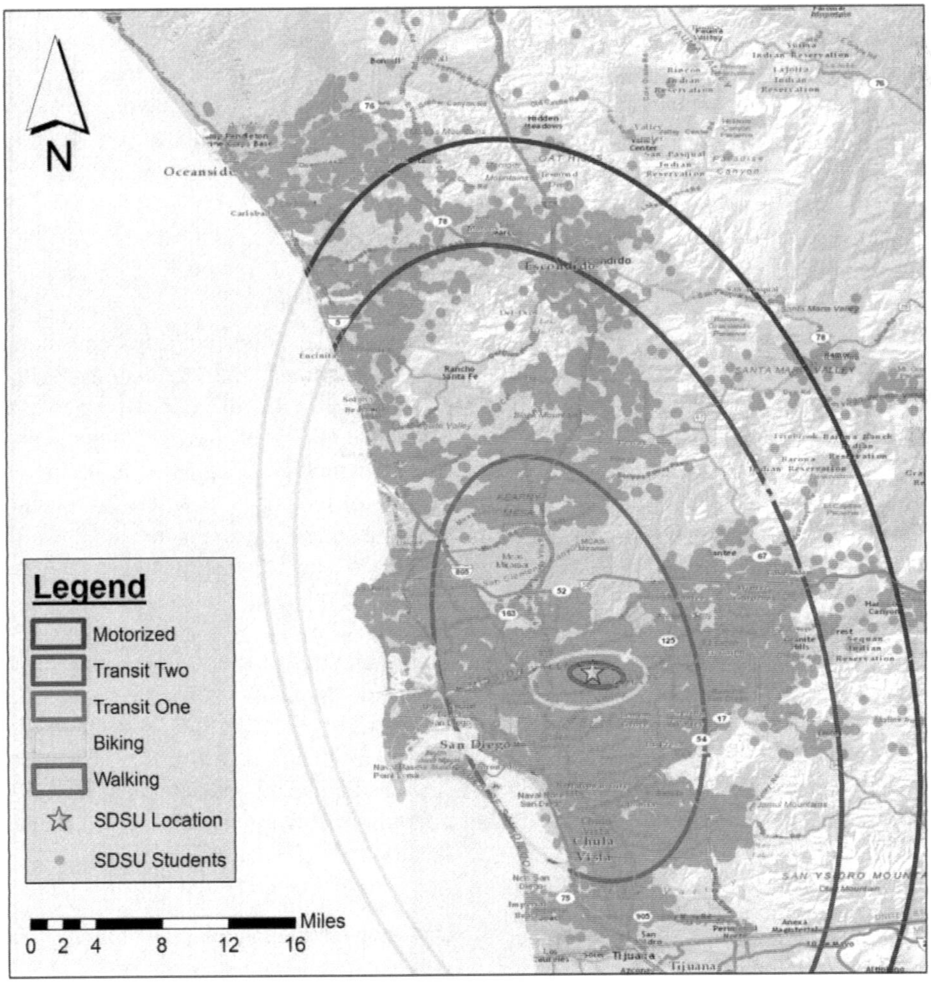

Figure 1. San Diego State University commute/policy zones.

most likely that more addresses are placed in outer-commute (mostly car-commute) zones. Therefore, if there is any miscalculation from this, it is possibly in the overstatement of our carbon emissions. To help moderate biases that may come from overestimating distance and emissions of the entire population, we apply the commute context calibration (CCC) method that works to isolate the appropriate mode split calculation to the commute context. For example, it does not overstate a walk share by misapplying it to the car-commute zone.

Sampling Rate and Weights. The sampling rate by CPZ was obtained by dividing the total number of survey respondents located in each CPZ by the total university-affiliated population for the academic year 2016–2017 located in that zone. After adjusting for students' reported address versus actual residence, researchers geolocated 20,845 out of 33,778 students living within the five CPZs. The five zones included a walking zone, a biking zone, two transit zones, and a motorized zone. Each represents a hierarchy of sustainable modes starting from walking to bicycling, to transit, and finally driving. The sampling weight was calculated based on geography and affiliation to better represent the entire survey population. It should be noted that we did not include gender as a weighting criterion because there was a minimal difference between raw and weighted results compared to affiliation and geographic criteria.

Based on the geocoded registrar data, the largest portions of students were living in the first trolley zone (43.5%), followed by the second trolley zone (31.7%), the walking zone (15.5%), the biking zone (6.6%), and finally the motorized zone (2.7%). The geocoded registrar data for faculty and staff were combined in GIS, so for the geographic calculation it was distributed evenly based on total population.

All CPZs for students, faculty, and staff had a survey-sampling rate of 2% or more. For students, the first trolley zone and the motorized zone had the highest sampling rate (4%), followed by the second trolley (3.9%), the biking (3.7%), and the walking zones (2.3%). Like student respondents, the first trolley zone had the highest sampling rate for both staff (19.6%) and faculty (10.6%).

Sampling weights were calculated separately for students, staff, and faculty by commuter zone to better correspond with the total university-affiliated population. The percentage of geocoded population was divided by the percentage of sample population to derive the sampling weight for each zone. The maximum sampling weight was for faculty in the motorized zone (4.05) and the lowest was for staff in the first trolley zone (0.85). The weighted mode share for each zone by students, staff, and faculty was used to calibrate the mode by zone for the final calculation of GHG emissions.

Calculating Emissions

Based on the inputs gathered from the travel survey and GIS analyses, emissions were calculated utilizing the University of New Hampshire's Campus Carbon Calculator v.9. The inputs include the number of people representing each affiliation, frequency of weekly travel, mode split percentages, and the trip distance average per mode. The calculator then provides fuel efficiency modifiers for automobile, bus, and light rail to calculate total miles and energy use per mode. It then uses fuel efficiency and emission factors to determine the million tons of carbon dioxide equivalent ($MTCO_2e$) from faculty/staff commuting and student commuting.

In total, 33,778 students were enrolled at SDSU during the academic year 2015–2016. There are roughly 29,234 undergraduate students and 15% live in college-owned and operated affiliated housing. These students (4,385) were excluded from the total calculation of commuting-related GHGs, as reflected in the survey.

Finally, 11.6% of the student body is part-time. For these students, based on the survey results, we assume that their average trip frequency is half that of a full-time student. Therefore, to account for these part-time students (3,410), only half of them are represented in the final calculation (1,705), leaving 27,688 as the final number of student commuters used to estimate the GHG emissions.

Results

Commute Mode Splits

Figure 2 reveals the weighted commuter mode for student, faculty, and staff for spring 2017. Overall, 83.2% of staff and 75.4% of faculty drive to campus alone compared to 71.2% of students who live off-campus. The SDSU campus is located in a suburban neighborhood, mostly surrounded by single-family homes with curvilinear, meandering street networks and hilly topography that discourages the use of non-motorized vehicles.

The carpooling rates are highest among faculty (7.8%), followed by students (7.2%), and staff (5.6%). The public transit use is highest among students (12.2%), followed by faculty (10.3%), and then staff (6.0%). Active transportation is also much more prevalent in the student population, with 9.4% of students walking, biking, or skating to campus, most likely following a proximity of off-campus residences, compared to 6.5% of faculty and 5.2% of staff.

Distance and Frequency

Survey results suggest that, on average, people commute 11.6 miles and take 23.1 minutes one-way to reach SDSU 4.63 times per week. The median one-way commute

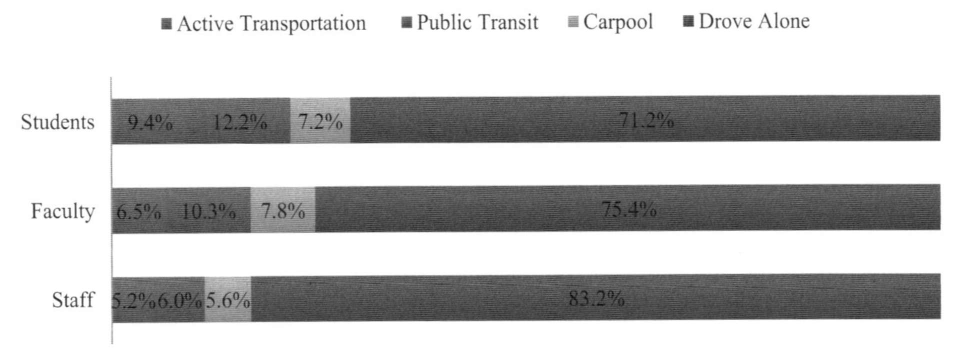

Figure 2. Spring 2017 SDSU commuter mode splits (weighted).

Table 1. Total Greenhouse Gas Emissions by Type of Commuters

Commute type	CO_2 (million tons)	CH_4 (million tons)	N_2O (million tons)	Total $MTCO_2e$
Auto	32,192.05	6.99	2.34	33,063.37
Bus	989.70	0.06	0.03	998.54
Light rail	1,776.56	0.02	0.03	1,787.51
				35,849.42

distance is 10 miles, 20 minutes, and five campus visits per week. The average commute mile per person is 23.2 miles per day and 46.2 minutes.

The people driving to SDSU alone or carpooling typically commute a little over 13 miles one-way to reach the SDSU campus. For bus commuters, the mean distance is around 8.2 miles, while trolley commuters live a little further on average by commuting 13.3 miles. For the people using rideshare, 7 miles is the average commuting distance.

The mean commute distance for skaters is 0.9 miles and for walkers 1.2 miles, while bikers commute 2.6 miles one-way on average to reach the SDSU campus. Most walkers and skaters do not travel more than 2 miles, while 78% of bike commuters do not commute longer than 4 miles to reach campus. Overall, 47% of bike commuters were women.

GHG Emissions

GHG emission from commuter transportation is one of the largest sources of pollution at SDSU. Campus commuting includes students, staff, and faculty traveling between their home and campus. For the academic year 2016–2017, automobile and public transit commuters to the SDSU campus contributed to the release of 35,849 $MTCO_2e$. Students were responsible for 81.4% of these emissions (29,184 $MTCO_2e$), while faculty and staff generated about 18.6% of the total emissions (6,666 $MTCO_2e$). Overall, SDSU staff were responsible for 10.8% of the total emission (3,862 $MTCO_2e$) and faculty contributed 7.8% (2,804 $MTCO_2e$).

Based on a total student commuting population of 27,688, each student, on average, is responsible for about 1.05 $MTCO_2e$ from commuting during the academic year. Each member of the faculty was responsible for 1.62 $MTCO_2e$ per year and staff had the highest rate of emissions per person at 1.81 $MTCO_2e$ per year. The higher per capita emission rate for faculty and staff can be explained by larger percentages of people driving longer distances to campus.

Table 1 highlights the total GHG emissions in $MTCO_2e$, while Table 2 compares the sources of three major GHG emissions—carbon dioxide (CO_2), methane (CH_4), and nitrous oxide (N_2O)—for relevant commuter types. People driving to campus in internal combustion engine vehicles were responsible for 92.1% of the emissions, while bus and light rail commuters were responsible for 2.8% and 5.1%, respectively. Commuters using non-motorized travel modes, such as walking, biking, or skating, did not contribute to any significant air pollution while creating health benefits and financial savings.

Health Cost of Transportation Emission

Commuters that drive and take public transit one fossil fuel-burning internal combustion engine vehicles release several additional pollutants, such as carbon monoxide

Table 2. Comparing Greenhouse Gas Emissions by Type of Commuters

Commuters	CO_2 (kg)	%	CH_4 (kg)	%	N_2O (kg)	%	Total
Auto	32,192,049.71	92.06%	6,990.80	0.02%	2,337.40	0.01%	92.09%
Bus	989,703.99	2.83%	55.20	0.00%	25.02	0.00%	2.83%
Light rail	1,776,557.85	5.08%	24.62	0.00%	34.70	0.00%	5.08%
Total	34,958,311.55	99.97%	7,070.62	0.02%	2,397.12	0.01%	100.00%

Table 3. Health Cost of SDSU Commuter Transportation Emission

Caltrans emission cost parameters			
Pollutant	CA urban area ($ per US ton)	SDSU commuting annual pollutant estimate (per US ton)	SDSU commuting health cost estimates
CO	$80.00	178.2	$14,256
NOx	$18,700.00	17.19	$321,458
PM10/2.5	$151,100.00	8.7	$1,313,824
VOCs	$1,305.00	15.29	$19,955
		Total	$1,669,493

Table 4. Estimated Cost Associated with Driving to Campus

	Annual total miles (2016–2017)	Commuting expenditure ($ 0.50 per mile)	Annual commuting hours	Annual commuting cost (at $13.65/hour)
Students	75,857,146	$37,928,573	4,286,371	$64,295,559
Faculty	7,242,915	$3,621,457	452,555	$6,788,326
Staff	10,206,606	$5,103,303	675,400	$10,131,000
Total	93,306,667	$46,653,334	5,414,326	$81,214,885

(CO), nitrogen oxide (NOx), particulate matter (PM), and volatile organic compounds (VOCs). These pollutants impact the quality of our air and health. According to the California Air Resources Board (CARB), more than 9,000 Californians are killed each year by automobile and truck pollution (*24*). Additional research suggests high exposure to tailpipe emissions increase the risk of dementia, cancer, and heart and respiratory diseases (*25, 26*).

Table 3 estimates the health costs from four types of SDSU commuter transportation emissions—CO, NOx, PM2.5, and VOCs—for the academic year 2016–2017. The emission factors for vehicle miles traveled are based on light-duty vehicle and truck estimates for 2011–2020, developed by CARB (*27*). The emission cost parameters come from the California Department of Transportation's (Caltrans) estimate for the California urban area in 2016 (*28*).

Using these assumptions, and based on the emissions estimates, the annual health costs (Table 2) from SDSU campus commuting-related non-GHG air pollution is estimated at $1.67 million per year.

Automobile Commuter Cost

Transportation costs are the second greatest household expenditure in the United States. According to a 2013 study by AAA, the average US household pays roughly $9,000 per year in car ownership-related costs such as gas, tolls, insurance, maintenance, and parking fees—about 60.8 cents per mile (*29*).

Table 4 estimates the costs associated with driving for students, faculty, and staff. During the academic year 2016–2017, SDSU auto commuters collectively drove 93.3 million vehicle miles and spent 5.4 million hours (233,333 days/639 years) inside automobiles just to get to campus. On a per capita basis, students spent 198 hours/8.2 days commuting to campus during the academic school year, while faculty spent 315 hours/13 days and staff spent 356 hours/15 days commuting to SDSU.

To estimate the true cost of auto commuting, we took a conservative estimate of $ 0.50 cents per mile to calculate the total cost for each affiliation, which is slightly below the Internal Revenue Service's estimate of $ 0.53.5 per mile for 2017 (*30*). The students drove a little over

76.8 million vehicle miles during the academic year 2016–2017, which translates to a cost of $40 million for all student commuters. For the staff the cost is around $5.1 million and the faculty spent around $3.6 million annually on driving. In total, the commuting expenditure for the SDSU commuters reaches a sum of $46.7 million per year. On a per capita basis, students annually spend $1,747 commuting to campus, while faculty and staff spend $2,519 and $2,687, respectively.

Another approach to calculate the cost of commuting is by understanding how much time is lost from auto commuting (time is money). According to Caltrans, for 2016 the value of time per person hour for people in automobiles is estimated at $13.65 (*28*). Every year, SDSU commuters spend collectively 5.4 million hours inside automobiles, which translates to roughly $81 million worth of time lost to auto commuting every year.

Parking Cost

To accommodate the many automobiles used, SDSU spends significant funds to build and maintain parking stalls. A simple back-of-the-envelope calculation illustrates the potential cost of parking. The average cost of an above-ground parking garage is $25,000 per parking space (*31*). The annual operation and maintenance cost ranges from $500 to $800 per space for a commercial parking garage (*32*). The cost of building 1,000 parking spaces is $25 million and annual maintenance is $ 0.5 million (at $500 per space).

Assuming two parking permits sold for each parking space used during the academic year, the annual (2016-17) revenue generated per structured parking space from semester parking passes sold at SDSU is about $660 ($165 × 2 semesters × 2 passes sold). Assuming a parking garage has a life span of 40 years, the annual cost and maintenance of one parking space is $1,125. We are choosing not to include revenue from summer passes, as most of the parking structures are empty during that time.

Based on these assumptions, SDSU is annually subsidizing about $465 per parking space ($1,125 annual cost/operation minus $660 from parking pass sales). For 1,000 parking stalls this would be $465,000 in annual subsidies per year. This is a very simple calculation to estimate the potential level of campus subsidies for parking. A deeper analysis of existing parking stalls by types—parking structure, curbside, paved surface—and annual maintenance cost and revenue from parking permits/payment with further economic analysis (inflation rate, and so forth) might provide more precise estimates of current subsidies for automobiles, but these figures seem in the reasonable range for policy discussions (*33*).

Policy Pathways

A series of policy strategies were developed to investigate what it would take to reach zero-carbon campus commuting by 2030. Building on previous policy testing for campus GHG-reduction plans (*34*), this section attempts to highlight how GHG emissions and pollution can be eliminated from SDSU's daily commuting patterns via four overarching policy strategies: (1) more on-campus student housing; (2) expanding public transit use; (3) electrifying vehicles; and (4) electrifying public transit.

According to SDSU's master plan, by 2030 there will be about 35,000 full-time equivalent (FTE) students (*35*). The increases in faculty and staff are based on a 2016 student ratio (1,820 faculty and 2,252 staff). Assuming business as usual (BAU) in mode share, trip frequency, and distance from the 2016–2017 academic year, emissions from campus commutes are expected to rise from 35,849 to 39,084 $MTCO_2e$ by 2030 following increased student, faculty, and staff populations. Each policy strategy (except policy strategy A) assumes 24,550 FTE student commuters following an increase in campus housing. All of the following scenarios sequentially build on each assumption.

Table 5 highlights the potential emission reduction from four major policy options (described below) based on a BAU projection in 2030.

Policy Pathway A: Changes in Student Housing. This policy scenario is based on the possibility of SDSU realizing a total of 10,000 beds on campus by 2030, as indicated in the 2007 SDSU Master Plan. Currently, 4,385 students

Table 5. Policy Pathways to Zero-Carbon Campus Commuting

Scenario	Emission reductions ($MTCO_2e$)	Emission reductions (percentage total reduction)
BAU 2030 (39,083.8 $MTCO_2e$)		
Policy pathway A: changes in student housing	6,392.80	16.40
Policy pathway B: expand public transit use	1,632.20	4.20
Policy pathway C: electric vehicles	26,445.90	67.70
Policy pathway D: electric public transit	4,612.90	11.80
Total	39,083.80	100.00

live in on-campus housing, leaving 30,615 FTE commuting students in 2030. Expanding the on-campus population by 5,615 students would reduce commuting-related emissions by approximately 6,392.8 $MTCO_2e$ per year (16.4%) by 2030 from the projected BAU scenario. This highlights the importance of university housing policy and land-use planning in the reduction of campus carbon footprint and expanding multimodal transport.

Policy Pathway B: Expand Public Transit Use by Lowering Cost of Transit Pass. This policy strategy aims to double student public transit mode share (from 4% to 8%) by 2030. In our survey of the students, we asked a series of willingness-to-pay questions; 42% of people in the survey indicated they would consider taking transit instead of driving if the cost of a \$154 transit pass were lowered to \$100 or less. We also asked similar questions about travel behavior shifts and acceptance of raising the cost of parking. This scenario assumes that 12% of those who drive alone to campus would switch to public transit if the cost of the student transit pass were lowered from \$154 per semester to just under \$100 per semester (to perhaps \$96). To reinforce this increase in transit ridership, we recommend SDSU simultaneously increase the student semester parking pass from \$165 to just over \$200 (to perhaps \$204). We recommend making these changes to "nudge" people toward more sustainable, healthier, and fiscally responsible behaviors by taking advantage of their cognitive aversion to loss (paying more for parking and less for transit) as discussed by Richard Thaler and other behavioral economists (*36*). Overall, this policy scenario improves transit ridership from 4% to 8% for bus, and 8% to 16% for light rail. This was found to reduce emissions by 1,632.2 $MTCO_2e$, or 4.2% of total commuter emissions.

Policy Pathway C: Electrify Vehicles. This policy strategy assumes all single-occupant vehicles and carpool commuting vehicles driven to campus by students, staff, and faculty will be electric vehicles by 2030 and powered by on-site or off-site renewable energy. Realizing fully electric auto commuting will potentially reduce emission by 26,445.9 $MTCO_2e$ (67.7%).

Policy Pathway D: Electrify Public Transit. This policy strategy assumes all buses and light rail systems become electric and powered by 100% renewable energy. Realizing fully electric public transit commuting will potentially reduce emissions by 4,612 $MTCO_2e$ (11.8%).

The combined effect of implementing these four policy options can bring direct campus commuter pollution down to zero. This of course assumes all electricity is generated from renewable sources, and does not account for fossil fuel emissions that may come from producing electric vehicles, buses, and so forth. It is also important to note that some of the implementation policies are outside the jurisdictional oversight of SDSU responsibility (such as electrifying the non-operation vehicle fleet), but as the oldest and largest educational institution in the region, there are plenty of opportunities to advocate for these changes while leading by example. SDSU is already working to expand on-campus and near-campus student housing and amenities. The opening of South Campus Plaza—a mixed-use medium-density development—is one recent success and SDSU should encourage land-use reform to allow greater density for campus-adjacent neighborhoods and better multimodal connectivity. The provision of a Universal Transit Pass (UTP)—unlimited access to public transit services for a mandatory fee—could be implemented quickly, while expanding solar generation and storage potential, and providing incentives for people to transition toward electric vehicles for their next purchase, can both be medium-term actions.

This policy-testing exercise shows it is possible to significantly reduce and possibly eliminate transport-related GHGs and air pollution. In sum, about 20.6% of these reductions come from changes in travel behavior (16.4% from 5,615 beds on campus, and 4.2% through expanding transit use), with almost 80% coming from changes in technology (67.7% from electrifying the private vehicle fleet, and 11.8% from electrifying the public transit fleet).

The City of San Diego CAP has set a goal to realize 100% renewable electricity generation by 2035 and converting all operational fleet to electric or hybrid vehicles (*2*). By reducing the demand for internal combustion engine vehicles, increasing access to public transit and shared mobility technology, greening campus streets, increasing housing near campus, and repowering everything through renewable energy, a zero-carbon transportation network is within reach. Benefits include reducing student auto traffic and congestion, saving students money, more curbside parking available for local residents in their neighborhoods, improving air quality, enhancing student engagement, creating a more vibrant campus, and providing better overall health outcomes throughout SDSU and its environs.

Conclusion and Discussions

This paper outlines the development of methods to calculate the annual transportation-related GHG emissions and economic and health costs of SDSU commuters and establishes policy pathways toward a zero-carbon future for campus commuting.

Specifically, this study contributes to the literature by providing new geospatial methods for calculating GHG emissions from commuters in a university population.

This involves modeling individuals' networked routes to campus from home, as opposed to using straight, crow-fly distances, on facilities based on the mode of commute. It also involves the innovative application of CPZs (walking zone, bicycling zone, and so forth) to better refine our measures of the travel behavior of a university community. These CPZs have proven useful in (1) quantifying emissions per mode for each zone and helping to weight the survey samples; (2) exploring and testing various policy scenarios; and (3) guiding policy discussions when working with stakeholders throughout the campus community. An appropriate term for this methodologic approach appears to be *calibrating within the commute context*, or CCC, and it seems to work reasonably well at accounting for the contextual factors that shape travel behavior. Through the analysis and discussions around these CPZs, strategies were developed to guide decarbonization policies from campus commuting.

For the academic year 2016–2017, automobile and public transit commuters to SDSU campus contributed to the release of 35,849 $MTCO_2e$, with 81.4% of total emissions generated by students (29,184 $MTCO_2e$) and 18.6% generated by faculty and staff (6,666 $MTCO_2e$) The biggest source of transportation-related emission is from auto commuters (92.1%), followed by bus commuters (2.8%), and light rail commuters (5.1%).

The economic and health costs of auto commuting annually are estimated to be $46.7 million in total SDSU auto commuting expenditure per year, $1.67 million in health costs from campus commuting-related non-GHG air pollution, and $81 million worth of time lost to automobile commuting every year, with auto commuters spending 5.4 million hours inside automobiles.

A collection of four synergistic policy options were tested toward realizing a pathway to zero-carbon campus commuting by 2030:

1. Land use and housing: changes in student housing by realizing the goal of 10,000 beds by 2030 will reduce emissions by 16.4%.
2. Shared mobility: expanding public transit use through more affordable transit passes will shift 12% of driving commuters to buses and light rail, which reduces emissions by 4.2%.
3. Electric vehicles: transitioning all single-occupant and carpool commuting vehicles to electric vehicles powered by on-site or off-site renewable energy will reduce emissions by 67.7%.
4. Electric public transit: electrifying all buses and light rail systems and powering them with 100% renewable energy will reduce emissions by 11.8%.

As per our survey, the most significant near-term actions that can be taken to reduce future parking demand and minimize pollution and carbon emissions, while saving students money and improving health outcomes, are to lower the price of the transit pass from $154 to just under $100 (to, say,$96), increase parking permits from $165 to just over $200 (to, say, $204) per semester, and increase student housing on or near campus. Providing a subsidized UTP with unlimited access showed a 10–30% reduction in automobile trips in some cases for eight US campuses and led to 400% increases in student ridership during the first years of program operation (*37*).

Part of this ideal implementation strategy would be to introduce a lower cost ($96 or less) UTP for incoming freshmen. Within four years all undergraduates will have the UTP to travel around the campus area and explore the San Diego Region using the trolley (light rail) or bus, which reduces the number of traveling vehicles and the level of congestion while improving the regional air quality and saving students money. The funding for the pass can be incorporated as part of the mandatory student registration fees. The survey result indicates that over 68.5% would support some level of fee for a transit pass.

Overall, there are encouraging trends as 15% of survey respondents are utilizing alternative and fuel-efficient vehicles, such as electric (2.2%), hybrid (7.7%), and biodiesel (5.1%) for campus commuting. These vehicles significantly reduce the release of GHG emissions. For example, zero-emission vehicles charged through renewable energy contribute zero air pollution during operation, although there are significant life-cycle impacts from extraction, processing, manufacturing, delivery, and disposal of these vehicles (*38*). The potential to power all our transportation systems via renewable energy and storage is becoming economically competitive in southern California.

This study also confirms previous findings that emissions are greatly influenced by home–work distances measured by both emissions per area and emissions per person. Moving outward from SDSU, each successive CPZ was responsible for greater emissions. This provides evidence to offer better and more affordable housing closer to campus for universities looking to reduce GHG emissions. Accordingly, this study also found faculty and staff to be responsible for greater emissions per individual than students. This was also found in the McGill Study (*17*). Often, the behavior of faculty and staff can be very different from that of students, which is why these groups must be examined separately.

In the United States the transportation sector is the biggest contributor to the release of carbon emissions, and the sector's responsibility in this area continues to grow. This research provides evidence that universities can improve and advocate for policies that lead to zero-carbon commuting. This starts with quantifying the economic, health, and environmental impacts of campus commuting and promoting more housing on or near campus to encourage active transportation and public transit use. The rapid deployment of renewable technologies can accelerate the decarbonization of campus auto commuters to improve regional air quality, enhance health outcomes, and maintain a livable climate for all.

Acknowledgments

The authors gratefully appreciate the support for this research by members of the San Diego State University Aztec community, especially Tom Abram, PE, CEM, LEED AP, Assistant Director, Campus Sustainability, at SDSU; Debbie Richeson; Sam St. Lifer, Courtney Armusewicz; Jessica Barlow, PhD; Geoff Chase, PhD; Sherry Ryan, PhD; Matthew Lauer; Laura Schinn, Cyndi Chie, Chad Teichman, and all the great SDSU Staff & Faculty who have helped with our data requests and surveys throughout the years.

Author Contributions

The authors confirm contribution to the paper as follows: study conception and design: BA, ARF, JM; data collection: BA, ARF, JM, EC; analysis and interpretation of results: BA, ARF, JM, EC; draft manuscript preparation: BA, ARF, JM. All authors reviewed the results and approved the final version of the manuscript.

References

1. California Air Resource Board. ARB Process for SB 375 Target Update. https://www.arb.ca.gov/board/books/2017/032317/17-3-7pres.pdf. Accessed April 10, 2017.
2. *City of San Diego Climate Action Plan*. City of San Diego, San Diego, CA, 2015.
3. *July 2017 Monthly Energy Review*. U.S. Energy Information Administration, Washington, D.C., 2017.
4. California Air Resources Board. California Greenhouse Gas Emission Inventory 2017 Edition. https://www.arb.ca.gov/cc/inventory/data/data.htm. Accessed July 29, 2017.
5. *Climate Action Plan for San Diego State University*. San Diego State University, San Diego, CA, 2017.
6. Ewing, R., G. Meakins, S. Hamidi, and A. C. Nelson. Relationship between Urban Sprawl and Physical Activity, Obesity, and Morbidity: Update and Refinement. *Health & Place*, Vol. 26, 2014, pp. 118–126.
7. Gilderbloom, J. I., W. W. Riggs, and W. L. Meares. Does Walkability Matter? An Examination of Walkability's Impact on Housing Values, Foreclosures and Crime. *Cities*, Vol. 42, 2015, pp. 13–24.
8. Sallis, J. F., L. D. Frank, B. E. Saelens, and M. K. Kraft. Active Transportation and Physical Activity. *Transportation Research Part A: Policy and Practice*, Vol. 38, No. 4, 2004.
9. Forsyth, A., M. Hearst, J. M. Oakes, and K. H. Schmitz. Design and Destinations: Factors Influencing Walking and Total Physical Activity. *Urban Studies*, Vol. 45, No. 9, 2008, pp. 1973–1996.
10. Heyman, J., and D. Ariely. Effort for Payment: A Tale of Two Markets. *Psychological Science*, Vol. 15, No. 11, 2004, pp. 787–793.
11. Amir, O., D. Ariely, A. Cooke, D. Dunning, N. Epley, U. Gneezy, B. Koszegi, D. Lichtenstein, N. Mazar, and S. Mullainathan. Psychology, Behavioral Economics, and Public Policy. *Marketing Letters*, Vol. 16, No. 3–4, 2005, pp. 443–454.
12. Deakin, E., A. Bechtel, A. Crabbe, M. Archer, S. Cairns, A. Kluter, K. Leung, and J. Ni. Parking Management and Downtown Land Development in Berkeley, California. *Transportation Research Record: Journal of the Transportation Research Board*, 2007. 1898: 124–129.
13. Appleyard, B. Sustainable and Healthy Travel Choices and the Built Environment. *Transportation Research Record: Journal of the Transportation Research Board*, 2012. 2303: 38–45.
14. Appleyard, B. New Methods to Measure the Built Environment for Human-Scale Travel Research: Individual Access Corridor (IAC) Analytics to Better Understand Sustainable Active Travel Choices. *Journal of Transport and Land Use*, Vol. 9, No. 2 , 2015.
15. Litman, T. *Land Use Impacts on Transport*. Victoria Transport Policy Institute, Victoria, BC, Canada, 2016.
16. Aldo, T.-R., M. Shirgoakar, E. A. Deakin, and W. W. Riggs. The Cost versus Price for Parking Spaces at Major Employment Centers: Findings from UC Berkeley. Presented at 94th Annual Meeting of the Transportation Research Board, Washington, D.C., 2015.
17. Mathez, A., K. Manaugh, V. Chakour, A. El-Geneidy, and M. Hatzopoulou. How Can We Alter Our Carbon Footprint? Estimating GHG Emissions Based on Travel Survey Information. *Transportation*, Vol. 40, No. 1, 2013, pp. 131–149.
18. Popovich, N. *Results of the 2013–14 Campus Travel Survey*. Publication UCD-ITS-RR-14-14. Institute of Transportation Studies, University of California, Berkeley, CA, 2014.
19. Climate Registry. http://www.climateregistry.org. Accessed August 2, 2016.
20. University of New Hampshire. Campus Carbon Calculator. http://www.sustainableunh.unh.edu/calculator. Accessed August 2, 2016.
21. Rice University. Climate Action Plan. http://rs.acupcc.org/site_media/uploads/cap/1211-cap.pdf. Accessed August 2, 2016.
22. Tulane University. 2014 Climate Action Plan. http://green.tulane.edu/PDFs/2014_Tulane_CAP_Final.pdf. Accessed August 2, 2016.
23. University of Maryland. Climate Action Plan. http://rs.acupcc.org/site_media/uploads/cap/278-cap.pdf. Accessed August 2, 2016.
24. California Air Resource Board. Fine Particle Air Pollution Responsible for 9,000 Premature Deaths in California Each Year. News Release. https://www.arb.ca.gov/newsrel/newsrelease.php?id = 149. Accessed July 29, 2017.
25. Krzyżanowski, M., B. Kuna-Dibbert, and J. Schneider. *Health Effects of Transport-Related Air Pollution*. WHO Regional Office Europe, Copenhagen, Denmark, 2005.

26. Samoli, E., R. W. Atkinson, A. Analitis, G. W. Fuller, D. C. Green, I. Mudway, H. R. Anderson, and F. J. Kelly. Associations of Short-Term Exposure to Traffic-Related Air Pollution with Cardiovascular and Respiratory Hospital Admissions in London, UK. *Occupational and Environmental Medicine*, Vol. 73, No. 5, 2016, pp. 300–307.
27. Methods to Find the Cost-Effectiveness of Funding Air Quality Projects: Emission Factor Tables. California Air Resources Board, Sacramento, CA, 2013.
28. California Department of Transportation, Division of Transportation Planning. *Life-Cycle Benefit-Cost Analysis Economic Parameters 2016*. http://www.dot.ca.gov/hq/tpp/offices/eab/benefit_cost/LCBCA-economic_parameters.html. Accessed July 29, 2017.
29. AAA. Cost of Owning and Operating Vehicle in U.S. Increases Nearly Two Percent According to AAA's 2013 "Your Driving Costs" Study. *AAA NewsRoom*. http://newsroom.aaa.com/2013/04/cost-of-owning-and-operating-vehicle-in-u-s-increases-nearly-two-percent-according-to-aaas-2013-your-driving-costs-study-archive. Accessed July 29, 2017.
30. Internal Revenue Service. Standard Mileage Rates for Business, Medical and Moving Announced, 2017. https://www.irs.gov/uac/2017-standard-mileage-rates-for-business-and-medical-and-moving-announced. Accessed July 29, 2017.
31. Shoup, D. Cutting the Cost of Parking. *ACCESS Magazine*, Vol. 1, No. 48, 2016.
32. Litman, T. *Transportation Cost and Benefit Analysis II: Parking Costs*. Victoria Transport Policy Institute, Victoria, BC, Canada, 2017.
33. Shoup, D. Roughly Right or Precisely Wrong. *ACCESS Magazine*, Vol. 1, No. 20, 2002.
34. Appleyard, B., J. Mckinstry, and A. R. Frost. Calculating the Campus Carbon Footprint: Measuring University Associated Greenhouse Gas Emissions from Transport. Presented at 96th Annual Meeting of the Transportation Research Board, Washington, D.C., 2017.
35. San Diego State University. Campus Master Plan: Campus Master Plan Elements. http://advancement.sdsu.edu/masterplan/2007/elements.html. Accessed July 29, 2017.
36. Thaler, R., and C. Sunstein. Nudge: Improving Decisions about Health, Wealth, and Happiness. *Constitutional Political Economy*, Vol. 19, No. 4, 2008, pp. 356–360.
37. Barata, E., L. Cruz, and J.-P. Ferreira. Parking at the UC Campus: Problems and Solutions. *Cities*, Vol. 28, No. 5, 2011, pp. 406–413.
38. Hawkins, T. R., B. Singh, G. Majeau-Bettez, and A. H. Strømman. Comparative Environmental Life Cycle Assessment of Conventional and Electric Vehicles. *Journal of Industrial Ecology*, Vol. 17, No. 1, 2013, pp. 53–64.

The Standing Committee on Transportation Energy (ADC70) peer-reviewed this paper (18-05580).

TRR
JOURNAL OF THE TRANSPORTATION RESEARCH BOARD

Article

Transportation Research Record
2018, Vol. 2672(24) 98–108

DOI: 10.1177/0361198118790375
journals.sagepub.com/home/trr
SAGE

Collaborative Geodesign for Alternative-Fuel Station Location using "Collablocation" Software

Michael Kuby[1], Keiron Bailey[2], Fangwu Wei[1], John Fowler[3], Daoqin Tong[1], Qing Zhong[2], Oscar Lopez[1], and William Sheaffer[4]

Abstract

Adoption of alternative-fuel vehicles requires a network of conveniently located fuel stations capable of serving the movement patterns of potential users. One hindrance to planning an effective network of stations is the difficulty of integrating the competing views of multiple stakeholders into an agreeable solution. This paper reports on a collaborative geodesign methodology applied to the problem of compressed natural gas fueling locations in the southwestern U.S.A. Geodesign is a multi-stakeholder spatial planning process that has gained currency in the GIS community but has not been applied to fuel infrastructure previously. We have developed an open-source interactive geovisual platform called Collablocation and a structured group process to facilitate interactive exploration of scenarios with multiple spatial data layers and to permit real-time computation and evaluation of network performance characteristics. For the pilot workshop we recruited expert stakeholders from industry, government, and local organizations. Breakout groups reached convergence on six locations on the I-10 and I-80, and at the Arizona–Mexico border, and several other less precise areas of need were identified. A post-workshop survey indicated high satisfaction with the technical features of the platform and the workshop design, and highlighted the ease of use and exchange of information.

Road transportation causes significant air pollution at global, regional, and local scales. Substituting alternative fuels for diesel and gasoline can help mitigate these environmental impacts (*1*). The paucity of conveniently located alternative-fuel stations, however, is a primary barrier to widespread adoption of alternative-fuel vehicles (AFVs) (*2*). This problem is recognized as a chicken-and-egg dilemma, in which lack of stations suppresses AFV adoption, which in turn reduces demand at stations, suppressing construction (*3*, *4*). Designing a starter network of stations that initiates the transition process is a major challenge. While market forces will gradually take over as the process unfolds, a large literature has emerged for planning an initial network to meet basic coverage requirements.

The typical approach to planning a network of alternative-fuel stations involves geographic information systems (GIS), operations research (OR), or both. GIS models for station location typically incorporate map layers relating to supply, demand, traffic, feasible sites, potential fleet adopters, incompatible land uses, etc. (*5*). Desirable and undesirable points, lines, and polygons can be spatially buffered, weighted, added, or subtracted using map algebra to form a suitability surface, and what-if scenarios can be evaluated quantitatively by several metrics (*6–8*). An exemplary GIS-based approach being used for planning stations is the California Hydrogen Infrastructure Tool which subtracts a map layer of coverage by existing stations from a demand surface to determine areas of high potential (*9*).

Rather than testing one configuration at a time, OR models for station location perform combinatorial optimization using exact or heuristic methods. Many models and objective functions have been proposed (*10*). OR models differ in how they represent the geographic

[1]School of Geographical Sciences and Urban Planning, Arizona State University, Tempe, AZ
[2]School of Geography and Development, University of Arizona, Tucson, AZ
[3]Department of Supply Chain Management, W. P. Carey School of Business, Arizona State University, Tempe, AZ
[4]Valley of the Sun Clean Cities Coalition, ASU Skysong Campus, Scottsdale, AZ

Corresponding Author:
Address correspondence to Keiron Bailey: kbailey@email.arizona.edu

demand for fuel, including by nodes (*11*, *12*), links (*13*), origin–destination (O-D) routes (*14*, *15*), and multi-stop tours (*16*, *17*). GIS and OR integrated into a spatial decision support system (SDSS) can form a powerful yet detailed modeling approach (*18–20*).

Nevertheless, most GIS, OR, and SDSS models are designed to be used by a single analyst or decision-maker. A successful transition to AFVs, however, requires a coordinated rollout of fuel production, fuel retailing, vehicle manufacturing, and supporting policies, institutions, and education. Researchers have referred to this web of essential elements as an alternative-fuels "ecosystem" (*21*). Melaina et al. (*22*) concluded from stakeholder interviews that the business case for alternative-fuel stations is highly dependent on coordination with automakers and fleets. Diverse industry stakeholders indeed have a stake in where and how many fuel stations are built.

The systematic goal of this research, therefore, is to develop a collaborative solution to this location problem given that multiple real-world factors constrain decision making. We build on recent interest in "geodesign"—a structured and collaborative approach to planning land uses—and extend it to locating systems of point facilities. Our first application of this approach is to truck stop locations for fueling heavy-duty trucks with compressed natural gas (CNG) in the U.S. Southwest. Trucking is a potential early market for natural gas transport because of price and maintenance advantages, high intensity of traffic, high fuel consumption, noise benefits, and reduction of particulates, nitrogen and sulfur oxides, and CO_2 compared with diesel (*14*, *23*, *24*). The second goal of this research is to evaluate the performance of the method used in a pilot workshop with the industry participants. Finally, the broader impact of the project is to introduce a new approach that can be improved and extended to different types of vehicles and fuels, geographic scales, regions, and stakeholders for generating more widely acceptable solutions to this thorny problem.

Methodology

While this problem suits a geospatial analytic approach, models based on literature do not always correspond to how decisions are made in the real world by the system designers, stakeholders, and users. Additional constraints and objectives may exist that are not evident to academic researchers. Fuel stations for trucking serve demand and face incompatibilities that involve a range of stakeholders and require viewing the problem from many different geographic scales from regional connectivity to local accessibility. A more flexible, collaborative, user-driven approach is called for that can tap into the "collective intelligence" of a group of experts.

Geodesign

Noted landscape theorist Carl Steinitz defines geodesign simply as "geography by design" (*25*). More specifically, "Geodesign is a design and planning method which tightly couples the creation of design proposals with impact simulations informed by geographic contexts" (*26*). Steinitz (*27*) emphasizes that good design and planning require both collaboration and geographic systems thinking. The term "geodesign" was coined around 2008 and has gained currency in recent years, though similar approaches were used previously (*27–29*). For our purposes, we define geodesign as the use of an interactive multi-user geographic platform to locate, simulate, and evaluate the performance of a network by multiple stakeholders in a workshop. The collaborative element involves simultaneous elicitation and integration of multiple stakeholder values into the decision support system, real-time evaluation of system designs using scientific performance metrics, and sharing of interim data and final solutions across participants, with discussion and feedback at each step. The workflow is what makes geodesign fundamentally different from other types of participatory GIS and structured public involvement (*30*).

Geodesign was originally developed to address multifaceted and/or contentious design problems at the scale of cities and regions, and several software packages are available. Geodesign Hub® (https://www.geodesignhub.com/) is free software designed for assigning land uses to land parcels. ESRI®'s GeoPlanner® (*30*) is a commercial product that allows users to work with points, lines, and polygons. We determined there was a need to develop a point-based, open-source package for multi-facility location problems, which we are calling Collablocation: a "Collaborative Laboratory for Location."

Specification of the Collablocation Geodesign Platform

The research team identified six essential functions for the collaborative geodesign software:

1. Display general-information base-map data and zoom in and out to any level of detail.
2. Visualize specific map layers important to the problem.
3. Point and click to add/relocate/remove facilities.
4. Compute in real time the performance or impact of the system design (consisting of the selected new facilities plus the existing facilities) using several system metrics.
5. Improve system designs over multiple iterations and stages.
6. Compare performance of system designs across groups and stages of the workshop, using tables, graphs, and maps.

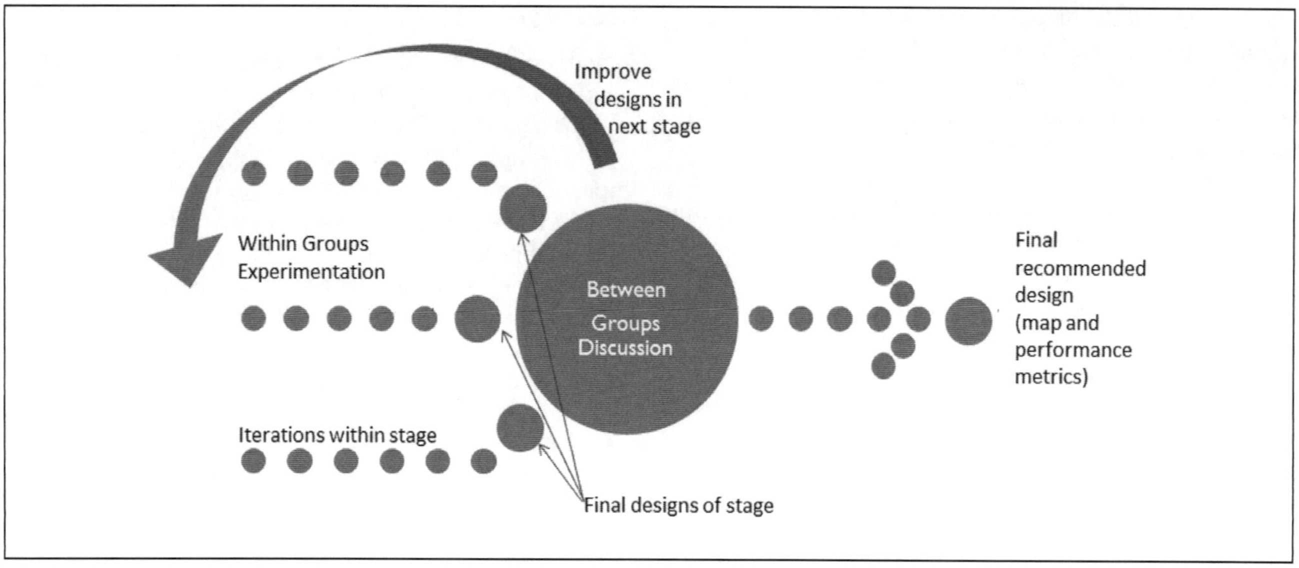

Figure 1. Basic structure of a geodesign workshop using Collablocation.

The first four functions are standard functionality in single-user SDSSs, while the latter two functions are essential for collaboration. Collaboration features include the ability to accept spatial inputs from multiple users simultaneously, display indicators for each group, import data from other groups, and synthesize group data. Rapid computation and ease of use are essential because technology that imposes an unnecessary overhead would slow collaboration and create a barrier to participation by industry experts, who are unlikely to be GIS users.

In addition to technology, successful geodesign requires a structured participation process. Workshop participants are divided into breakout groups, and the workshop as a whole is divided into rounds or stages (Figure 1). In each stage, groups go through multiple iterations of a network design, testing the performance of each iteration before finishing a final design for that stage. Breakout groups then come together for a between-groups discussion to compare and contrast the strategies, locations, and performance of their final designs. Groups then "go back to the drawing board" to improve previous designs in a subsequent stage. They can pull up any set of stations developed by any group to use as a basis for further editing and testing. This process is repeated several times to converge towards a final recommended solution.

Collablocation Software Design

Collablocation (https://collablocation.shinyapps.io/home) was developed iteratively through the software development life cycle process, including definition of requirements, design and development, testing, and implementation (*31*). Taking prototype feedback from the team, students, faculty, and the Valley of the Sun Clean Cities Coalition (VSCCC), the final version of the platform included seven tabs (Figure 2, top):

1. *Main* enables concurrent users/groups to interact with the map by using the right-panel visualization and analysis functions.
2. *Selected New CNG Stations* presents information about a group's selected truck stops individually, including name, address, geographic coordinates, and nearby traffic volume and truck fleets.
3. *Group Performance Measures by Stage* provides a vertical comparison of performance measures within a group by stages (see below for the specific metrics).
4. *Comparing to Other Groups* provides table and graph comparisons of performance measures between groups.
5. *Spatial Comparison* enables groups to compare their selections on a base map by visualizing similarity/dissimilarity of locations of selected stations between groups.
6. *About* introduces the Collablocation platform, team, and potential applications.
7. *Help & Data* provides instructions for use and explanations about data.

The Collablocation platform is written in R, a popular scripting language. Several R packages play important roles in platform design and development, data preprocessing, and real-time analysis. Shiny was used to structure the web-based interface. Leaflet is integrated with Shiny for interactive mapping. The platform is

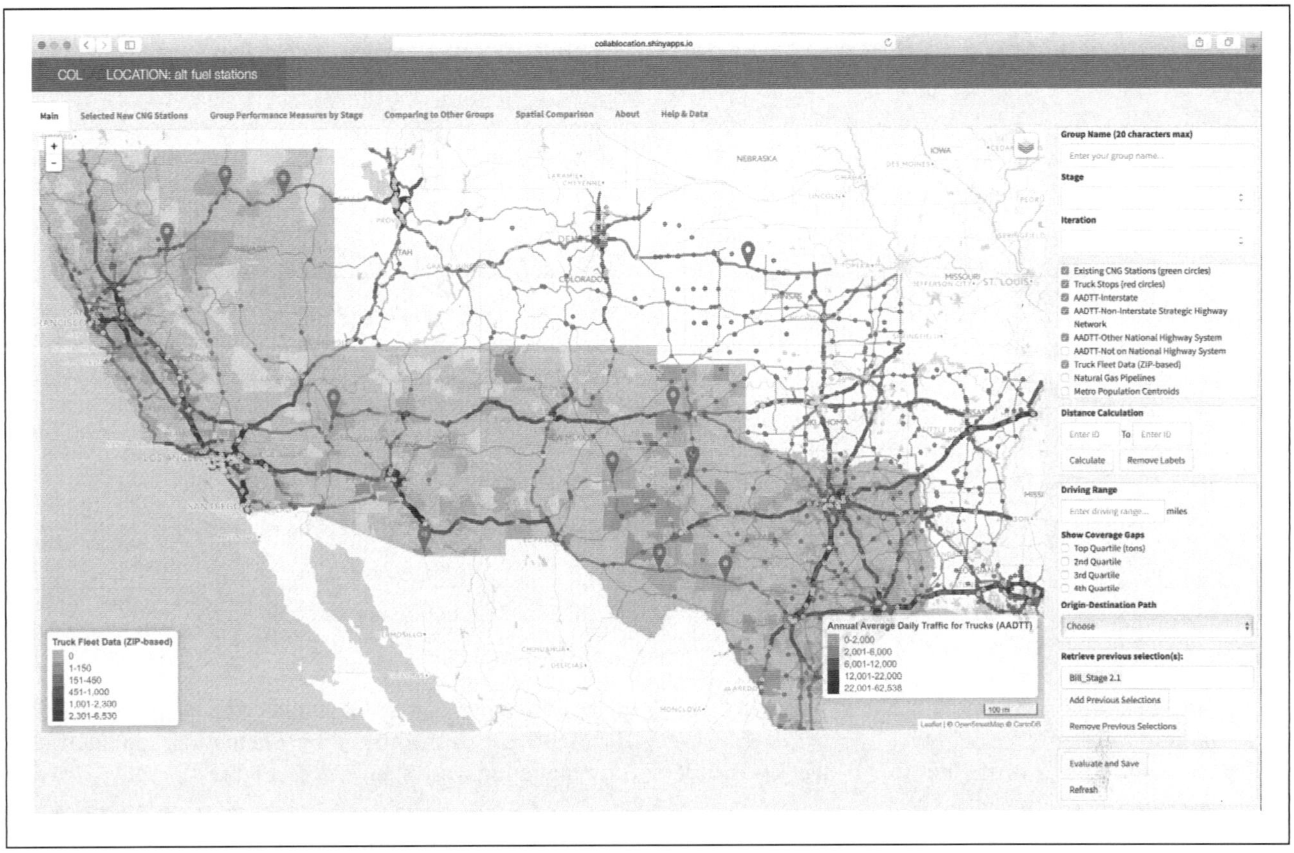

Figure 2. Collablocation user interface *Main* tab with 11 stations chosen. The seven screen tabs are displayed across the top, while the visualization and analysis settings are found in the right-side panel.

hosted on a cloud server by using Shinyapps.io service. Pre-processed input data are stored on the server for visualization and data analysis. Real-time processing group selections and results are saved on remote storage web service MongoLab using MongoDB, a type of NoSQL database.

Collablocation source code and non-proprietary data are available as free, open-source software under the same General Public License governing R and Shiny, at https://github.com/collablocation. Other organizations can modify and update the data, features, and functions for their own purposes.

Case Study: CNG Truck Stops in the Southwest United States

Study Area

The focus of the initial Collablocation case study is the five-state region of California, Nevada, Arizona, New Mexico, and Texas. When considering long-distance trucking, however, these states cannot be viewed in isolation. The application therefore included the surrounding states of Utah, Colorado, Kansas, Oklahoma, Arkansas, and Louisiana as origins and destinations of freight feeding into and out of the core region and areas where supporting stations are also needed. The Southwest was chosen because of severe air pollution in many cities and the importance of its seaports and U.S.–Mexico border ports as sources of freight. The Southwest's geographic extent necessitates coordination of multiple stations to refuel long shipments. In addition, the team's established relationships with alternative-fuel logistics, supply, and policy stakeholders in the Southwest helped recruit workshop participants.

Data

The most basic function of the software—geovisualization—requires three types of spatial data layers that can be selected and displayed on the map:

1. Point layers including existing CNG stations, diesel truck stops, and metropolitan population centroids
2. Polyline layers including annual average daily truck traffic (AADTT) flows, natural gas

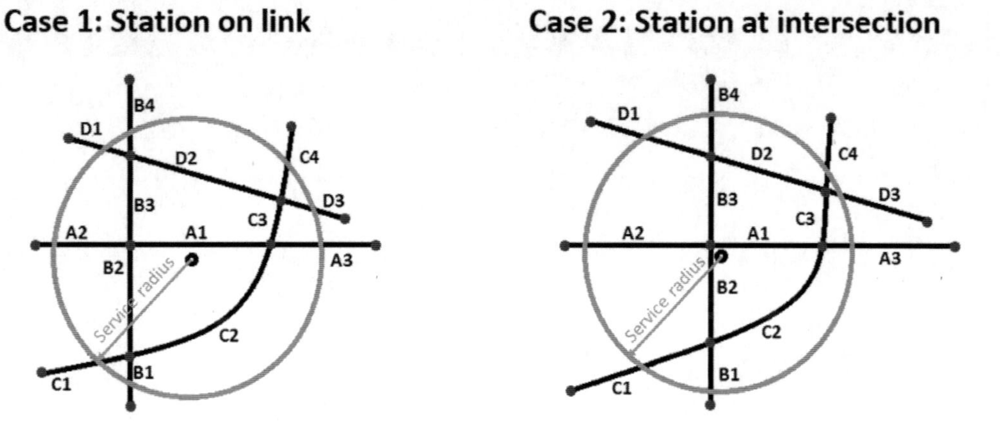

$$AADTT = A1 + \alpha(B_{max} + C_{max} + D_{max}) \qquad AADTT = \frac{A1 + A2 + B2 + B3}{2} + \alpha(C_{max} + D_{max})$$

Figure 3. AADTT calculation for Case 1 (station between intersections) and Case 2 (station at intersection).

pipelines, and O-D paths between major metropolitan areas in the Southwest
3. A polygon layer of the number of trucks at fleet headquarters within zip codes

Existing CNG stations available to large trucks as of October 31, 2016 were downloaded from NREL's Station Locator (https://www.afdc.energy.gov/locator/stations/). Potential new CNG truck stops (candidate sites) are limited to diesel truck stops from AllStays (http://www.allstays.com). Data used in the 2017 pilot workshop are described below; see the Help & Data tab on the 2018 platform for subsequent updates.

The platform provides several performance metrics to evaluate group station selections in real time. For land-use planning, these metrics are called impact simulations or evaluation maps and are based on models of subsystems for hydrology, green infrastructure, residential use, and transportation. For CNG truck stops infrastructure planning, three performance submodels were included.

Annual Average Daily Truck Traffic (AADTT). In Collablocation, AADTT is a rough estimate of the number of individual trucks passing near a candidate station daily. Without O-D flow data at a detailed geographical scale, it is impossible to calculate this exactly. The following assumptions, however, help avoid double-counting of trucks that travel on several arcs in the area, and discount traffic that must detour to get to the candidate station. Only interstate and non-interstate strategic highways within three miles of a station are considered, consistent with willingness to detour in other studies of commercial vehicles (*32*, *33*). Two cases are defined based on whether the candidate station is at a highway intersection (Figure 3):

- Case 1: If a station is in the middle of one arc, its AADTT is counted fully.
- Case 2: If a station is at a highway intersection, the AADTT of all four arcs are summed and divided by two.

For both cases, the AADTT of other highways in the three-mile buffer that do not directly pass the station are discounted by a factor α, which is applied to the segment of each continuous highway with the maximum traffic volume. α accounts for willingness to detour for fuel and for the unknown fraction of traffic from non-adjacent highways that is already counted on the arcs directly facing the station. For these reasons, $0 \leq \alpha \leq 1$, and we chose 0.5 as the midpoint of this range. Although admittedly a rough estimate, this AADTT calculation is a reproducible method using readily available data based on the simple notion that more trucks on nearby road segments translates to more refueling.

Number of Fleet Trucks. Data on Class 8 truck fleets registered to owners within specified research area zip codes were purchased from RL Polk, a subsidiary of IHS Markit, for the five core states only. For each potential new CNG station, the size of local truck fleet excludes very large rental fleets, which are more likely to be distributed nationally across many locations.

Paths Covered. While the AADTT and truck fleet metrics are pre-calculated for individual candidate sites and summed over all chosen stations, the path coverage analysis is done in real time and depends on the combination of new *and existing* stations. Given the stations chosen and the user-defined driving range, this metric analyzes whether trucks can complete round trips between major metropolitan origins and destinations. Results from this analysis are reported in terms of number of paths, tons, and ton-miles that can be covered and are based on data for about 100 major O-D pairs from the Freight Analysis Framework (FAF). For each O-D pair, the fastest travel-time path over the FAF road network was generated based on speed limits. The FAF estimates freight flows among major metropolitan areas, as well as a "Rest of State" zone. For states with at least one FAF metropolitan area, the Rest of State flows were omitted because their origin/destination cannot be accurately estimated. To ensure rapid computation, the 2017 version eliminated short and low-volume pairs while still representing 90% of tons and 84% of ton-miles. In the 2018 update, we include all 323 metropolitan O-D pairs with no detectable increase in computation time.

Consistent with the AADTT estimation, any existing or new station within three miles of a path is considered to be able to refuel it. A round trip is considered "covered" or "feasible" if the distance between adjacent stations along the path is always less than or equal to the driving range, and the distance between either endpoint and the first station is less than or equal to half of the driving range, which ensures a trip is repeatable (*14*, *15*). After a solution is evaluated using the Refresh and Save button, users can visualize uncovered gaps between stations relative to the fastest route for any O-D pair.

Pilot Workshop

Stakeholder Participants

The pilot workshop included 11 stakeholders from five states: Arizona (6), Texas (2), and New Mexico, California, and Colorado (1 each). Participants represented a range of organizations: Clean Cities coordinators (4), fuel retailing companies (2), universities (2), state government agencies (1), utilities (1), and truck manufacturers (1). We offered $175 travel stipends to out-of-state participants. Over 70 stakeholders were invited individually in order to secure these 11 hands-on participants, although another 10 to 15 expressed strong interest.

Workshop Process

The design of the collaborative geodesign workshop was based on the structured group process method developed and used by research team members in previous applications such as: collaborative highway corridor location (*34*); collaborative electric powerline placement (*35*); archaeological site modeling for highway agencies (*36*); and context-sensitive large bridge design (*37*). These group methods are all based on the value forum approach that elicits, evaluates, and applies decision criteria from small expert groups (*38*). They are enriched by arranging the value forum around an interactive geovisual and/or geospatial system designed to promote value convergence, that is, to identify solutions with stronger agreement and/or higher objective performance. In this case, the value convergence was intended to be captured in a set of more strongly agreed locational choices for a given set of station locations.

The workshop lasted five hours and was intended to be structured in four phases, beginning with an initial "blue skies" visioning using only a general reference map, and proceeding to three stages of initial experimentation, subsequent refinement, and final convergence using all data layers and performance metrics. Each group worked independently at first, then a structured cross-group solution share and discussion was hosted in which breakout groups explained and compared the reasoning behind their strategies, both generally and regarding particular locations. Participants self-organized into three groups of three or four participants each.

Results

Location Results

Stage 0: Blue Skies Visioning. In the blue skies stage, breakout groups used a limited version of the tool lacking data layers, performance metrics, and existing and candidate sites. The goal of this stage was to uncover guiding or tacit criteria that affect decision making without any influence from the features built into the tool. Using the base map only, the first objective was to choose where and how many facilities there should be for the study area. Groups varied in the number of stations they included, from 18 to 34 to 133. All groups spaced their stations at intervals but not strictly equidistant. Most stations were placed along interstate highways at cities or major highway intersections (or both).

Stage 1: Initial Constrained Design. In Stage 1, groups used the full platform with visualization layers and ability to evaluate performance metrics for network configurations. Participants were able to see where existing stations are located and add new stations at diesel truck stops. The goal for this stage was to locate 30 new stations. Groups varied greatly in how many iterations they created and how many stations they included, from a

Table 1. Pilot Workshop Results for Final Iterations of Stages 1 and 2, by Breakout Group

GROUP_STAGE	Number of new stations located	AADTT Total	AADTT Average	Class 8 fleet trucks in zip code Total	Class 8 fleet trucks in zip code Average	Covered O-D paths: Driving range assumed (miles)	Covered O-D paths: Number of paths covered (out of 98)	Covered O-D paths: Kilotons covered	Covered O-D paths: Ton-miles covered
Stage 1 (goal of 30 stations)									
F&L_Stage 1.10	29	276,388	9,531	11,359	392	350	87	254,064 (97.6%)	92,008 (93.3%)
Bill_Stage 1.5	28	160,194	5,721	6,939	248	250	78	248,107 (95.3%)	86,821 (88.0%)
Group B_Stage 1.2	20	85,957	4,298	6,809	340	100	11	73,758 (28.3%)	12,036 (12.2%)
Stage 2 (goal of 10 stations)									
F&L_Stage 2.12	11	66,068	6,006	8,608	783	350	83	251,290 (96.5%)	91,135 (92.4%)
Bill_Stage 2.3	11	35,394	3,218	3,870	352	275	58	201,904 (77.5%)	61,771 (62.6%)
Group B_Stage 2.1	10	58,522	5,852	4,356	436	100	11	73,758 (28.3%)	12,036 (12.2%)

low of two iterations (with a maximum of 20 stations) to a high of 10 iterations (with a maximum of 29 stations). Table 1 shows the performance metrics for the final iteration of each group at the end of stages 1 and 2. Note that groups assumed different driving ranges, so the covered O-D path results are not comparable.

Stage 2: Final Convergence. After the blue skies and initial configuration stages, there was not enough time for an intermediate refinement stage, so groups were instructed to recommend their 10 highest-priority stations in Stage 2. Figure 4 shows locations chosen by multiple groups, and the between-groups discussion developed consensus on these priorities:

- Two stations on I-80 between the San Francisco Bay Area and Salt Lake City
- Two stations on I-10 between El Paso and San Antonio
- Two stations near Arizona–Mexico border crossings on I-8 and I-19, at Yuma and Nogales

While time ran out before stakeholders could converge on the other four locations, the geodesign process nevertheless identified priority areas, especially between Dallas and Albuquerque and between Denver and Kansas City.

Participant Survey Results

At the end of the workshop a six-question survey using a nine-point Likert scale (1-Very Unsatisfied, 3-Unsatisfied, 5-Neutral, 7-Satisfied, and 9-Very Satisfied) was administered anonymously to assess the utility of the platform and workshop design. The mean "strength-of-agreement" (SOA) scores ranged from 7.8 to 8.7:

1. How satisfied are you overall with the workshop design? ($\bar{X}$ = 8.4)
2. How satisfied are you overall with the solution from the iterated constrained version [which turned out to be the final Stage 2]? ($\bar{X}$ = 7.8)
3. How satisfied are you that the list of factors generated during the workshop is comprehensive (meaning, no significant factors are missing)? ($\bar{X}$ = 8.1)
4. How satisfied are you with the level of knowledge exchange the workshop promoted between YOU and OTHER PARTICIPANTS (not including the researchers and organizers)? ($\bar{X}$ = 8.7)
5. How satisfied are you with the EASE OF USE of the geodesign platform? ($\bar{X}$ = 8.2)
6. How satisfied are you with the TECHNICAL ANALYTIC capacity of the geodesign platform? ($\bar{X}$ = 8.1)

The mean SOA scores of 8.2 for "ease of use" and 8.1 for "analytic capacity" are very encouraging. The SOA score of 8.7 for the promotion of knowledge exchange indicates a high effectiveness in this design criterion.

The lowest (but still good) SOA score of 7.8 was for the solution to the 10-station final stage likely reflects insufficient time to complete the exercise. Several participants assured us that another few minutes would have been reasonable to finalize the 10-station network. The research team, however, terminated the exercise on schedule despite the imminent convergence because one factor in encouraging repeat participation in stakeholder meetings generally is strict observance of announced time limits.

Written commentary boxes asked participants what they found most and least useful.

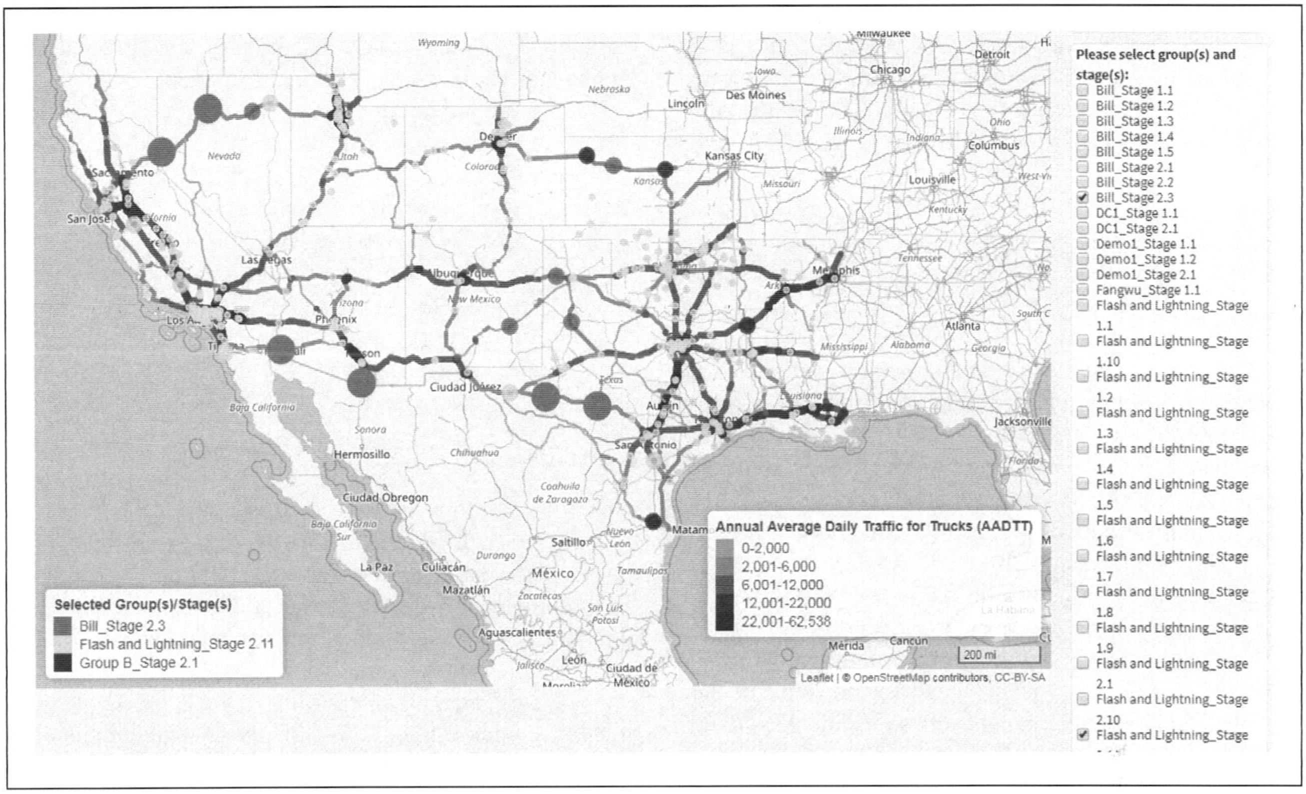

Figure 4. *Spatial Comparison* tab for final stage and iteration for three breakout groups. Large gray circles indicate stations chosen by multiple groups.

Most useful:

- "Sessions of small groups working with the tool was useful for exchange of ideas"
- "Opportunity to actually use the tool ourselves and design a system"
- "Collaborative efforts of a diverse group of participants"
- "Interactive and discussion between participants and facilitators"
- "The model has tremendous capability! Keep going!"
- "The collegiality, 'geodesign' conceptual and technical functionality and hands on"
- "The geodesign platform was beyond phenomenal"
- "The detail in the platform is high, and the tool is extremely useful and has multiple applications. Nice work!"
- "The overall discussion of the participants"
- "The sharing with industry knowledge"
- "Tool and new input from group"

Least useful:

- "The blue sky session was not as useful when we could not visualize the constraints we wanted to design with"
- "Set up of room"
- "Ultimately, a commercial focus on application would make the model more useful"
- "Interchange between ASU and participants"

In terms of the features of the platform being easy for all to use and to stimulate further contributions and enhancements, these comments demonstrate a measure of success.

Suggestions for Improvement

Participants provided the following specific suggestions for improving the platform and/or workshop process:

1. "As a strategy to reconcile different points on the map, perhaps have industry reps (both station developers and CNG fleet users) rank the criteria in order of importance to them to prioritize which criteria to drive final decisions. Ultimately, as planners we may focus on certain criteria but if the end users think it's not important, their opinions should probably outweigh ours!"
2. "Add additional reps from the industry to include dispatchers and fleet managers."

3. "More specifics in data layers as follows: 1) local fleet traffic (e.g., refuse, local haul, transit); 2) local gas supply – volume and pressure; 3) mitigating factors – future growth, incentives ($), and regulatory (unknowns)."

We certainly agree with the general concern behind Point 1 regarding the need to consider how user groups rank the criteria. Academic research that fails to identify, connect with, and solve the real-world problem is undesirable for all concerned. It is unclear, however, what exactly this anonymous comment is proposing. While the research team did choose which performance metrics to incorporate into the tool, we did not weight or rank the criteria; a formal weighting of objectives would take the process out of the realm of geodesign and into the realm of multiobjective optimization or analytic hierarchy processes. Because of their expressed concern about how to "reconcile different points," we suspect this comment also stems from a lack of time in the final stage to negotiate over the differences among the groups' final solutions. In any case, Point 1's call for more participation by CNG fleet users is echoed by Point 2, and we hope to include more trucking industry participants in subsequent forums.

Point 3 highlights other layers that workshop discussions highlighted as important factors missing from the database. The first suggestion—local fleet traffic—highlights an essential characteristic of fuel stations; even if a location was motivated by a particular class of users (e.g., long-haul truckers), the station can be used by other users. Passenger cars and fleets of vans, taxis, or buses can provide additional demand for a station, thus supporting profitability. The second suggestion regarding a station's CNG flow rate stems from the fact that fueling a large truck with CNG already takes a discouragingly long time for a driver who wants to get back on the road. Low flow rate adds significantly to the refueling time, and drivers share such information via social media. Workshop participants requested that the flow rate at existing stations be added to both the NREL Station Locator and Collablocation. The third suggestion—state incentives for station development—was also discussed in the between-groups discussion, for example, if a company were deciding between locations on either side of the Arizona-California border.

Discussion

Recruiting real-world participants for a one-day commitment to an unfamiliar process (at least in the fuel station setting) proved a major challenge. Partnering with VSCCC was crucial to the success of the initial workshop. VSCCC hosted the meeting, helped identify and convince industry stakeholders to participate, lent the workshop a "real-world" legitimacy, and provided a familiar location, organization, and sponsor.

One of the advantages of a geodesign process is that workshop participants are free to bring omitted factors into their decision making and indeed to emphasize them. The tool provides results in terms of AADTT, fleet coverage, and O-D coverage gaps, but participants can choose to focus on those metrics to varying degrees. For instance, in the pilot workshop, participants also shared information about the need for closer spacing of stations in mountainous areas.

Conclusion

Collaborative, multi-stakeholder planning and design of transportation infrastructure is a demanding undertaking. Because stakeholders have non-aligned and sometimes competing objectives, the goal of a collaborative platform is to serve as medium for information elicitation, sharing, analysis, and development of features or keys of agreement. Collaborative geodesign appears a promising methodology for involving stakeholders in transportation problems such as alternative-fuel corridor planning. In addition to promoting learning and information sharing, it can help streamline the slow process of reconciling the needs of multiple stakeholders by resolving dilemmas, reducing uncertainty, and facilitating consensus. The Collablocation platform performed surprisingly well for new software. The evaluation scores and the written commentary illustrate several strong features. The workshop design was also highly rated by participants, although we learned important lessons such as the importance of starting with a smaller number of stations, allotting a full day for a workshop, and agreeing on a common driving range assumption prior to the final stage.

Geodesign cannot be expected to generate or coerce "consensus" or complete agreement on solutions (*34–37*). In a relatively small expert group such as the one that piloted the platform, a strong measure of convergence was observed during the final stage on six specific locations and two other general areas. With a little more time, it may have been possible to obtain agreement on a set of ten high-priority locations. The lack of unanimous final agreement, however, is not necessarily problematic. In a paper called "The Trouble with a Strong Concept Fully Worked Out," Steinitz (*39*) argues that completeness is not essential because change is constant and even the collective intelligence of a group cannot know, let alone control, everything. In this sense, the more precise convergence on the I-10 and I-80, and Arizona–Mexico border stations and the less precise identification of needs in West Texas and Kansas may accurately convey the uncertainty surrounding the collective recommendations from the workshop.

How well geodesign works in larger forums is an open question. The core problems with collaborative geospatial methods are in both scale and scope (*40*); scale meaning non-linear or even diminishing returns to larger group sizes, and scope meaning the range of tacit and overt stakeholder objectives (i.e., levels of agreement among stakeholders about purpose, need, feasibility, desirability, and other attributes).

Although the pilot application for this platform was CNG truck stops in the Southwest, the platform is readily adaptable to other vehicles and fuels, from DC fast chargers for electric vehicles to hydrogen stations for fuel-cell vehicles. The data can be easily replaced by spatial layers for other fuels, vehicles, regions, and geographic scales. Our hope is that the open-source Collablocation software will be downloaded and modified by other researchers for other collaborative geodesign projects for other types of combinatorial, multistakeholder facility location problems.

Acknowledgments

The authors would like to thank: the Arizona Board of Regents for funding through the Research Innovation Fund; ASU and UA for matching funds; Larry Head (UA) and Ed Smaglik (NAU) for leading the ABOR-RIF project; and most of all the workshop participants, whose dedication and expertise is the real key to this geodesign approach.

Author Contributions

The authors confirm contribution to the paper as follows: study conception and design: M. Kuby, D. Tong; data collection: M. Kuby, K. Bailey, F. Wei, Q. Zhong; analysis and interpretation of results: M. Kuby, K. Bailey, D. Tong, F. Wei, Q. Zhong; draft manuscript preparation: M. Kuby, K. Bailey. All authors reviewed the results and approved the final version of the manuscript.

References

1. Sperling, D., and D. Gordon. *Two Billion Cars: Driving Toward Sustainability*. Oxford University Press, Oxford, UK, 2009.
2. Melendez, M. *Transitioning to a Hydrogen Future: Learning from the Alternative Fuels Experience*. Publication NREL/TP-540-39423. National Renewable Energy Laboratory, Golden, CO., 2006.
3. Melaina, M. Initiating Hydrogen Infrastructures: Preliminary Analysis of a Sufficient Number of Initial Hydrogen Stations in the US. *International Journal of Hydrogen Energy*, Vol. 28, No. 7, 2003, pp. 743–755. https://doi.org/10.1016/S0360-3199(02)00240-9.
4. Nicholas, M., S. Handy, and D. Sperling. Using Geographic Information Systems to Evaluate Siting and Networks of Hydrogen Stations. *Transportation Research Record: Journal of the Transportation Research Board*, 2004. 1880: 126–134.
5. Ko, J., T. H. T. Gim, and R. Guensler. Locating Refuelling Stations for Alternative Fuel Vehicles: A Review on Models and Applications. *Transport Reviews*, Vol. 37, No. 5, 2017, pp. 551–570. https://doi.org/10.1080/01441647.2016.1273274.
6. Ji, W., M. Nicholas, and G. Tal. Electric Vehicle Fast Charger Planning for Metropolitan Planning Organizations: Adapting to Changing Markets and Vehicle Technology. *Transportation Research Record: Journal of the Transportation Research Board*, 2015. 2502: 134–143.
7. Wood, E., S. Raghavan, C. Rames, J. Eichman, and M. Melaina. *Regional Charging Infrastructure for Plug-In Electric Vehicles: A Case Study of Massachusetts*. National Renewable Energy Laboratory, Golden, CO., 2017.
8. National Renewable Energy Lab. *Hydrogen Demand and Resource Analysis Tool (HyDRA)*. https://maps.nrel.gov/hydra/. Accessed October 27, 2017.
9. California Air Resources Board. *Hydrogen Fueling Infrastructure Assessments*. https://www.arb.ca.gov/msprog/zevprog/hydrogen/h2fueling.htm. Accessed July 24, 2017.
10. Jing, W., Y. Yan, I. Kim, and M. Sarvi. Electric vehicles: A Review of Network Modelling and Future Research Needs. *Advances in Mechanical Engineering*, Vol. 8, No. 1, 2016, pp. 1–8.
11. Honma, Y., and O. Kurita. A Mathematical Model on the Optimal Number of Hydrogen Stations with Respect to the Diffusion of Fuel Cell Vehicles. *Journal of the Operations Research Society of Japan*, Vol. 51, No. 2, 2008, pp. 166–190. https://doi.org/10.15807/jorsj.51.166.
12. Brey, J. J., R. Brey, A. F. Carazo, M. J. Ruiz-Montero, and M. Tejada. Incorporating Refuelling Behaviour and Drivers' Preferences in the Design of Alternative Fuels Infrastructure in a City. *Transportation Research Part C: Emerging Technologies*, Vol. 65, 2016, pp. 144–155.
13. Lin, Z., J. Ogden, Y. Fan, and C. W. Chen. The Fuel-Travel-Back Approach to Hydrogen Station Siting. *International Journal of Hydrogen Energy*, Vol. 33, No. 12, 2008, pp. 3096–3101. https://doi.org/10.1016/j.ijhydene.2008.01.040.
14. Kuby, M., I. Capar, and J. G. Kim. Efficient and Equitable Transnational Infrastructure Planning for Natural Gas Trucking in the European Union. *European Journal of Operational Research*, Vol. 257, No. 3, 2017, pp. 979–991. https://doi.org/10.1016/j.ejor.2016.08.017.
15. Kim, J. G., and M. Kuby. A Network Transformation Heuristic Approach for the Deviation Flow Refueling Location Model. *Computers & Operations Research*, Vol. 40, No. 4, 2013, pp. 1122–1131. https://doi.org/10.1016/j.cor.2012.10.021.
16. Tu, W., Q. Li, Z. Fang, S. L. Shaw, B. Zhou, and X. Chang. Optimizing the Locations of Electric Taxi Charging Stations: A Spatial–Temporal Demand Coverage Approach. *Transportation Research Part C: Emerging Technologies*, Vol. 65, 2015, pp. 172–189. https://doi.org/10.1016/j.trc.2015.10.004.

17. Kang, J. E., and W. Recker. Strategic Hydrogen Refueling Station Locations with Scheduling and Routing Considerations of Individual Vehicles. *Transportation Science*, Vol. 49, No. 4, 2016, pp. 767–783. https://doi.org/10.1287/trsc.2014.0519.
18. Kuby, M., L. Lines, R. Schultz, Z. Xie, J. G. Kim, and S. Lim. Optimization of Hydrogen Stations in Florida Using the Flow-Refueling Location Model. *International Journal of Hydrogen Energy*, Vol. 34, No. 15, 2009, pp. 6045–6064. https://doi.org/10.1016/j.ijhydene.2009.05.050.
19. Sathaye, N., and S. Kelley. An Approach for the Optimal Planning of Electric Vehicle Infrastructure for Highway Corridors. *Transportation Research Part E: Logistics and Transportation Review*, Vol. 59, 2013, pp. 15–33. https://doi.org/10.1016/j.tre.2013.08.003.
20. Stephens-Romero, S. D., T. M. Brown, J. E. Kang, W. W. Recker, and G. S. Samuelsen. Systematic Planning to Optimize Investments in Hydrogen Infrastructure Deployment. *International Journal of Hydrogen Energy*, Vol. 35, No. 10, 2010, pp. 4652–4667. https://doi.org/10.1016/j.ijhydene.2010.02.024.
21. Lu, C., K. Rong, J. You, and Y. Shi. Business Ecosystem and Stakeholders' Role Transformation: Evidence from Chinese Emerging Electric Vehicle Industry. *Expert Systems with Applications*, Vol. 41, No. 10, 2014, pp. 4579–4595. https://doi.org/10.1016/j.eswa.2014.01.026.
22. Melaina, M., M. Muratori, J. McLaren, and P. Schwabe. Investing in Alternative Fuel Infrastructure: Insights for California from Stakeholder Interviews. Presented at 96th Annual Meeting of the Transportation Research Board, Washington, D.C., 2017.
23. Jaffe, A. M., R. Dominguez-Faus, L. Allen, K Medlock, N. Parker, D. Scheitrum, A. Burke, H. Zhao, and Y. Fan. *Exploring the Role of Natural Gas in U.S. Trucking*. No. UCD-ITS-RR-15-05. University of California at Davis, Institute for Transportation Studies, Davis, CA, 2015.
24. Osorio-Tejada, J. L., E. Llera-Sastresa, and S. Scarpellini. Liquefied Natural Gas: Could It Be a Reliable Option for Road Freight Transport in the EU? *Renewable and Sustainable Energy Reviews*, Vol. 71, May 2017, pp. 785–795. https://doi.org/10.1016/j.rser.2016.12.104.
25. Steinitz, C. *A Framework for Geodesign: Changing Geography by Design*. ESRI Press, Redlands, CA, 2012.
26. Flaxman, M. *Quoted in "Designing Geodesign: Summit on New Field that Couples GIS and Design."* ESRI ArcUser News. 2010. http://www.esri.com/news/arcuser/0410/geodesign.html. Accessed July 10, 2017.
27. Paar, P., and J. Rekittke. Wheeling a Trojan Horse to Teach MLA Students Geoinformation Methods. *Proceedings of the 13th Digital Landscape Architecture Conference (DLA 2012)*, Wichmann, Heidelberg, Germany, 2012, pp. 32–40. https://researchbank.rmit.edu.au/view/rmit:38407.
28. Goodchild, M. Towards Geodesign: Repurposing Cartography and GIS. *Cartographic Perspectives*, No. 66, 2010, pp. 1–16. https://doi.org/10.14714/CP66.93.
29. Batty, M. Defining Geodesign. *Environment and Planning B: Planning and Design*, Vol. 40, No. 1, 2013, pp. 1–2.
30. Srivasta, P., and M. Coast. *New GeoPlanner App Enables Smart Planning with Geodesign. ArcWatch: GIS News, Views, and Insights*. ESRI, July 2014. http://www.esri.com/esri-news/arcwatch/0714/new-geoplanner-app-enables-smart-planning-with-geodesign. Accessed October 27, 2017.
31. Larman, C., and V. R. Basili. Iterative and Incremental Developments. A Brief History. *Computer*, Vol. 36, 2003, pp. 47–56. https://doi.org/10.1109/MC.2003.1204375.
32. Kelley, S., and M. Kuby. Decentralized Refueling of Compressed Natural Gas (CNG) Fleet Vehicles in Southern California. *Energy Policy*, Vol. 109, 2017, pp. 350–359. https://doi.org/10.1016/j.enpol.2017.07.017.
33. Suzuki, Y. A Generic Model of Motor-Carrier Fuel Optimization. *Naval Research Logistics*, Vol. 55, No. 8, 2008, pp. 737–746.
34. Grossardt, T., K. Bailey, and J. Brumm. Analytic Minimum Impedance Surface: Geographic Information System Based Corridor Planning Methodology. *Transportation Research Record: Journal of the Transportation Research Board*, 2001. 1768: 224–232.
35. Jewell, W., R. Gill, T. Grossardt, and K. Bailey. A New Method for Public Involvement in Electric Power Transmission Line Routing. *Transactions of the Institute of Electrical and Electronics Engineers Power Delivery*, Vol. 24, No. 4, 2009, pp. 2240–2247.
36. Ripy, J., T. Grossardt, M. Shouse, P. Mink, K. Bailey, and C. Shields. Expert System Archaeological Predictive Model. *Transportation Research Record: Journal of the Transportation Research Board*, 2014. 2403: 37–44.
37. Bailey, K., J. Williams, L. Toole, T. Grossardt, and B. Bryant. Structured Public Involvement in Context-Sensitive Large Bridge Design Using Casewise Visual Evaluation: Case Study of Section 2 of Ohio River Bridges Project. *Transportation Research Record: Journal of the Transportation Research Board*, 2007. 2028: 85–92.
38. Keeney, R., D. Von Winterfeldt, and T. Eppel. Eliciting Public Values for Complex Policy Decisions. *Journal of Management Science*, Vol. 36, No. 9, 1990, pp. 1011–1030. https://doi.org/10.1287/mnsc.36.9.1011.
39. Steinitz, C. The Trouble with a Strong Concept, Fully Worked Out. *Landscape Architecture*, Vol. 69, 1979, p. 565.
40. Bailey, K., and T. Grossardt. Towards Structured Public Involvement: Justice, Geography and Collaborative Geospatial/Geovisual Decision Support Systems. *Annals of the Association of American Geographers*, Vol. 100, No. 1, 2010, pp. 57–86. https://doi.org/10.1080/00045600903364259.

The Standing Committee on Alternative Transportation Fuels and Technologies (ADC80) peer-reviewed this paper (18-02899).

TRR
JOURNAL OF THE TRANSPORTATION RESEARCH BOARD

Article

Transportation Research Record
2018, Vol. 2672(24) 109–121

DOI: 10.1177/0361198118792329
journals.sagepub.com/home/trr
SAGE

Quantitative Evaluation of MD/HD Vehicle Electrification using Statistical Data

Zhiming Gao[1], Zhenhong Lin[1], Stacy Cagle Davis[1], and Alicia K. Birky[2]

Abstract

This paper presents a wide-ranging analysis of Class 3-8 commercial vehicle electrification by means of developing a framework tool which uses a quantitative method of estimating electric vehicle energy consumption and appropriate charging considerations. The Fleet DNA composite statistics on real-world driving behavior is used to evaluate feasible or market-ready battery electric vehicle (BEV) technologies in medium- and heavy-duty (MD/HD) applications. In the paper, ten representative Class 3-8 commercial vehicle electrifications have been evaluated as a function of various service coverages, including applications in port drayage tractors, refuse trucks, delivery trucks, buses, and bucket trucks. The results indicate significant energy savings and fuel cost savings across all MD/HD vehicle electrifications. The majority of MD BEVs, with the exception of Class 3 bucket trucks, achieve better than a 5-year payback with 50–75% service coverage. For HD BEVs, with the exception of the Class 8 port drayage tractors, the 90% service coverage results in a 10-year or longer payback time, while the 50% service coverage yields a 7–8 year payback. Class 8 port drayage tractors should achieve payback in no more than a 3.5 years with 50–75% service coverage. Thus, the analysis indicates a highly feasible potential for Class 3-6 MD vehicles to be electrified, and feasible opportunities for electrification in Class 7-8 HD short-distance applications.

The United States is the world's largest user of oil and refined petroleum products (*1*). Much of this demand is contributed by the transportation sector, which has been responsible for about 65% of U.S. carbon dioxide emissions over the last 20 years (*2*). Transportation vehicle electrification offers substantial benefits for local air quality, greenhouse gas emissions, and petroleum use (*3*). Electrification is particularly vital for medium-duty (MD) and heavy-duty (HD) vehicles which account for nearly 20% of the nation's transportation fuel consumption (*4*). In fact, battery electric vehicle (BEV) technology is preferably suited for many MD and HD applications with well-defined routes, frequent stops, long idle times, and lower average speeds. Many U.S. nationwide fleets have shown significant interest and made commitments to electrify part of their fleet, even with significant cost penalties, in order to benefit the community, environment, and economy (*5*, *6*). Significant growth is expected in the future MD and HD electric vehicle market (*7*).

However, compared with light-duty (LD) BEVs, there are few MD and HD BEVs on the road to date. This is in part because not all LD BEV technologies can be transferred directly to MD and HD applications because of requirements for higher power and the durability to withstand harsh working environments (*8*). Moreover, the MD and HD vehicle market covers a wide diversity of commercial vocations, operational patterns, and duty cycles (*3*). Table 1 summarizes the existing MD/HD vehicle electrification technologies and readiness. Because of differences in duty cycles, one truck electrification model cannot fit many diverse truck applications. As a result, MD and HD vehicle electrification options are still in limited commercial availability, although electrification technologies have been widely applied to transit buses, school buses, MD and HD trucks. At present, five manufacturers offer commercial battery electric transit buses for sale in the U.S., including Proterra, BYD, New Flyer, GreenPower, and Nova (*8*). These commercial transit electric buses are being tested in New York (*9*), Chicago (*10*), St. Louis (*11*), California (*12*), and so forth. Like transit buses, school buses operate in urban and suburban environments. Lion, Motiv Power

[1]Oak Ridge National Laboratory, Oak Ridge, TN
[2]Energetics Incorporated, Columbia, MD

Corresponding Author:
Address correspondence to Zhiming Gao: gaoz@ornl.gov

Table 1. MD/HD Vehicle Electrification Technologies and Readiness (*7, 8, 16–18*)

Type	Vehicle	OEM	Battery (kWh)	Range (miles)	Charging (kW or hr)
Transit bus[a]	30–60 ft city bus	BYD Motors	197–591	120–200	80–200
	35–60 ft city bus	New Flyer	100–480	120	–
	35–40 ft city bus	Proterra	79–660	47–350	0.2–5 hrs
	40 ft city bus	Nova	–	15	–
	40 ft city bus	GreenPower	320	>185	–
	Shuttle bus	Balqon	–	100–150	–
	Shuttle bus	Phoenix	–	100	–
	Shuttle bus	Motiv	–	65–100	60
School bus[b]	School bus	Lion	90/105/130	50–100	–
	School bus	Motiv	85/106/127	65–100	25
	School bus	TransPower	108	75	1.5–2.0 hrs
	School bus	Adomani	210–400	>175	–
MD truck[b]	Delivery van	AMP	60	60	–
	Delivery van	Motiv	85/106/127	65–100	25
	MD cargo van	Zenith Motors	52–62	90	6
	Class 5 delivery truck	BYD Motors	145	155	100
	Class 6 delivery truck	BYD Motors	175	124	100
	MD work truck	Motiv	85/106/127	65–100	–
	MD terminal truck	Orange EV	80–160	–	–
	MD utility truck	Phoenix	105	100	2.5–3 hrs
HD truck[c]	Class 7 Semi-tractor	Cummins	140 kWh	100	1 hr
	Class 8 tractor	TransPower	–	–	–
	Class 8 refuse truck	Motiv	170–212	50–80	45
	Class 8 tractor	BYD Motors	188	92	80
	Class 8 drayage	Balqon	140–250	150	–

[a]Commercially available.
[b]Limited commercial availability.
[c]Demonstration phase.

systems, TransPower, and Adomani have demonstrated their battery all-electric school bus with 80–400 kWh battery (*13–16*). California Highway Patrol's electric school bus demonstration projects show the significant potential for fuel cost savings (*13, 14*). However, the large school bus manufacturers in the U.S., such as Bluebird, IC Bus, and Thomas Built, are not involved in electric bus production because of a number of factors, including safety certification (*8*). School buses are not yet as commercially available as transit buses.

Class 3-6 MD vehicle electrification could benefit from technology transferred from LD applications to a larger degree than Class 7-8 HD trucks and buses, because MD trucks and vans require less power and drive fewer miles. The electric vehicles in this category are being used in optimal duty cycles such as local delivery and food distribution (*15–17*). Several manufacturers produce electric trucks/vans for food/beverage, parcel delivery, utility and maintenance work, and other vocational applications, with gross vehicle weight (GVW) ranging from 10,000 to 26,000 pounds (*17*). High-profile fleets, such as Frito-Lay, Staples, and Coca-Cola, have made big investments in E-trucks that haul as much as 16,000 pounds, with a range up to 100 miles on a single charge (*15*). It is expected that widespread penetration into the market place will occur in the next 5–10 years (*8*). HD vehicle electrification is currently in the field demonstration phase by the manufacturers including Balqon, BYD, Electric Vehicles International, Motiv Power System, TransPower, ZeroTruck, and others (*3, 7*). There are also 12 other Class 8 BEVs in varying stages of prototype manufacture (*7*), including Cummins and Tesla (*18*). The major efforts of truck electrification in Class 8 have been started with vocations that have optimal duty cycles, like short-haul for drayage operations, local delivery, and urban refuse trucks (*8, 18*). However, there are not commercial HD battery electric trucks available at the moment because of the large gap of battery cost and fast charging technologies.

MD/HD vehicle electrification is still in very early stages of development, despite high expectations for their future market. Significant research is needed to understand (1) BEV feasibility and priority across types and duties (or vocations) of MD/HD vehicles; (2) cost effectiveness of electrification across types and duties of these commercial vehicles; and (3) the high potential of MD/HD electrification technologies and market opportunities to align stakeholder needs and policy considerations with qualitative and quantitative analysis planning. An in-depth analysis is particularly essential to identify the challenges and opportunities of MD/HD electrification over a wide range of driving patterns related to MD/HD vocations. Therefore, this study will apply a quantitative method with electric vehicle energy consumption and appropriate charging assumptions to conduct a comprehensive priority analysis of commercial vehicle electrification. A wide range of vehicle types and operations used in the method comes from Fleet DNA statistical data, which are collected by National Renewable Energy Laboratory (NREL), covering a substantial number of day-to-day real routes performed by MD and HD vehicles (*19*). To keep the problem tractable, the energy consumption analysis was restricted to use the modified vehicle tractive energy methodology using the statistical drive data and component efficiency for describing vehicle performance. In addition, MD/HD electrification cost effectiveness is also addressed in the developed tool.

Driving Data, Methodology, and Assumption

This section describes drive data, vehicle energy methodology for tractive energy, brake forces, and vehicle component efficiencies, as well as the assumptions for modeling MD/HD conventional and BEVs.

Driving Data

Fleet DNA data provide a wide range of operation statistics for many of today's Class 3-8 MD and HD vehicles, covering transit buses, school buses, delivery vehicles, regional tractors, utility trucks, refuse trucks, bucket trucks, cabs, sleeper-cab trucks, and many others (*19*). The datasets include extensive vehicle driving statistical information, such as aggregated route distance, average speed, stop events per mile, deceleration events per mile, drive duration, stop duration, deceleration duration, load and grade statistics, and many other drive-cycle metrics and travel features. The Fleet DNA data also provide specific operating energy consumption, allowing more customized results to be produced from the Fleet DNA data formation. To keep the current study in reasonable order, ten representative Class 3-8 MD/HD vehicles with optimal driving cycle features were selected in the current electrification analysis. Table 2 summarizes the driving information of these vehicles.

The ten Class 3-8 MD/HD vocations cover a widely distinguished range of driving patterns. Naturally, Class 7-8 vehicles have longer daily travel mileage, and their maximum daily travel mileage is much greater than the average daily travel mileage. Figure 1 displays two examples of the driving statistics (i.e., Class 7 food delivery tractors and Class 8 port drayage tractors), showing a long-tailed distribution in both the cases. The

Table 2. Ten Representative Class 3-8 MD/HD Vehicle Driving Behavior Summary

Drive characteristics summary	Class 8 port drayage tractor	Class 8 refuse pickup refuse truck	Class 7 food delivery tractor	Class 7 mass transit city transit bus	Class 6 school bus school bus	Class 5 linen delivery step van	Class 5 food delivery straight truck	Class 4 parcel delivery step van	Class 3 food delivery straight truck	Class 3 telecom bucket truck
Average drive distance (mile/day)	119	73	164	102	61	66	40	55	37	26
Annual travel mileage[a]	43386	26566	60001	37284	22112	23992	14478	19911	13551	9652
Max drive distance (mile/day)	612	273	568	268	181	141	81	132	79	93
Average drive time (hr/day)[b]	4.29	3.52	4.36	5.05	2.09	2.42	1.18	2.79	1.12	1.09
Max drive time (hr/day)[b]	13.80	8.78	11.52	13.80	4.52	4.21	2.05	4.56	2.14	2.61
Average vehicle on time (hr/day)[c]	9.06	7.33	7.26	9.11	2.96	6.18	2.98	7.06	1.60	3.10
Max vehicle on time (hr/day)[c]	23.96	13.21	15.69	22.35	7.73	12.63	18.16	11.38	3.29	8.52
Average drive speed (mph)[b]	27.70	20.68	37.71	20.23	23.54	27.18	33.52	19.57	33.19	24.35
Max drive speed (mph)[b]	76.50	69.59	71.87	69.07	79.24	69.90	70.66	70.84	70.34	74.63
Average vehicle on speed (mph)[c]	14.11	9.46	18.93	11.36	17.74	11.87	18.23	8.30	22.84	10.75
Average stops per mile	1.02	4.01	0.49	1.72	1.40	1.49	0.78	3.33	0.81	1.74
Max stops per mile	13.37	16.75	8.45	4.76	3.86	3.02	3.04	6.58	3.03	3.13
Average stops per day[d]	7.10	3.00	10.68	3.32	2.00	13.96	8.58	9.42	8.89	7.74
Max stops per day[d]	28.00	3.00	23.00	14.00	2.00	30.00	16.00	28.00	17.00	20.00

[a] 1 year = 365 days.
[b] Vehicle speed >0.
[c] Vehicle speed ≥ 0.
[d] Stop or idling time >5 minutes.

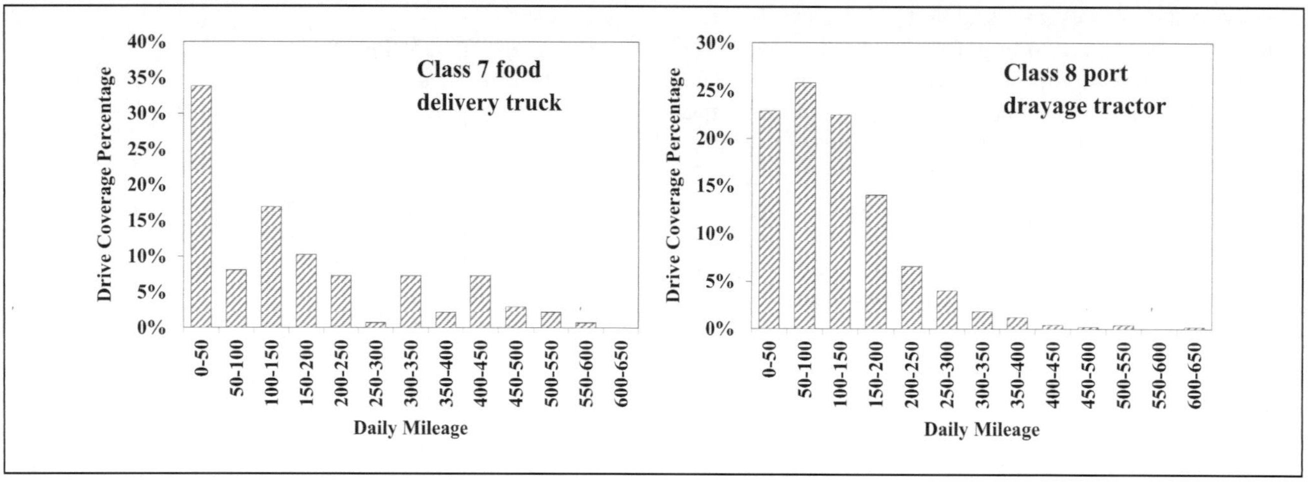

Figure 1. Driving statistics of Class 7 food delivery and Class 8 port drayage tractors.

observation illustrates that the percentage of aggressive daily travel mileage is actually quite limited. This insight illustrates that it is not necessary for MD/HD electric vehicles to cover all-days service conducted by conventional vehicles in order to achieve cost effectiveness and reasonable battery size.

In addition, the Fleet DNA data also shows the stop or idling events of MD/HD vocations. The stop or idling events with a minimum 5-minute stop typically occurs at the fleet garage, depot, or vocation stop sites after completing a major operation or a trip, as are defined as trip count in the Fleet DNA database. The trip count per day can actually be considered as the opportunities of BEVs' daily en-route charging, and is discussed later in this paper.

Vehicle Energy Methodology

An equation of typical tractive power demand is used to account for vehicle forward acceleration, rotational inertia, aerodynamic loss, rolling resistance loss, and road grade. For any instant of time, the tractive power demand is described as (20–22):

$$W_{\text{tract}} = m \cdot V \cdot dV/dt + \frac{1}{2}\rho \cdot C_d \cdot A_f \cdot V^3 + m \cdot g \cdot C_{\text{rr}} \cdot V + m \cdot g \cdot V \cdot \sin(\theta) \quad (1)$$

where W_{tract} is vehicle tractive power; V is velocity; ρ is air density; C_d is the aerodynamic drag coefficient; A_f is the projected frontal area; C_{rr} is the rolling resistance coefficient; θ is road gradient; g is gravity; t is time; m is GVW, including passengers and key components for powertrain and drivetrain systems (e.g., engine, clutch/torque converter, gearbox, final drive, wheel, chassis, generator, battery, mechanical and electrical accessory, as well as motor and high-voltage battery for hybrid powertrain).

For a conventional vehicle, the tractive power is positive if the vehicle is being actively propelled; and the tractive power becomes negative during braking. Braking represents a dissipative force that depletes the energy that is stored as vehicle kinetic and potential energy. According to Fleet DNA data, a vehicle propels energy over a daily operating time of ΔT, estimated in the modified equation listed below:

$$E_{\text{prop}} = \int_{\Delta T} \left[m \cdot V \cdot dV/dt + \frac{1}{2}\rho \cdot C_d \cdot A_f \cdot V^3 + m \cdot g \cdot C_{rr} \cdot V + m \cdot g \cdot V \cdot sin(\theta) \right] dt + E_{\text{brk}} \quad (2a)$$

Then the equation is further formatted as:

$$E_{\text{prop}} = [m \cdot (\bar{E}_k(a>0) + \bar{E}_k(a<0)) + \rho \cdot C_d \cdot A_f \cdot \bar{E}_{\text{Cd}} + m \cdot C_{\text{rr}} \cdot \bar{E}_{\text{Crr}} + m \cdot (\bar{E}_p(\theta>0) + \bar{E}_p(\theta<0))] \cdot \Delta T + E_{\text{brk}} \quad (2b)$$

where a represents acceleration; $\bar{E}_k(a>0)$ and $\bar{E}_k(a<0)$ are averaged specific positive/negative kinetic energy in unit of kW/kg; $\bar{E}_{\text{Cd}}$ is averaged specific aerodynamic energy in unit of kJ-m/kg; $\bar{E}_{\text{Crr}}$ is averaged specific rolling resistance energy in unit of kW/kg; $\bar{E}_p(\theta>0)$ are $\bar{E}_p(\theta<0)$ are averaged specific positive/negative potential energy in units of kW/kg, respectively. All these statistical energy terms are provided by the Fleet DNA database. E_{brk} represents braking energy loss for all vehicle deceleration events, which is used for compensating the negative tractive energy encompassed in the derived propel energy shown in the first term at the right side of Equation 2*a* and *b*. E_{brk} is evaluated based on the statistical data (i.e., $\bar{V}_{drive}$: average drive speed; δ_{decel}: average deceleration event duration; n_{decel}: deceleration events per mile; and d_{dist}: driving distance) with the assumption of a linear deceleration stop of $V = \alpha_{decel} \cdot (\delta_{decel} - t)$, where $\alpha_{decel} = \bar{V}_{drive}/\delta_{decel}$, then it is derived from Equation 1 without considering elevation changes:

$$E_{brk} = \left(\frac{1}{2}m \cdot a_{decel}^2 \cdot \delta_{decel}^2 + \frac{1}{8}\rho \cdot C_d \cdot A_f \cdot a_{decel}^3 \cdot \delta_{decel}^4 + \frac{1}{2}m \cdot g \cdot C_{rr} \cdot a_{decel} \cdot \delta_{decel}^2\right) \cdot n_{decel} \cdot d_{dist} \quad (3)$$

When a conventional vehicle is being actively propelled, the vehicle engine provides power to the wheels. Tractive power becomes zero when the vehicle is at idling or braking, but the engine still has to run to satisfy accessory loads. Thus, mechanical power output from the engine is derived from drivetrain component efficiencies, as described below:

$$E_{eng}^c = E_{acc}^c + E_{prop}^c/\eta_{wh} \cdot \eta_{fd} \cdot \eta_{gb} \cdot \eta_{cl} \quad (4a)$$

$$E_{fuel}^c = E_{eng}^c/\eta_{eng} \quad (4b)$$

where η_{wh} is wheel efficiency; η_{fd} is final drive efficiency; η_{gb} is gearbox efficiency; η_{cl} is clutch efficiency; E_{acc}^c is overall energy for accessory load of a conventional vehicle, simply expressed as $\int_{\Delta T} W_{acc}^c dt$, and W_{acc}^c is considered as constant value; the superscript *c* represents conventional vehicle.

In an electric vehicle, electric power output from the battery is derived based on the efficiencies of electric components (i.e., motor and battery) and related drivetrain components (e.g., final drive and wheel), shown in Equation (5). However, an electric vehicle is capable of converting vehicle kinetic energy into a storable form of battery energy during braking if satisfying the thresholds of vehicle speed and acceleration. The current electric vehicle model assumes braking energy regeneration is allowed to occur once vehicle speed is not below the given value (i.e., v_{lb}), as is described as Equation 6; otherwise, braking energy regeneration does not occur in order to avoid a kinetic energy regeneration that is too low. The battery electrical energy consumption is described as:

$$E_{batt}^e = E_{acc}^e + E_{prop}^e/\eta_{wh} \cdot \eta_{fd} \cdot \eta_{mot} \cdot \eta_{batt} - E_{regen}^e \quad (5)$$

where η_{mot} is motor efficiency; η_{batt} is battery efficiency; η_{chgr} is charger efficiency; E_{acc}^e is overall energy for electric accessory load, simply expressed in $\int_{\Delta T} W_{acc}^e dt$, where W_{acc}^e is considered as constant value; the superscript *e* represents electric vehicle. E_{regen}^e is potential electrical energy generated from braking energy regeneration, estimated as below:

$$E_{regen}^e = \left(E_{brk} - E_{brk,lb}\right)/\eta_{wh} \cdot \eta_{fd} \cdot \eta_{mot} \cdot \eta_{batt} \quad (6a)$$

$$E_{brk\_lb} = \left(\frac{1}{2}m \cdot V_{lb}^2 + \frac{1}{8}\rho \cdot C_d \cdot A_f \cdot V_{lb}^3 \cdot \delta_{lb} + \frac{1}{2}m \cdot g \cdot C_{rr} \cdot V_{lb} \cdot \delta_{lb}\right) \cdot n_{decel} \cdot d_{dist} \quad (6b)$$

where V_{lb} is the low boundary for vehicle speed in converting kinetic energy into electric energy; δ_{lb} is deceleration time from low boundary speed to full stop, estimated by $V_{lb} = \alpha_{decel} \cdot \delta_{lb}$; E_{brk_lb} represents braking energy loss resulting from too low vehicle speed in recovering kinetic energy during vehicle deceleration events.

Unlike for conventional vehicles, it is important for electric vehicle drivers to deliberate en-route charging activities. To simplify the process, trip count per day, n_{trip}, in Fleet DNA is considered as electric vehicle (EV) en-route charging slots for each service day. However, the charging process should be limited by charger power level, battery capacity and battery energy consumption, shown in Equation 7. In the simulations, different charging power levels and times are required:

$$E_{chg}^e = \min\left(E_{batt}^e/n_{trip}, \eta_{chgr} \cdot \int_{\delta_{chg}} W_{chgr}^e dt, (1 - SOC_{lb}) \cdot C_{batt}\right) \quad (7)$$

where η_{chgr} is charger efficiency, which is considered to be constant; W_{chgr}^e is charging power; δ_{chg} is charging time; SOC_{lb} is the low-boundary SOC for safety and battery protection, which is assumed to be 10%; C_{batt} is battery capacity.

Also, an EV battery typically leads to a significant weight penalty as a result of low battery energy density. Thus, the BEV weight penalty is considered in the GVW of the above tractive energy equations, and given as:

$$m_{extra} = m_{batt} + m_{mot} + m_{cov/inv} - m_{eng} - m_{cpl} - m_{gb} - m_{cat} \quad (8)$$

where m_{batt}, m_{mot}, $m_{cov/inv}$, m_{eng}, m_{cpl}, m_{gb} and m_{cat} are the weight of battery, motor, converter/inverter, engine, clutch, gearbox, and aftertreatment system, respectively. These component weights are estimated based on their peak powers except battery. Battery weight is determined based on battery capacity. The detailed mass estimations have been addressed in previous works (*22*).

Vehicle and Charging Assumptions

The ten studied conventional MD/HD vehicle configurations were constructed based on an ORNL database and open literature. The considered components include engine, clutch, gearbox, final drive, wheel, and chassis. Table 3 details these modeled MD/HD applications. The frontal area, rolling resistance, and aerodynamic drag coefficient were estimated based on the ORNL MD/HD vehicle database. Also, appropriate constant average efficiencies for each powertrain and drivetrain components were assumed (*22–25*). Constant values for the accessories power consumption were adopted correspondingly. The considered GVWs of the MD/HD vehicles are based on the statistical data provided in FLEET DNA.

To ensure the methodology and assumptions provide reasonable predictions for MD/HD conventional vehicles, a Class 7 transit bus was selected for the validation. The predicted simulations were compared with one-year real driving data recorded by ORNL in three 2005 Optima buses (*24*). The results show an overall 4.74 mpg in the current simulation vs. 4.38 mpg of the measurement, indicating an 8% over-predicted error, implying the modified methodology and assumptions for conventional vehicles achieve a reasonable accuracy.

For electric vehicles, the major components considered are battery, motor, final drive, wheel, and chassis. Similarly constant average efficiencies for motor and battery components were assumed.(*3*, *22*). Table 3 lists the electric component features. Other drivetrain components and chassis parameters are considered as the same as those used in the conventional vehicle. Similarly, a constant value of accessory load is considered in the modeled electric vehicle, but the accessory load in the electric vehicles is assumed 50% less than a conventional vehicle. The identical assumption has been widely used in open literature due to lower accessory energy consumption in EV (*22–25*). The BEV mass weight is estimated based on the comparable conventional vehicle weight plus BEV mass penalty predicted using Equation 8.

Because of the very limited MD/HD electric vehicle data available, it is difficult to calibrate the EV model directly. Thus, the study utilized the previous well-

Table 3. Key Parameters Used for Modeled Conventional and Battery Electric Vehicles

Vehicle type	Class 8 port drayage tractor	Class 8 refuse truck	Class 7 transit city bus	Class 7 food delivery tractor	Class 6 school bus	Class 5 linen delivery van	Class 5 food delivery truck	Class 4 parcel delivery van	Class 3 bucket truck	Class 3 food delivery truck
C_d	0.65	0.82	0.79	0.65	0.80	0.55	0.55	0.55	0.50	0.46
C_{rr}	0.0080	0.0098	0.0098	0.0080	0.0090	0.0095	0.0095	0.0095	0.0110	0.0120
Frontal area (m^2)	8.5	8	8.2	8.2	8.2	6	6	6	4	5
Veh weight (kg)	34545	23823	13000	12727	11818	8377	8208	7273	5909	5909
Eng eff	0.36	0.32	0.32	0.328	0.32	0.32	0.32	0.32	0.32	0.32
Clutch eff	0.86	0.86	0.86	0.86	0.86	0.86	0.86	0.86	0.86	0.86
Gear box eff	0.92	0.92	0.92	0.92	0.92	0.92	0.92	0.92	0.92	0.92
Final drive eff	0.98	0.98	0.98	0.98	0.98	0.98	0.98	0.98	0.98	0.98
Wheel eff	0.99	0.99	0.99	0.99	0.99	0.99	0.99	0.99	0.99	0.99
Motor eff	0.88	0.88	0.88	0.88	0.88	0.88	0.88	0.88	0.88	0.88
Battery eff	0.98	0.98	0.98	0.98	0.98	0.98	0.98	0.98	0.98	0.98
Conv_Acc (kw)	4	6	12	8	6	2	5	2	4	4
EV_Acc (kw)	2	3	6	4	3	1	2.5	1	2	2

Table 4. Charging Assumption for the Modeled Vehicles

Charging assumption	Charging event	Charging time (s)
Class 8 drayage tractor	Data*	1800
Class 8 refuse truck	2	600
Class 7 food delivery tractor	Data*	300
Class 7 city bus	Data*	300
Class 6 school bus	1	7200
Class 5 linen delivery van	Data*	300
Class 5 food delivery truck	Data*	300
Class 4 parcel delivery van	Data*	300
Class 3 food delivery truck	Data*	300
Class 3 telecom truck	0	0

*Based on stop times >300 s provided in Fleet DNA data.

calibrated electric bus model based on the same 2005 Optima bus, as well as ORNL second-by-second driving data, to validate the current BEV model using statistical data. The results show 2.16 kWh/mile using the statistical data vs. 2.17 kWh/mile using ORNL s-by-s driving data, meaning less than 1% error, again implying that these BEV methodologies and assumptions are reasonable.

For BEV charging in the simulations, all MD/HD battery electric vehicles are assumed to be fully charged every morning before beginning daily service. En-route charging is considered during the period of vehicle daily route operation. However, the en-route charging strongly depends on MD/HD vocations and driving patterns. For example, Class 8 port drayage tractors frequently upload, drive, and download cargoes between airport/seaport and warehouse, and stop/waiting time at airport/seaport and warehouse represent fairly good opportunities for recharging; school buses typically travel one or two routes in the morning and again in the afternoon, and there is adequate time for recharging before serving afternoon activities. Similarly, transit bus and delivery trucks and vans typically experience significant time at bus depots, marketplace delivery stops, and so on. These are all good opportunities for battery recharging through appropriate regular charging or wireless charging technologies during the slots. However, utility trucks usually require immediate travel to any unscheduled site for emergency service, so en-route recharging for utility trucks during their travel is not appropriate. Consequently, Table 4 lists the charging assumptions for the modeled vehicles after considering their actual drive patterns. Except school bus, drayage, refuse, and utility trucks, a vehicle stop of more than 5 minutes provided by Fleet DNA is considered to be a charging event. Also, as shown in Table 1, charging power technologies vary with different BEV manufacturers and MD/HD applications, covering 0.2–5 hrs charging time. To simplify the current electrification studies and analyses, a fast recharging technology is considered, which enables 45 minutes to complete full battery charging, but also limits the maximum charging power to not more than 480 kW. Thus, the considered charging power is between 25 and 480 kW, depending on battery size and vehicle type. The charger efficiency is assumed as 97% for charger power with 80 kW or higher; otherwise it is 85%. The considered charger efficiency is the overall energy efficiency of the charging procedure from grid electrical energy at charging facility to BEV battery. The efficiency does not account for grid distribution losses.

Results and Discussion

Impact of BEV Service Coverage

As discussed above, the most extreme service days covered by MD/HD conventional vehicles often account for a very small percentage of total vehicle miles. Therefore, excessive battery capacity and cost penalties can be avoided very well with little to no impact on service provision for fleets if electric vehicles are not required to cover all service-days in the dataset. Instead, conventional vehicles are used to satisfy such extreme service days. Therefore, the study focus is on evaluating battery capacity need for 50–90% service coverage. The service coverage of less than 50% is not considered because it may limit substantial BEV applications as a result of inadequate service opportunities. Service coverage represents the percentage of all operating days in a year. For example, 50–90% service coverage means the service covers 50–90% of all operating days in a year. The results are plotted in Figure 2, showing that Class 7-8 HD vehicles require considerably larger battery capacity than Class 3-6 MD vehicles. For 90% service coverage, the battery requirement of Class 7-8 vehicles is 250-600 kWh, as is less than 60 kWh for Class 3-6 vehicles. The lower service coverage enables a vehicle to have reduced battery capacity, particularly for Class 7-8 vehicles. When Class 7-8 HD BEV service coverage is reduced down to 50%, the battery capacity requirement of Class 8 port drayage tractors, Class 8 refuse trucks, Class 7 food delivery tractors, and Class 7 city buses is only 20–57% of the battery needed for their 90% service coverage. This significant reduction of battery cost indicates a considerable potential of Class 7-8 HD electrification applications for short-distance service. For Class 3-6 MD vehicle, the similar service coverage reduction does not lead to a battery reduction as large as Class 7-8 HD vehicles. This is because the battery size used in Class 3-6 MD electrification is usually a smaller capacity than Class 7-8 MD vehicles (see Figure 2*a*).

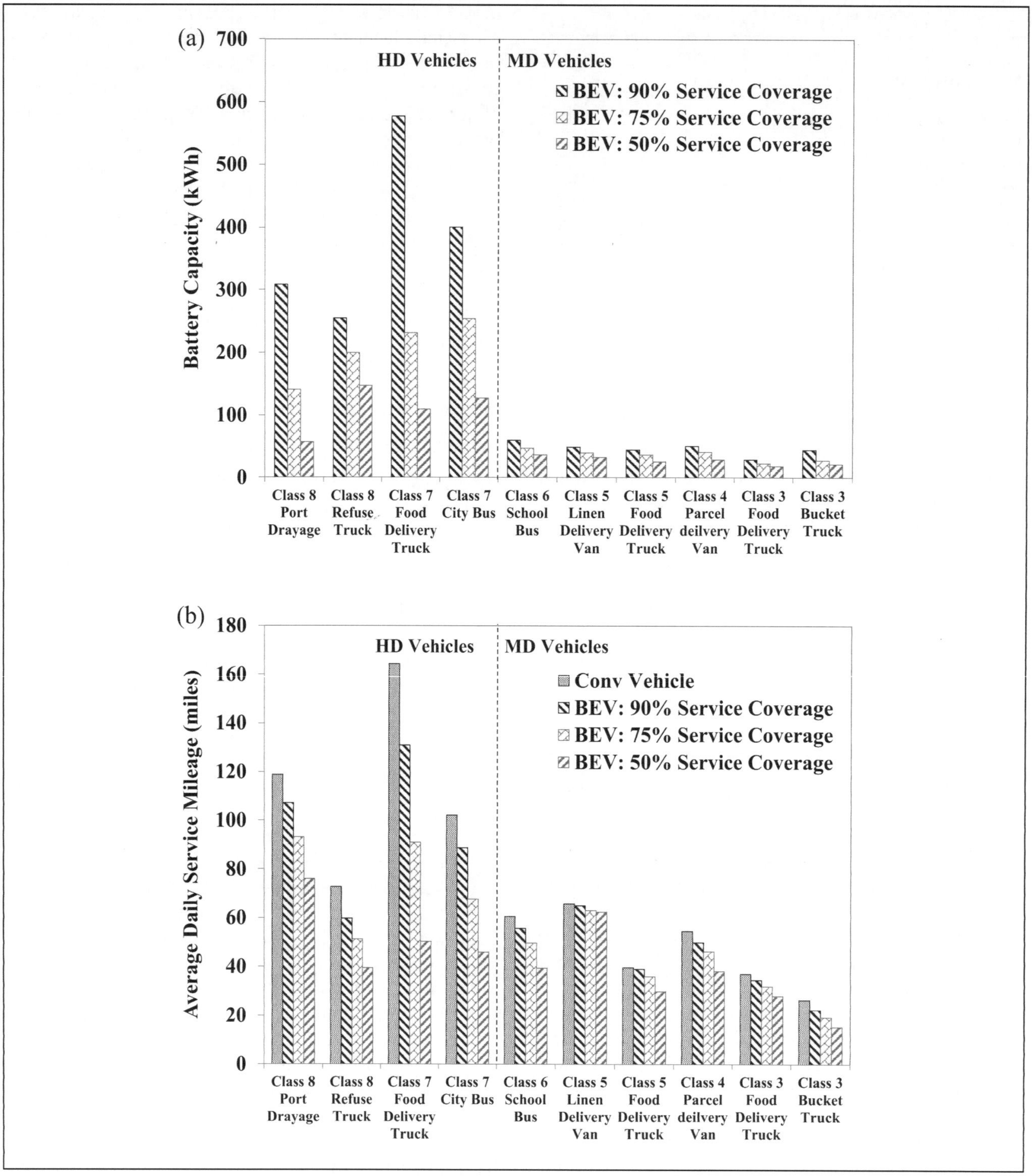

Figure 2. Effect of MD/HD vehicle electrification for 50–90% service coverage on battery capacity requirement and daily service mileage. En-route charging is considered with 45 minutes of full battery charging, except bucket trucks. (*a*) battery capacity; (*b*) daily service mileage.

On the other hand, for the MD/HD electric vehicles with 50–90% service coverage, the smaller battery capacity leads to less average daily service mileage, shown in Figure 2*b*. Corresponding to substantial battery capacity reduction to HD BEVs with less service coverage, the daily service mileage of the HD electric vehicles is also considerably reduced. For Class 3-6 MD vehicles, their average daily service mileage does not decrease very much, although their service days are decreased (see Figure 3). Figure 3 shows totals of 245 and 136 service days for conventional MD and HD food delivery trucks, respectively. With the range of 50, 75, 90% service coverages, the Class 5 food delivery electric trucks cover 123, 184, and 220 days of service, respectively; and Class 7 food delivery electric trucks can cover 68, 102, and 122 days of service, respectively. For HD applications, the uncovered services are caused typically by aggressive daily driving mileage (i.e., >450 miles) or longer driving per stop charging (i.e., >75 miles/stop charging). A similar phenomenon occurred with MD BEVs. Therefore, it is important for MD/HD vehicle electrification to appropriately alter EV driving patterns for fitting well-planned short route and unique en-route charging activities in order to achieve successful MD/HD vehicle electrification with reasonable battery size.

Figure 4*a* reveals the energy consumption comparison between conventional and BEVs. The results show all the BEVs achieve 11–34% and 68–80% energy consumption savings compared with conventional vehicle engine-out mechanical energy and fuel energy, respectively. The significant energy savings is mainly attributable to BEV braking energy recovery, energy reduction at idling, and accessary load reduction, as have been well addressed in previous works (*3*, *22*). To well understand the energy consumption savings related to BEVs, BEV battery electric energy consumption was converted into diesel-equivalent fuel economy. The BEV diesel-equivalent fuel economy is calculated based on the assumption that diesel is used as fuel in a power plant with 45% efficiency, and the produced electricity is delivered to a charging station with 95% distribution efficiency. Thus, compared with the identical conventional vehicles, the projected equivalent fuel economy improvement is 34–77% in the BEV scenario with 75% service coverage (see the details of Figure 4*c*). Figure 4*b* displays the energy cost savings of MD/HD BEVs in unit of $/mile. The BEVs save 55–69% of the energy cost compared with the identical diesel vehicles. In the calculations, the used diesel fuel price is $2.577 per gallon, which is the average U.S. market price of diesel as of February 2017 (*26*). The price for electricity is 9.4 cents per kWh, which is the average U.S. market cost of electricity for December 2016 (*27*). In general, both MD and HD vehicle electrifications achieve significant energy cost savings. A similar conclusion has been reported by Gao et al. (*3*).

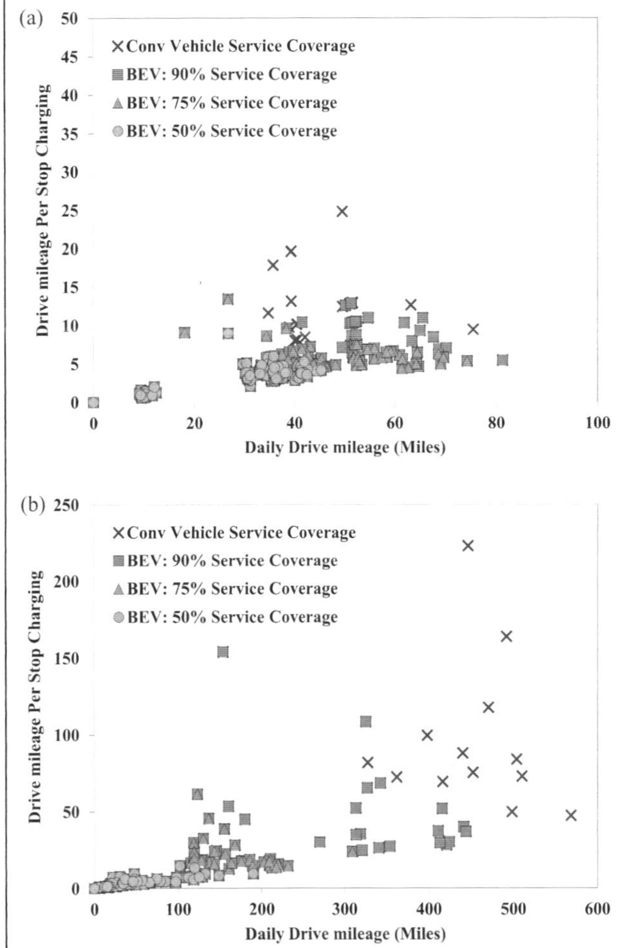

Figure 3. The service days and driving mileage per charging stop of Class 5 and Class 7 food delivery electric trucks with 50–90% service coverage. En-route charging is considered with 45 minutes of full battery charging. Each distributed point represents a day of driving. (*a*) Class 5 food delivery truck; (*b*) Class 7 food delivery truck.

Discussion on Payload Limit and Payback Period

EV batteries typically lead to significant weight penalties because of low battery energy densities. It is important to understand how much the mass penalty could be increased as a result of electrification. Figure 5*a* shows the BEV mass penalty for MD/HD vehicles as a function of different service coverages. Clearly the mass penalty is significant for HD applications with 90% service coverage. Class 7 food delivery truck electrification with 90% service coverage requires ~600 kWh battery capacity and results in a 3,500 kg mass penalty. The mass penalty certainly reduces the delivery truck payload, meaning more BEVs are needed than conventional vehicles to complete the food delivery. However, with less service coverage leading to less battery need, the mass penalty is considerably decreased. Therefore, future studies are merited for the optimization among service coverage, battery size, and payload reduction. MD vehicle electrification causes no more than 160 kg mass penalty, which should not impact vehicle payload. With less service coverage, the mass penalty could even be eliminated. The

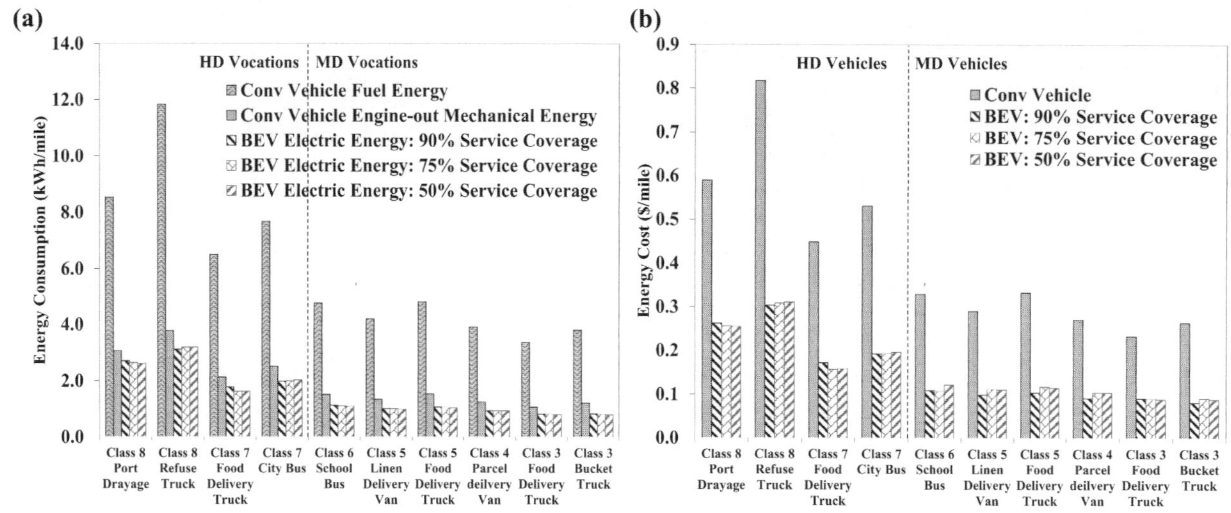

(c)

MD/HD Electrification: 75% service coverage	Eng-out Energy (kWh/mile)	Fuel Economy (mpg)	Battery Energy (kWh/mile)	Equivalent Fuel Economy (mpg)
Vehicle	Conventional vehicle		BEV	
Class 8 Port Drayage	3.1	4.4	2.7	5.8
Class 8 Refuse Truck	3.8	3.2	3.2	4.9
Class 7 Food Delivery Truck	2.1	5.7	1.6	9.5
Class 7 City Bus	2.5	4.9	2.0	7.7
Class 6 School Bus	1.5	7.8	1.1	13.9
Class 5 Linen Delivery Van	1.3	8.9	1.0	13.4
Class 5 Food Delivery Truck	1.5	7.8	1.1	12.8
Class 4 Parcel delivery Van	1.3	9.5	0.9	14.3
Class 3 Food Delivery Truck	1.1	11.1	0.8	16.5
Class 3 Bucket Truck	1.2	9.8	0.8	16.4

Figure 4. Energy consumption, energy cost and fuel economy of MD/HD conventional vehicles and electric vehicles with 50–90% service coverage. En-route charging is considered with 45 minutes of full battery charging except bucket truck. (*a*) Class 5 food delivery truck; (*b*) Class 7 food delivery truck; (*c*) diesel equivalent fuel economy.

reason is mainly because of the smaller battery size of MD BEVs (see Figure 2*a*).

The earlier simulations show that MD/HD vehicle electrification leads to energy savings, but significant battery capacity requirement leads to high BEV purchase cost. Thus, it is particularly interesting to understand the payback period for these MD/HD vehicle electrifications. In the cost-effectiveness analysis, the 2016 U.S. DOE projected prices for Li-ion batteries, electric motors, and power electronics (*28*, *29*) were used, while open literature cost data (*30*, *31*) are used for conventional power devices like engines, clutches, transmissions, and aftertreatment systems. Similar to Equation 8, the cost penalty for MD/HD electrification is estimated based on the cost variance between E-Powertrain cost and conventional powertrain cost. The cost penalty is further used with annual travel mileage related to various service coverage scenarios and energy cost savings shown in Figure 2 to determine the potential payback period of each MD/HD BEV. The annual travel mileage of various service coverage BEVs is based on the corresponding reduced daily service mileage for a year of 365 day operation; in other words, the BEVs with less service coverage are designed to provide year-around short-distance service. The results are plotted in Figure 5*b*. Although higher service coverage leads to a longer payback period, the majority of MD BEVs except Class 3 bucket trucks achieve less than a 5-year payback period, particularly true for the 50–75% service coverage. For HD BEVs, except Class 8 port drayage tractors, the 90% service coverage results in a 10-year or even longer payback period, but the 50% service coverage just requires 7–8 years. For Class 8 port drayage tractor, it should achieve no more than a 3.5-year payback period for 50–75% service

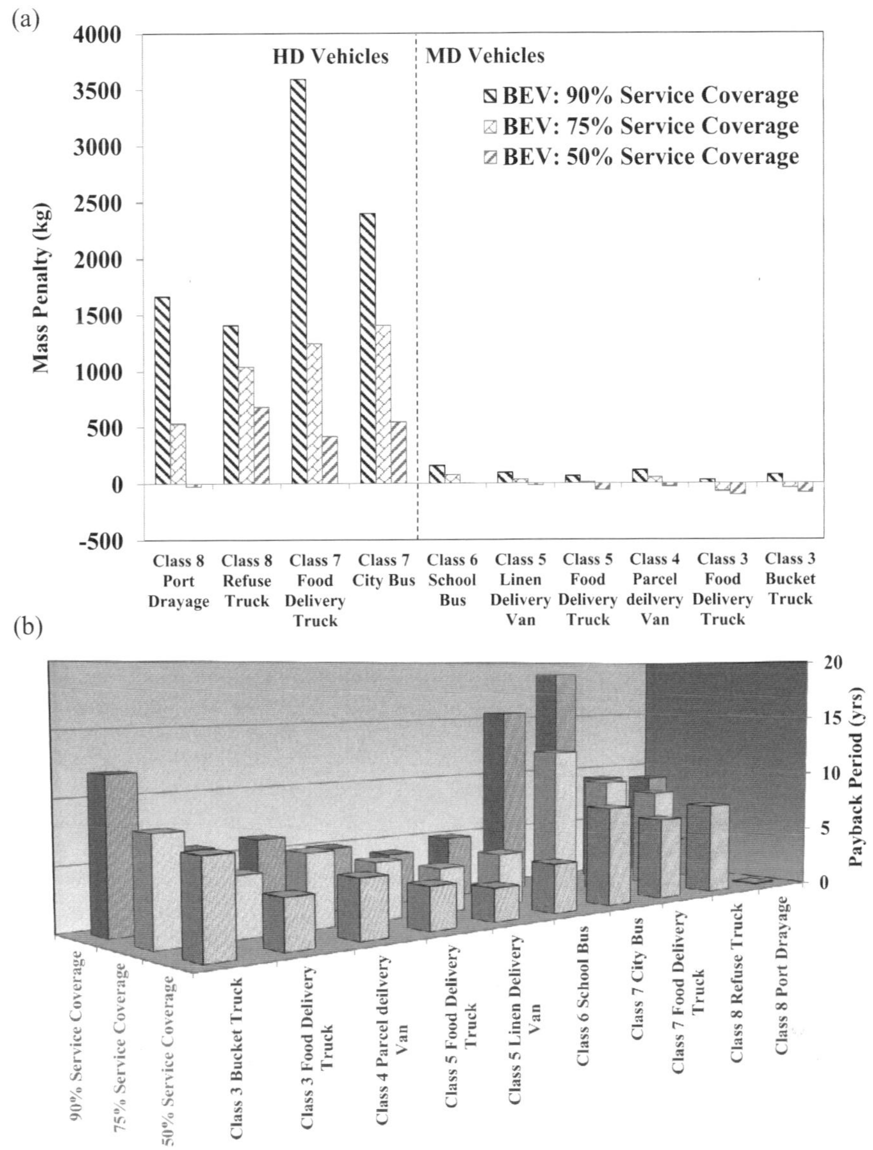

Figure 5. BEV mass penalty and payback period of MD/HD electrification as a function of various service coverages. En-route charging is considered with 45 minutes of full battery charging, except bucket trucks. (*a*) BEV mass penalty; (*b*) BEV payback period.

coverage. One explanation could be because of the unique driving feature of port drayage tractors, which frequently drive from ports to ports or warehouses. It is clear that there is good potential for Class 8 port drayage tractor electrification based on cost effectiveness.

All the cost-effective analyses do not consider the cost related to repair, maintenance, and battery disposal. In addition, all the above analyses do not consider the impact of climate (i.e., cold and hot weather) and ambient temperature on BEV driving range and cost effectiveness. Such impact is important and deserves study in future work. However, a comprehensive discussion in this paper is beyond the scope of the current study.

Conclusion

A wide range of analysis of Class 3-8 commercial vehicle electrification has been conducted by means of developing a framework tool which adopts a quantitative method of estimating electric vehicle energy consumption and appropriate charging considerations using the Fleet DNA composite statistics on driving behavior. The developed tool provides insights into BEV feasibility and cost effectiveness across types and duties of MD/HD vehicles, and identifies high-potential MD/HD electrification technologies and market opportunities. In the paper, the ten representative Class 3-8 commercial vehicle electrifications have been evaluated and reported as a function of various service coverages, including applications in port drayage tractors, refuse trucks, delivery trucks, buses, and bucket trucks.

The results demonstrate Class 7-8 HD vehicles require considerably larger battery capacity compared with Class 3-6 MD vehicles. For 90% service coverage, Class 7-8 BEVs need a 250–600 kWh battery compared with less than 60 kWh for Class 3-6 vehicle. The mass penalty is also significant for the HD applications with 90% service coverage and could impact vehicle payload. However, less service coverage enables a meaningful reduction of battery capacity and mass penalty for Class 7-8 vehicles, but not for Class 3-6 MD vehicles. This indicates a high potential of Class 7-8 HD electrification applications for short-distance service with lower battery capacity needs. Class 3-6 MD vehicles seem more feasible for vehicle electrification, with reasonable battery size and greater service coverage.

Moreover, MD/HD vehicle electrification results in significant energy savings and fuel cost savings. The projected diesel-equivalent fuel economy of BEVs represents a 34–80% improvement and saves 55–69% in fuel energy cost compared with the comparable diesel vehicles. In considering payback, the cost-effectiveness analysis illustrates that higher service coverage leads to longer payback years because of higher battery capacity requirements. The majority of MD BEVs, except Class 3 bucket trucks, achieve better than a 5-year payback, particularly for the 50–75% service coverage. For HD BEVs, except Class 8 port drayage tractors, the 90% service coverage results in a 10-year or longer payback period while the 50% service coverage yields a 7–8 year payback. Class 8 port drayage tractors should achieve payback in no more than 3.5 years for 50–75% service coverage.

Overall, MD vehicle electrification provides more feasible opportunities with less battery cost and payback period, and similar energy savings. However, various MD/HD vehicles have different operations and drive patterns, which may lead to different cost and energy saving scenarios in terms of electrification. In particular, this requires a further optimization study with evolving battery and infrastructure technologies.

Acknowledgments

This project was sponsored by Rachael Nealer of the U.S. Department of Energy's VTO program. The authors would like to recognize Adam Duran of NREL for his helps in Fleet DNA data. We also appreciate the organizers and reviewers for their support and unpaid worker time.

This manuscript has been authored by UT-Battelle, LLC under Contract No. DE-AC05-00OR22725 with the U.S. Department of Energy. The United States Government retains and the publisher, by accepting the article for publication, acknowledges that the United States Government retains a nonexclusive, paid-up, irrevocable, worldwide license to publish or reproduce the published form of this manuscript, or allow others to do so, for United States Government purposes. The Department of Energy will provide public access to these results of federally sponsored research in accordance with the DOE Public Access Plan.

Author Contributions

The authors confirm contribution to the paper as follows: The corresponding author: Zhiming Gao; Study conception and design: Zhiming Gao, Zhenhong Lin; Data collection: Zhiming Gao, Alicia Birky; Analysis and interpretation of results: Zhiming Gao, Zhenhong Lin, Alicia Birky; Draft manuscript preparation: Zhiming Gao, Zhenhong Lin, Alicia Birky, Stacy Davis. All authors reviewed the results and approved the final version of the manuscript.

References

1. CIA. *The World Factbook*. 2014. https://www.cia.gov/library/publications/the-world-factbook/rankorder/2246rank.html. Accessed July 28, 2017.
2. EPA. *Inventory of U.S. Greenhouse Gas Emissions and Sinks: 1990-2014*. 2014. https://www3.epa.gov/climate-change/Downloads/ghgemissions/US-GHG-Inventory-2016-Chapter-Executive-Summary.pdf. Accessed July 28, 2017.
3. Gao, Z., Z. Lin, and O. Franzese. The Energy Consumption and Cost Savings of Truck Electrification for Heavy Duty Vocational Applications. *Transportation Research Record: Journal of the Transportation Research Board*, 2017. 2628: 99–109.
4. Daw, C., Z. Gao, D. Smith, T. LaClair, J. Pihl, and K. Edwards. Simulated Fuel Economy and Emissions Performance during City and Interstate Driving for a Heavy-Duty Hybrid Truck. *SAE International Journal of Commercial Vehicles*, Vol. 6, No. 1, 2013, pp. 161–182.
5. Chen, Y., and J. Borken-Kleefeld. Real-Driving Emissions from Cars and Light Commercial Vehicles–Results from 13 Years Remote Sensing at Zurich/CH. *Atmospheric Environment*, Vol. 88, 2014, pp. 157–164.
6. Toon, J. *Diesel or Electric? Study Offers Advice for Owners of Urban Delivery Truck Fleets*. 2013. http://www.news.gatech.edu/2013/09/25/diesel-or-electric-study-offers-advice-owners-urban-delivery-truck-fleets. Accessed July 28, 2017.
7. Calstart. *Electric Truck & Bus Grid Integration Opportunities, Challenges & Recommendations*. 2015. http://www.westcoastelectricfleets.com/portfolio-items/electric-truck-bus-grid-integration-opportunities-challenges-and-recommendations/. Accessed July 28, 2017.
8. California Environmental Protection Agency, Air Resources Board. *Technology Assessment: Medium- and Heavy- Duty Battery Electric Trucks and Buses*. 2015. https://www.arb.ca.gov/msprog/tech/techreport/bev_tech_report.pdf. Accessed July 28, 2017.
9. MTA. *MTA Tests Electric Buses for Use on NY Streets*. 2017. http://www.mta.info/news-nyct-bus/2017/04/26/mta-tests-electric-buses-use-ny-streets. Accessed July 28, 2017.

10. Edelstein, S. *Chicago Transit Authority to Add Dozens of Electric Buses after Successful Tests*. 2016. http://www.greencarreports.com/news/1102130_chicago-transit-authority-to-add-dozens-of-electric-buses-after-successful-tests. Accessed July 28, 2017.
11. Gray, B. *Metro Begins Test Driving Electric Bus on Downtown St. Louis Route*. 2017. http://www.stltoday.com/business/local/metro-begins-test-driving-electric-bus-on-downtown-st-louis/article_e4a529e2-8a18-5a76-aae7-6c145c99d4a8.html. Accessed July 28, 2017.
12. NREL. *Foothill Transit Electric Bus Evaluation*. https://www.nrel.gov/transportation/fleettest-electric-foothill.html. Accessed July 28, 2017.
13. TransPower. *Economical Electric School Bus (EESB) Final Project Report*. http://www.transpowerusa.com/downloads/TransPower-Economical-Electric-School-Bus-Final-Report.pdf. Accessed July 28, 2017.
14. Adomani. *Help yellow Go Green*. http://www.adomanielectric.com/phoenix.zhtml?c = 254475&p = irol-products. Accessed July 28, 2017.
15. Lyden, S. *The State of All-Electric Trucks in the U.S. Medium Duty Market*. 2014. http://zerotruck.com/wp-content/downloads/GRN_medium.pdf. Accessed July 28, 2017.
16. California HVIP. *Hybrid and Zero-Emission Vehicles*. https://www.californiahvip.org/docs/HVIP_EligibleVehicles.pdf. Accessed July 28, 2017.
17. Birky, A. K., M. Laughlin, K. Tartaglia, R. Price, and Z. Lin. *Transportation Electrification beyond Light Duty: Technology and Market Assessment*. 2017. https://info.ornl.gov/sites/publications/Files/Dev296070.pdf. Accessed November 12, 2017.
18. Muller, J. *Cummins Beats Tesla To The Punch, Unveiling Heavy Duty Electric Truck*. https://www.forbes.com/sites/joannmuller/2017/08/29/take-that-tesla-diesel-engine-giant-cummins-unveils-heavy-duty-truck-powered-by-electricity/#7195f52678f1. Accessed October 30, 2017.
19. NREL. *Fleet DNA: Commercial Fleet Vehicle Operating Data*. https://www.nrel.gov/transportation/fleettest-fleet-dna.html. Accessed July 28, 2017.
20. Gao, Z., D. E. Smith, C. S. Daw, K. D. Edwards, B. C. Kaul, N. Domingo, J. E. Parks, and P. T. Jones. The Evaluation of Developing Vehicle Technologies on the Fuel Economy of Long-Haul Trucks. *Energy Conversion and Management*, Vol. 106, 2015, pp. 766–781.
21. Gao, Z., S. J. Curran, J. E. Parks, II, D. E. Smith, R. M. Wagner, C. S. Daw, K. D. Edwards, and J. F. Thomas. Drive Cycle Simulation of High Efficiency Combustions on Fuel Economy and Exhaust Properties in Light-Duty Vehicles. *Applied Energy*, Vol. 157, 2015, pp. 762–776.
22. Gao, Z., Z. Lin, T. J. LaClair, C. Liu, J. M. Li, A. K. Birky, and J. Ward. Battery Capacity and Recharging Needs for Electric Buses in City Transit Service. *Energy*, Vol. 122, 2017, pp. 588–600.
23. Gao, Z., M. Y. Kim, J. S. Choi, C. S. Daw, J. E. Parks, and D. E. Smith. Cold-Start Emissions Control in Hybrid Vehicles Equipped with a Passive Adsorber for Hydrocarbons and Nitrogen Oxides. *Proceedings of the Institution of Mechanical Engineers, Part D: Journal of Automobile Engineering*, Vol. 226, No. 10, 2012, pp. 1396–1407.
24. Gao, Z., C. S. Daw, D. E. Smith, T. J. LaClair, J. E. Parks, II, and P. T. Jones. Comparison of Parallel and Series Hybrid Power Trains for Transit Bus Applications. *Transportation Research Record: Journal of the Transportation Research Board*, 2016. 2570: 97–106.
25. Gao, Z., T. J. LaClair, D. E. Smith, and C. S. Daw. Exploring Fuel-Saving Potential of Long-Haul Truck Hybridization. *Transportation Research Record: Journal of the Transportation Research Board*, 2015. 2502: 99–107.
26. EIA. *Gasoline and Diesel Fuel Update*. http://www.eia.gov/petroleum/gasdiesel/. Accessed July 28, 2017.
27. EIA. *Electric Power Monthly*. http://www.eia.gov/electricity/monthly/epm_table_grapher.cfm?t = epmt_5_6_a. Accessed July 28, 2017.
28. Howell, D., B. Cunningham, T. Duong, and P. Faguy. *Overview of the DOE VTO: Advanced Battery R&D Program*. 2016. https://energy.gov/sites/prod/files/2016/06/f32/es000_howell_2016_o_web.pdf. Accessed July 28, 2017.
29. Rogers, S., and S. Boyd. *Overview of the DOE VTO: Electric Drive Technologies R&D Program*. 2016. https://energy.gov/sites/prod/files/2016/06/f32/edt000_rogers_2016_o_web.pdf. Accessed July 28, 2017.
30. U.S. Department of Transportation, National highway Traffic Safety Administration. *Commercial Medium- and Heavy-Duty truck Fuel Efficiency Technology Cost Study*. DOT HS 812177. June 2015.
31. Brown, A. *Assessment of Fuel Economy Technologies Medium- & Heavy-Duty Vehicles: An Overview of the National Academies Study*. http://cta.ornl.gov/TRBenergy/trb_documents/2011_presentations/TRB_Andrew_Brown_Jr_376.pdf. Accessed July 28, 2017.

The Standing Committee on Alternative Transportation Fuels and Technologies (ADC80) peer-reviewed this paper (18-04193).

Article

TRR
JOURNAL OF THE TRANSPORTATION RESEARCH BOARD

Transportation Research Record
2018, Vol. 2672(24) 122–133
© National Academy of Sciences: Transportation Research Board 2018
Article reuse guidelines: sagepub.com/journals-permissions
DOI: 10.1177/0361198118773194
journals.sagepub.com/home/trr
SAGE

Electrification of Transport Refrigeration Units for Temperature-Sensitive Freight: U.S. Environmental Protection Agency Region 10 Technical Assistance Case Study

John Thornton[1], John MacArthur[2], and Husam Barham[3]

Abstract

Transport refrigeration units (TRUs) powered by integral diesel engines provide necessary temperature control for temperature-sensitive freight. TRU engines on trucks or trailers run while parked for temperature control (known as idling)—commonly for 40 to 60% of engine run time while at a home base distribution center. TRU engine idling is a sizable source of fuel consumption, causing air pollution, negative health effects, noise, and unnecessary cost. Electrification is a viable alternative to engine idling in TRUs parked at distribution centers, cold storage warehouses, truck stops, public rest areas, packing houses, terminals and other goods-movement facilities by adopting plug-in electric transport refrigeration units (eTRUs) with grid-supplied electricity. While electrification promises environmental and economic advantages over engine idling, adoption of eTRUs with grid-supplied electricity in the United States is slow. This paper presents results from a U.S. Environmental Protection Agency technical assistance pilot project to address market and behavior barriers of electrification of transport refrigeration to reduce TRU idling. The project findings revealed fleets often underestimate idling time and cost, while being skeptical about the long-term savings and benefits of electrification. The paper introduces a technical assistance approach to work with businesses to reduce operating costs while also reducing greenhouse gas emissions and toxic air pollution of temperature-controlled freight.

The U.S. economy relies heavily on the freight transportation system (*1*). The freight industry consists of 10 million diesel-engine trucks on the roads, moving over 12 billion tons of goods (*2, 3*). Idling by a truck's main engine and secondary engine for transport refrigeration, both on the side of the road and in goods-movement facilities, consumes almost a billion gallons of fuel annually, with fuel cost of more than US$3 billion and resulting in emission of millions of tons of CO_2 and other air pollutants (*3–5*).

In response, developing alternative idling-reduction technologies have been the focus of research efforts and academic studies in the United States in the past decade (*4, 6*). In addition, over 30 states have implemented idling-reduction-related policies (*7, 8*). But so far, little has been accomplished in reducing idling time or reducing emissions during idling (*3*).

Electrification of transport refrigeration units (eTRUs), is a promising but underutilized alternative to engine idling. Operating eTRU from grid-supplied electricity is a cleaner, quieter, and a more economical alternative to diesel-engine idling to run transport refrigeration units (TRUs) (*9–11*).

As part of a U.S. Environmental Protection Agency (EPA) funded technical assistance project to reduce pollution, Portland State University Transportation Research Education Center, Forth, and CleanFuture, Inc. provided technical assistance and conducted a field study to learn more about refrigerated fleets' TRU engine-idling behavior and the potential of electrification adoption (*12*). To create awareness and offer technical support on TRU idle reduction and electrification

[1]CleanFuture, Inc., Tigard, OR
[2]Transportation Research and Education Center (TREC), Portland State University, Portland, OR
[3]Department of Engineering & Technology Management, Portland State University, Portland, OR

Corresponding Author:
Address correspondence to John MacArthur: macarthur@pdx.edu

benefits, TRU operating patterns were characterized to create a baseline TRU operating profile for fleet participants. This baseline was used to help businesses understand their TRU fleet-idling behavior and operating practices, and how electrification can add value and reduce cost (*12*).

This project looked to answer these questions: What are the engine-idling patterns among refrigerated fleets? How does a shift to TRU electrification contribute positively to reduce engine idling and its side effects? What are the prospects and main barriers to achieve such a shift? This paper represents the findings of this project.

Literature Review

This section reviews engine idling, addressing some of its types, side effects, and the related government laws. Then, a review of electrification of transport refrigeration is presented, detailing how it works, what benefits it has over engine idling, and the challenges hindering its adoption.

Engine Idling

One well-known form of engine idling is truck engine idling, which can be defined as the condition in which a truck's main engine is running while the truck itself is parked (*13*). Truck drivers keep the truck engine running for many reasons, such as supplying power to the main cab or trailer, engine performance, or behavior reasons (*4*). A study by North American Council for Freight Efficiency estimated that 8% of all fuel consumed by trucks is burnt in engine idling (*14*).

TRUs maintain the temperature of temperature-sensitive goods while in transit, yet another form of engine idling occurs in TRUs while parked. A TRU is a refrigeration system powered by an auxiliary internal combustion engine designed to control the environment of temperature-sensitive products transported in refrigerated trucks and refrigerated trailers. TRUs may be capable of heating and cooling. Traditionally, TRUs are powered by a dedicated diesel engine. The TRU engine must run whenever temperature-sensitive goods are inside a truck or trailer, even when parked at a distribution center, or other goods-movement facilities (*9, 13*).

Engine idling has several negative side effects (*15*). First, the diesel engine is a main source of air emissions with direct environmental and health impacts. Diesel emissions contain nitrogen oxides, particulate matter, and carbon monoxide, which can have negative impacts on health, quality of life, and also produce CO_2 that contributes to climate change (*4, 13, 16–18*). For example, the Oregon Department of Environmental Quality estimates that diesel fuel accounts for only 29% of all transportation fuel consumed in Oregon. However, diesel transportation emissions account for 49% of nitrogen oxide and 60% of fine particulate matter (PM; 2.5 and 10) (*19*). In another example, in New York City it is estimated that emissions from trucks engines causes over 2,000 deaths, 4,800 asthma emergency room visits, and 1,500 hospitalizations for respiratory and cardiovascular disease annually (*5, 13*).

Engine idling consumes a great amount of diesel resulting in an estimation of billions of dollars in higher fuel cost for fleets (*3, 14*). In addition, parked diesel engines generate noise at transportation centers and in neighborhoods and businesses while loading and unloading goods (*4*).

A literature review of several publications related to idling reveals that the focus is either on diesel-engine idling in general, or on truck diesel-engine performance, with little focus on TRU diesel engines (*3, 4, 6, 14, 20–22*). More field studies are needed to determine TRU idling patterns and locations where TRU idling is most prevalent, and to determine if there are any specific negative impacts related to TRU idling aside from the general negative impacts of diesel-engine idling.

Over 30 states have policies, regulations, or administrative codes to reduce engine idling, at the state, county, and city levels, which mainly set time limits on idling. These time limits differ by state, county, and city locations, and also by vehicle size, use, and engine type (*8, 21, 23, 24*). TRUs are generally exempt from idling regulations because of the necessary function to control cargo temperature. One exception is in California where strict standards were enacted calling for ultra-low emissions from TRUs (*17*). California Air Resources Board has stringent in-use performance standards for particulate matter from TRUs operating in California (*25*). California is moving toward regulations to limit and then prohibit fossil fuel use in TRUs at distribution centers and other facilities in California (*26–28*). The initial concept of the proposed regulation proposes to limit the time a transport refrigeration system powered by a fossil-fueled internal combustion engine can operate at any facility, and then gradually decrease the time limit to transition to zero-emission transport refrigeration.

Electrification of Transport Refrigeration

A cleaner and more economical alternative to TRU diesel-engine idling is electrification. Using electric power along any part of the freight supply-chain is a promising approach to reduce diesel-fueled TRU idling through the use of grid-connected eTRUs (*9, 11, 17*). Under this alternative, when a TRU is at a distribution center, or other goods-movement facility, the TRU can be powered

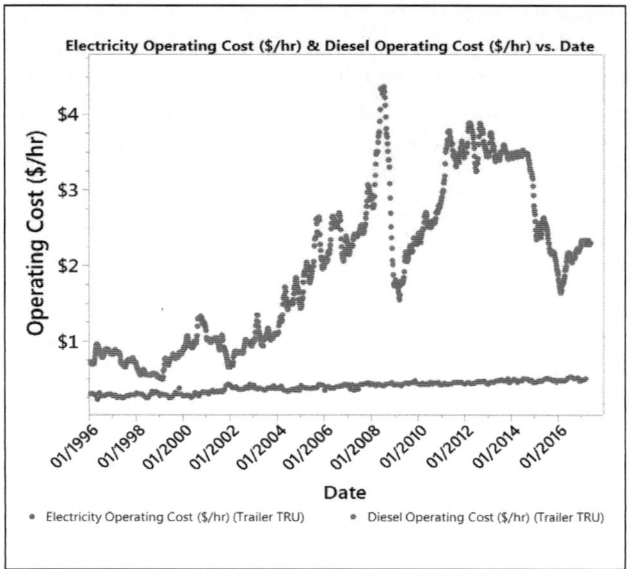

Figure 1. Hourly TRU operating cost in Oregon by fuel source (Data: Energy Information Administration).

from the electric grid while loading/unloading, waiting for a place at the dock, or while staged in preparation to leave the distribution center. In the refrigerated transport industry, this is known as electric standby refrigeration or may also be termed shore power, both terms for using electrified parking spaces (EPS), instead of the diesel engine for temperature control (*9, 17, 29*).

TRU electrification has many advantages as it is cleaner, cheaper, and quieter than diesel operation. Local air quality is improved with electrification as it is cleaner than diesel-engine emissions (*9, 17*). Furthermore, eTRU has economic advantages with lower operating costs and electricity prices are stable and low compared with diesel prices (Figure 1), which is more expensive and prone to fluctuating prices (*30, 31*). With fewer electrical components, eTRU has a lower maintenance cost compared with diesel engines in TRUs, plus it increases the lifespan of the diesel engine by reducing its utilization (*32, 33*). Studies have shown eTRUs on grid-supplied electricity to reduce hourly operating costs by US$1.41 to US$2.68 per engine run hour (*11*). eTRUs operating on electricity are quieter than diesel engines and hence reduce the noise side effect of idling (*29*).

Several eTRU commercial solutions are available, ranging from hybrid electric or electric standby TRUs that can run on grid-supplied electricity when stationary or on diesel while mobile (*34–36*). For stationary storage, engineless eTRUs are available to be plugged into the grid (*37*). A hybrid electric/electric standby TRU has a higher initial cost at acquisition compared with a diesel-only TRU, however the operating cost savings using electricity can easily make it less expensive on a long-run life-cycle cost basis due to fuel savings and reduced maintenance (*32*).

Electrification as an alternative to diesel-engine idling has been applied successfully in other transportation supply-chain elements, including ships in ports, truck roadside rest stop areas, and freight trains (*17*).

Slow Adoption Challenges

Plug-in eTRUs have been adopted successfully in Europe for decades and have been commercially available for well over a decadein the United States (*9*). Yet, refrigerated fleets have been slow to adopt eTRUs, particularly to use EPS to plug-in eTRUs to run from grid-supplied electricity. It is estimated that sales of electric-capable TRUs is between 11% and 25% of new TRU sales, however few such TRUs have ever been plugged-in (*10, 21*). Even though several initiatives, incentives, and pilot projects were conducted to encourage the adoption of eTRU and highlight its environmental and economic benefits, few fleets have adopted the practice of plugging in TRUs (*10, 32, 38*). Few locations have the electric infrastructure for electrified idle reduction in trucks or TRUs. Out of about 6,500 truck stops and travel plazas in the United States, only 113 offer truck-stop electrification (*9, 39*).

Several factors contribute to the slow adoption of electrification. Below is a summary of the main reasons identified by several studies (*3, 9, 10, 20, 32*):

- Lack of awareness – managers of fleets have a notion that the initial infrastructure cost and the operational cost of eTRU is high.
- Slow TRU fleet changeover – TRUs typically go through a 7 to 10-year replacement cycle; effectively slowing the adoption of eTRU.
- Lack of standardization – the current shore power solutions vary and require different expertise and handling.
- Return on investment – the general notion that the cost savings for electricity have limited value when diesel prices are low, although price analysis shows that, in recent decades, even when diesel prices were at their lowest levels they were still higher than electricity prices.

Case Study

U.S. EPA's Pollution Prevention Program (P2) provides technical assistance and training to businesses on source reduction techniques to reduce pollution. Food manufacturing and the supply chain is a national emphasis area under this program, as the food manufacturing sector is a major user of energy and other resources (*12*). Under

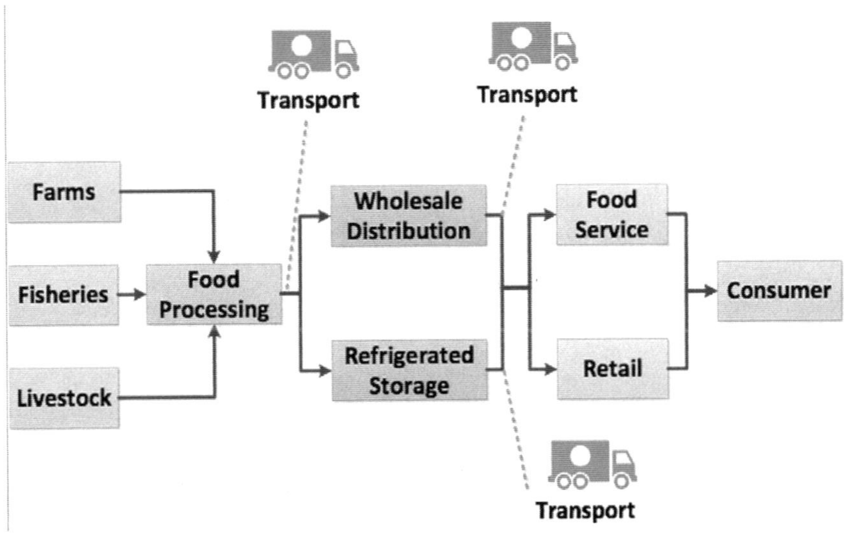

Figure 2. Food distribution supply chain and transport.

this program, a technical assistance project targeting food supply-chain-related pollution, with a focus on reducing pollution in transport refrigeration (Figure 2), was initiated in 2015. The Pollution Prevention Technical Assistance for Idle Reduction and Electrification of Transport Refrigeration project(called "the P2 project" hereafter) served as a pilot project to address market and behavior barriers to electrification of transport refrigeration. This project focused on EPA Region 10 (Pacific Northwest area), mainly with businesses in Oregon, yet also provided analyses of fleets operating in other areas of the United States, with the goals of creating awareness about TRU engine idling impacts and alternatives, and to provide businesses with technical assistance to implement technology (electrification) and behavioral solutions (best practices).

The project incorporated the following program elements:

- Make specific technical assistance available to businesses seeking information about source reduction opportunities, including funding for experts to provide on-site technical advice to businesses and to assist in the development of source reduction plans;
- Target assistance to businesses for whom lack of information is an impediment to source reduction; and
- Provide training in source reduction techniques. Such training may be provided through local engineering schools or other appropriate means.

Figure 3 represents the project plan and approach to achieve the project's goals.

Fleets Participating in Technical Assistance

The project explored the operations of 19 companies representing 38 distinct operating fleets. Return-to-base fleets at home distribution centers with local or regional deliveries were the main focus, although five long-haul fleet carriers' terminal operations were studied. The technical assistance approach followed general steps with participating fleets (not necessarily in the same order or identical for all fleets).

- Presenting P2 project goals to management with main message on cost reduction opportunities; secondarily on why TRU idle reduction is important, how electrification is a promising alternative, and why it makes sense for them to adopt such technology and best practices.
- Conduct field visits to distribution centers to review operating practices and equipment.
- Gain access to telematics data (more details about telematics devices are provided in next section) if such a system is installed, or install telematics devices if a fleet does not have any installed.
- Establish an electronic connection to the fleet's telematics data, download the data, and process them. Analyze operating patterns within the data for idling and to identify other behavioral practices for improvement. See Figure 3 for a schematic diagram on this process.
- Follow-up meeting with management to share results and recommendations for improvement based on an analysis of their fleet operations and other recommended best practices. Included is an

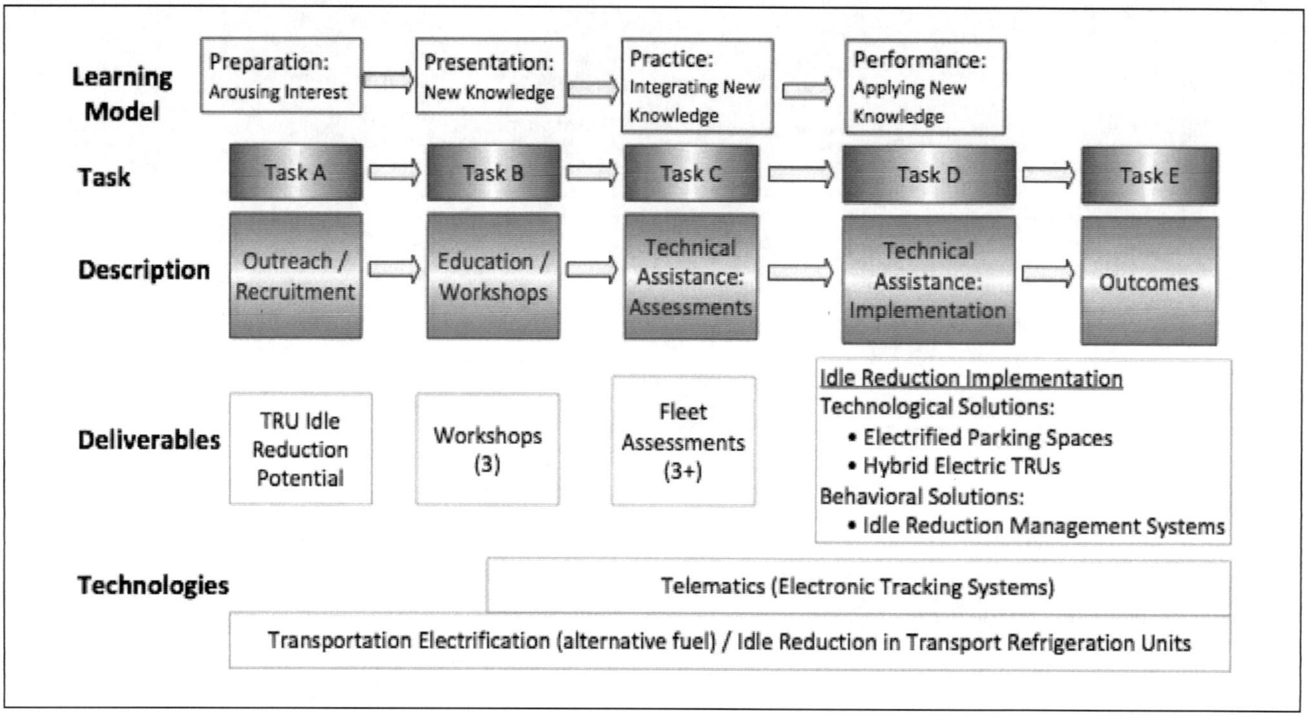

Figure 3. Learning model process approach with P2 project deliverables.

economic analysis offering financial justification on how electrified TRUs could be a cost-effective alternative to their patterns of idling.

- Offer technical support on adopting and implementing technology and best practices: analysis, implementation planning, system design, and best practices.
- Provide technical assistance to implement eTRU idle reduction technologies (EPS and eTRU) and best operating practices, including TRU idle management systems.

Table 1 represents fleets that received technical assistance and the status of the implementation when the project concluded in June 2017. Outputs included high-level scoping studies for 16 businesses, and comprehensive technical studies for 11 businesses representing 24 facility locations with telematics data using ReeferAnalytics and a third-party telematics system to develop a baseline for stationary TRU engine run time at respective home locations. Another three technical studies were undertaken with telematics data from yet another telematics system.

Methodology

To determine baseline TRU engine-idling patterns among fleet participants, the project extracted data from telematics devices attached to fleets' TRUs. A descriptive analysis was conducted on the data to understand the baseline TRU operating patterns (including stationary TRU engine operating time). Finally, the results were analyzed and presented to fleet management.

Data Collection

TRU telematics devices are third-party communication and GPS devices attached to TRUs on refrigerated trucks or trailers. The project used both basic and advanced telematics devices attached to the TRU. Advanced telematics devices have communication and control capabilities with TRU electronic control units to access engine and refrigeration system operating parameters, and also have sensors to monitor and collect information periodically, and when certain TRU events occur, like door open/close, reefer on/off, fuel level change, and so on. Cellular data networks or satellite connections provided real-time reporting and control. Advanced telematics allow for remote monitoring and control of the reefer unit temperatures: reacting and responding to alarms and alerts, optimized logistics and management of cold chain assets and resources, and so on. Basic telematics lack electronic control units (ECU) only (*40*).

There are different commercial telematics systems in the market. These telematics systems vary by vendor and

Table 1. Businesses Receiving Field P2 Technical Assistance

ID#	Type of Business	Activity	Status
P2-001	Food distribution	Implementation Assistance	Electrified parking spaces (EPS) operational, with usage monitoring system.
P2-002	Food distribution	Technical Assistance, Assessment	Electrified parking spaces being installed, for completion July 2017.
P2-003	Foodservice distribution	Technical Assistance, Assessment	Installing electrified parking spaces for completion August 2017. Acquiring electric standby TRUs.
P2-004	Food manufacturing	Technical Assistance, Assessment	Acquiring used trailers with electric standby capability. Installation of electrified parking spaces to be determined.
P2-005	Foodservice distribution	Technical Assistance, Assessment	Installing electrified parking spaces for planned completion October 2017.
P2-006	Foodservice distribution	Implementation Assistance	Planned upgrade to electrified parking spaces already installed and expansion in 2018.
P2-007	Food distribution	Technical Assistance	Electrified parking spaces and new TRUs are in 2018 capital plan at California distribution center. Grant funding being sought for EPS and demonstration of all-electric TRUs.
P2-008	Carrier, LTL refrigerated	Technical Assistance, Assessment	Acquired 12 new TRUs with electric standby. TBD on electrified parking spaces.
P2-009	Food distribution	Technical Assistance, Assessment	Only conducted an electrification assessment.
P2-010	Food distribution	Technical Assistance, Assessment	Acquiring electric standby TRUs, but will likely wait to acquire more eTRU to have enough to justify EPS investment.
P2-011	Food manufacturing	Technical Assistance, Assessment	Stationary storage study results initiated a study on cold storage facility needs (for instance, evaluation of adequate cold storage needs to add permanent cold storage space vs. use of electric standby refrigerated trailers or electric refrigerated shipping containers).
P2-012	Food distribution	Technical Assistance, Assessment	Moving forward on EPS once formal grant awarded and approved to install electrified parking spaces at California distribution center.
P2-013	Food manufacturing	Technical Assistance, Assessment	Analysis of stationary diesel storage trailers under review. Business is evaluating construction to expand cooler space vs. continued stationary storage with electric trailers/containers.
P2-014	Food distribution	Technical Assistance, Assessment	This location is acquiring electric standby TRUs, but is waiting to acquire enough eTRUs to justify EPS investment.
P2-015	Foodservice distribution	Technical Assistance Assessment at 1 Local/Total of 10 Facilities	Technical study results justified including EPS in new distribution center under construction. Evaluating cost to add EPS at other 9 distribution center locations. Applied for grant funding at 1 location, interested in funding for other locations.
P2-016	Foodservice distribution	Implementation Technical Assistance	EPS for trailers and trucks implemented.
P2-017 (a–f)	Carrier	Technical Assistance Outreach at Local Terminal Led to Corporate Engagement and Evaluation at 4 of 15 Terminals.	This Carrier has about 300 hybrid electric TRUs with plug-in capability. The company is evaluating the addition of EPS idle reduction at freight terminals, including incentive funding for a pilot at 1 to 4 terminals.

features, and usually are bundled with an enterprise software providing the ability to monitor, control, and get reports on fleet behavior. Fleets usually implement TRU telematics devices to leverage its reporting features mainly designed around temperature reporting, temperature alerts to ensure compliance, asset utilization, fuel utilization, and fuel theft monitoring (*41–45*).

Most fleet telematics devices offer the ability to closely monitor TRU during operations at certain distribution centers and other locations of interest by creating "geofences." A geofence is a virtual geographic boundary defined by GPS coordinates that enables software to trigger a response when a TRU enters or leaves an area. The P2 project closely monitored TRU operations inside certain

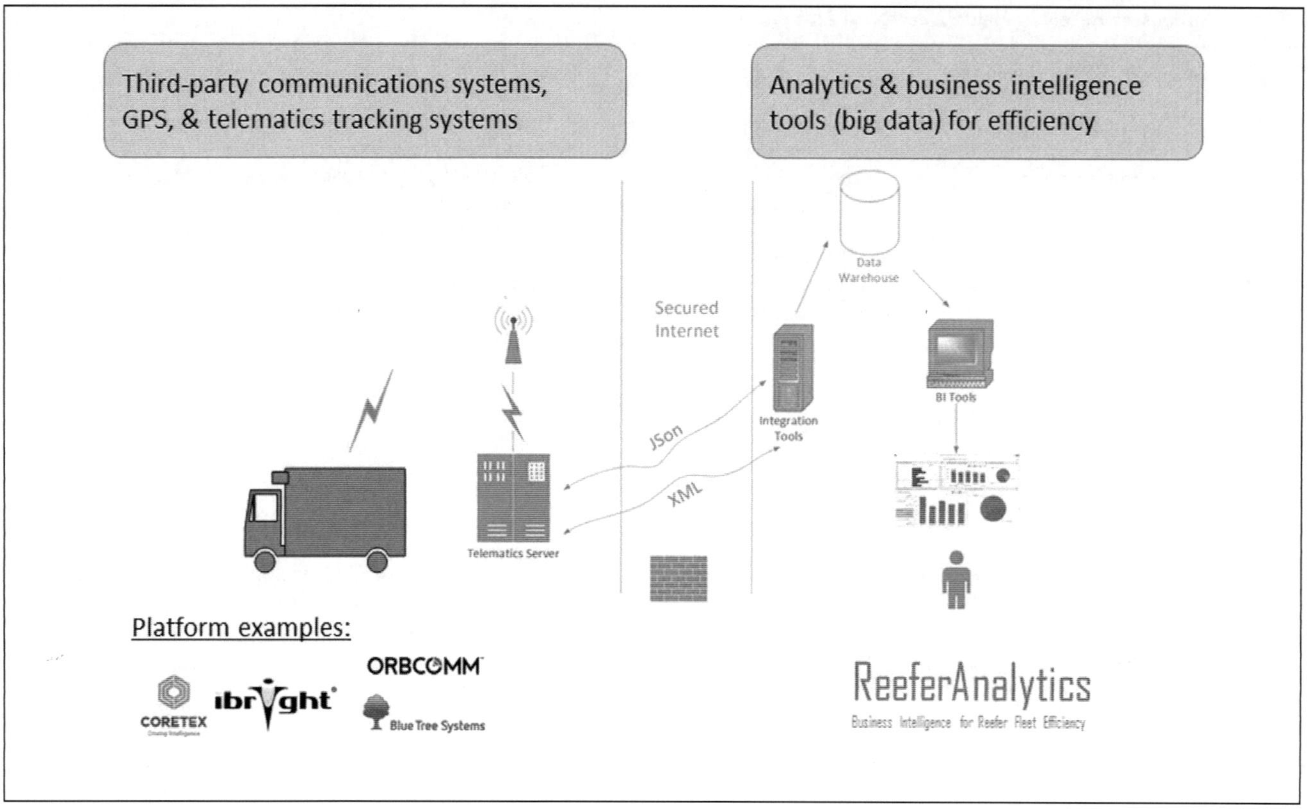

Figure 4. Big data technology architecture used in the P2 project.

geofences, however there was no special reporting or attention on TRU operations while outside the geofenced areas.

As explained earlier, 38 fleets participated in this study. Several fleets had telematics devices and systems implemented in their fleets. For fleets without telematics devices the project team installed them for data acquisition purposes. The ibright® Refrigeration system by Coretex was used by nine fleets and Blue Tree Systems was used by two fleets participating in the project.

To investigate idling behaviors and patterns, CleanFuture, Inc., one of the project partners, developed a big data solution, ReeferAnalytics. This system was used to extract data from various telematics sources, especially data related to fleet behavior inside geofences, to cleanse and reformat the data to be suitable for idling analyses, and to store it. Following this, ReeferAnalytics visualized the descriptive analyses to identify TRU idling patterns. See Figure 4 for a schematic.

Analysis and Results

Data was acquired, analyzed, and visualized for all fleets participating in the technical assistance project. An example analysis at a food distribution fleet is shown in Figure 5. TRUs at this food distributor were determined to accumulate about 58% of annual engine hours at the home distribution center.

A total cost of operations (TCO) hourly rate was determined for TRU diesel-engine hours, and then compared with total cost of operations for TRUs if running on electricity. The benefits of TRU electrification are translated into environmental and cost savings and compared with the opportunity cost of doing nothing over time. For this analysis, a TCO formula was used with cost per run hour for operations and maintenance costs. Operating cost per operating hour included diesel fuel costs and electrical energy costs at each site, and maintenance costs were based on fleet-wide historical maintenance costs whenever possible combined with fleet-wide engine hours. The organization's actual cost of supplied diesel fuel and electricity for operating costs, accompanied with actual diesel maintenance costs, or maintenance rates for diesel and electric hours as defined in the service agreement for each facility were used when available. Whenever data elements were not provided by an organization, we substituted with industry standard data. For this analysis, for the sake of simplicity, a discount rate, depreciation, or residual values were not used in working with fleet end-users.

Hazardous materials reduced is calculated by determining tailpipe emission reductions of nitrogen oxides (NOx), reactive organic gases, and particulate matter using the emissions factors for 25 hp Tier 4 Final engines

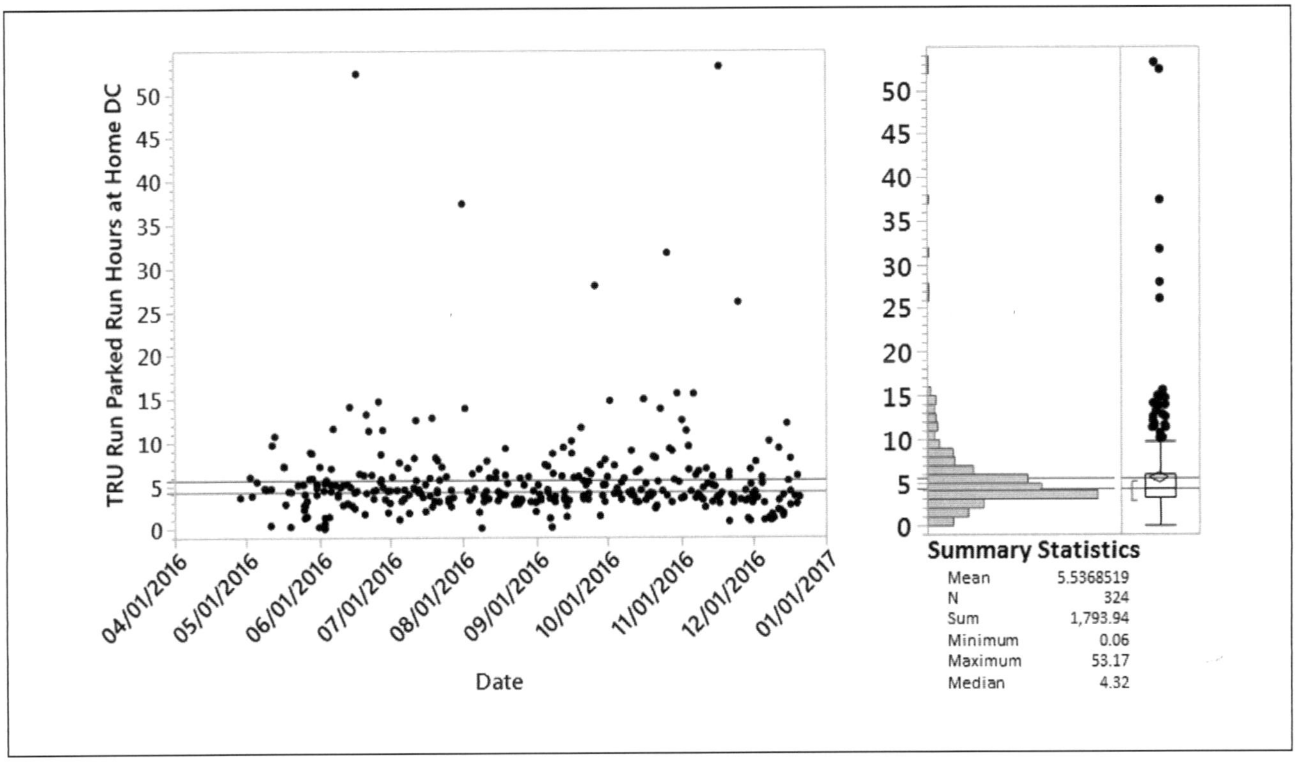

Figure 5. TRU idling behavior at a distribution center. Figure 5*a* shows elapsed trailer TRU engine hours at the distribution center during the measurement period. Figure 5*b* summarizes trailer TRU run events in a histogram.

based on annual TRU engine-idling hours reduced (*46*). TRU annual idling hours were determined through technical assistance activities. Hazardous materials reduction is conservatively calculated by using Tier 4 Final emissions factors as a simplified assumption, however since the TRU idle reduction occurs in fleets with a mix of Tier 1, Tier 2, Tier 4 Interim and Tier 4 Final engines the actual hazardous material reductions are greater. Carbon dioxide reduction is calculated from diesel fuel avoided based on annual TRU diesel-engine idling hours (*46, 47*).

Cost savings were calculated based on annual idling hours reduced and operating cost savings per hour. Operating cost savings per hour were determined by using the fleet's off-road diesel price multiplied by the TRU diesel consumption rates. Actual TRU diesel consumption rates were used when enough operating data was obtained from participating companies, otherwise a diesel consumption rate of 1 gal/h was used. The cost of electricity on a TRU run hour basis is subtracted from the diesel fuel cost savings. Typical maintenance cost savings for running on diesel compared with electricity is added to the hourly savings, typically US$ 0.20 to US$ 0.30 ph (although some fleet's diesel run rate savings are greater, between US$ 0.60 to US$1.00 ph). Hourly operating cost savings and annualized cost savings did not differ appreciably for refrigerated fleets in different climate zones in the United States. Variability in TRU idling was mainly attributed to fleet operating practices, which had a greater impact on savings than regional differences in ambient temperatures across the United States.

Findings

Analysis of the data and the field studies of the fleets revealed several findings about engine idling and the challenges facing electrification adoption in Region 10 as an alternative to engine idling.

- Most fleets' management has a broad underestimation of the average time spent idling by TRU units in their fleets.
- Most fleets' management thought the initial cost of adopting electrification did not justify the savings made by electrification, although they did not know what the initial cost was or what the savings would be.
- Few of the fleets' management knew that electricity is cheaper than diesel to power TRUs. Most fleets mistakenly believed that the low diesel prices during the project period made it uneconomical to use electricity. Moreover, businesses do not realize

Table 2. Fleet Level Outcomes Implemented During Project Period

Source	P2 activity	Sector	Haz. materials reduced (lb)	$MTCO_2e$ reduced	Savings (US$)
P2-001	Replaced diesel TRU idling with grid electricity (eTRU)	Food distribution	4,578	324	$71,081
P2-016	Replaced diesel TRU idling with grid electricity (eTRU)	Foodservice distribution	4,288	523	$60,570
P2-020	Reduced diesel TRU idling through behavior change	Grocery distribution	1,259	154	$16,224
Total			10,125	1,001	$147,875

Note: $MTCO_2e$ = metric tons of carbon dioxide equivalent

that electricity prices are stable and predictable in comparison with diesel's fluctuating prices.
- Fleets operating in the proximity of commercial and residential areas are prone to noise complaints from neighbors.

These findings align with findings from previous studies, as explained in the literature review of this paper.

Project Outcomes and Outputs

Working with fleets under this project has yielded outputs and outcomes to reduce TRU idling. The primary outputs are:

- Increasing awareness about idling problems and electrification as an alternative to idling: Fleets' management were educated about idling, its negative effects, recent related legislation, and how electrification is a better economic and environment alternative.
- Offering technical assistant on electrification: Fleets' management and operation staff were offered technical support on how to implement electrification on site, and help with implementation and best practices.
- Based on the work of this project, outcomes included several fleets deciding to replace diesel TRUs idling with eTRUs and EPS to run eTRUs while parked. Three fleets implemented EPS in their distribution centers during the 20-month project period (outcomes are presented in Table 2).

Other fleets are planning future technology implementation by adopting the specification requirements of electric-capable TRUs, or adoption of idle reduction policies and best practices. Anticipated outcomes from additional implementation over the next 3 years are shown in Table 3. Three-year projected outcomes were determined using the estimated annual hours of TRU idle reduction and multiplying this by the anticipated implementation period over the next 3 years, which is then multiplied by a probability of implementation score.

Conclusions

The P2 project offers a different approach to promoting a cleaner and more economically feasible alternative energy technology. Unlike other projects and programs that typically promote cleaner technologies by offering grants or incentives to replace older vehicles or equipment, P2 offered education and technical assistance to reduce idling, resulting in clean technology adoption and behavior change through the implementation of best practices.

By promoting a promising alternative – the use of plug-in eTRUs – the P2 project addressed an important source of pollution in the food supply-chain: the TRU engine-idling phenomenon, with its negative environmental, health, and noise impacts, and economic cost. eTRU technology is a cleaner, healthier, quieter, and more economically viable option compared with traditional diesel-only TRUs. However, a lack of awareness in fleets – also found in the literature review – is a barrier to eTRU adoption.

The P2 project targeted this issue by offering education and technical support to fleets to break this barrier. This effort included (a) studying and analyzing idling patterns of 38 fleets operating in Region 10, (b) sharing the findings with the fleets' management along with feasibility studies to show how adopting electrification provides economic benefits, as well as positive environmental and health impacts on fleet operations, and (c) showing the return on investment and how the initial cost of electrification can be covered by the resulting savings with a shore payback period.

This project found that, usually, fleets' management underestimated average idling hours and the corresponding cost. Also, most did not realize how electricity is better than diesel in relation to cost, price stability, potential return on investment and environmental benefits. After working with the fleets, most are now interested in electrification as an alternative to idling, and three fleets have already started/have plans to adopt electrification.

Table 3. Anticipated Outcomes Over Next 3 Years

Source	Sector	Anticipated implementation period (years)	Probability	Haz.materials reduced (lb)	$MTCO_2e$ reduced	Savings (US$)
P2-001	Food distribution	3	1	14,842	1,811	$209,664
P2-002	Food delivery	3	1	596	73	$9,477
P2-003	Foodservice distribution	2.75	1	10,495	1,280	$379,922
P2-004	Food manufacturing	0.5	0.1	253	31	$4,021
P2-005	Food distribution	2.5	1	5,301	647	$75,348
P2-006	Foodservice distribution	2.25	0.8	3,816	466	$52,229
P2-007	Food distribution	1.5	0.9	15,266	1,862	$202,176
P2-008	Carrier	2	0.75	2,735	334	$44,675
P2-009	Food distribution	2	0.5	618	75	$8,900
P2-010	Food distribution	1	0.5	1,153	141	$16,386
P2-011	Food manufacturing	2	0.5	495	60	$7,032
P2-012	Food distribution	2.5	1	6,998	854	$87,126
P2-013	Food manufacturing	2	0.1	264	32	$3,355
P2-014	Food distribution	1	0.25	610	74	$8,611
P2-015a	Foodservice distribution	2	0.75	7,492	914	$182,150
P2-015b	Foodservice distribution	2	0.75	6,181	754	$141,048
P2-015c	Foodservice distribution	2	0.8	6,678	815	$142,950
P2-015d	Foodservice distribution	1.5	0.1	42	5	$936
P2-015e	Foodservice distribution	1	0.75	1,022	125	$23,277
P2-015f	Foodservice distribution	1	0.75	3,096	378	$75,774
P2-015g	Foodservice distribution	1	0.75	2,751	336	$67,031
P2-015h	Foodservice distribution	1	0.75	648	79	$15,852
P2-015i	Foodservice distribution	1	0.75	2,127	260	$48,631
P2-015j	Foodservice distribution	1	0.75	3,222	393	$79,029
P2-015k	Foodservice distribution	2.5	0.8	6,078	742	$130,098
P2-016	Grocery distribution	1	1	4,288	523	$60,570
P2-017a	Carrier	1	0.5	2,474	302	$34,944
P2-017b	Carrier	2.5	0.5	16,491	2,012	$232,960
P2-017c	Carrier	2	0.5	9,894	1,207	$139,776
P2-017d	Carrier	2	0.5	6,596	805	$93,184
P2-018	Foodservice distribution	2	1	26,385	3,219	$372,736
P2-019	Grocery distribution	1	0.5	57,717	7,042	$815,360
P2-020	Grocery distribution	3	1	51,916	6,334	$669,232
P2-021	Food manufacturing	1	0.1	524	64	$6,664
P2-022	Seafood	2.25	0.8	6,303	769	$75,684
P2-023	Food distribution	2.75	0.8	17,621	2,150	$219,373
P2-024	Foodservice distribution	2	0.8	11,140	1,359	$143,598
P2-025	Carrier	0 (low interest)	0	-	-	$ 0
Total				**455,124**	**38,044**	**$4,879,779**

Note: $MTCO_2e$ = metric tons of carbon dioxide equivalent

Recommendations

The following recommendations are based on this project work and findings.

- Promoting alternative technologies by creating awareness and facilitating engagement with the targeted sector, is an effective approach that in many cases offers better value than funding for equipment incentives alone.
- The eTRU is a viable alternative to diesel-engine based TRUs, but a lack of awareness about its benefits is hindering adoption.
- Fleet management support is an important key to adopt eTRUs, therefore, working with fleet management and communicating the economic benefits of electrification are vital to achieve widespread acceptance and adoption of this technology.
- Using big–data-based solutions to capture and analyze the huge amount of data being generated electronically from telematics devices saves time and leads to more robust and comprehensive results (*48*).

This was a pilot project targeting one region in the United States. To achieve better results and widespread acceptance of electrification, similar projects must be conducted in other major regions of the country. The literature review revealed other reasons for the slow adoption of electrification besides lack of management awareness and support, those reasons must also be analyzed and addressed. Finally, most available studies about engine idling and its negative effects are focused on engine idling in general or on truck engine idling. Therefore, more specialized studies targeting TRU idling are needed to better understand its extent and impact.

Acknowledgments

This research was funded in part by the U.S. EPA Region 10 through the Pollution Prevention Program. Additional support was provided by Forth and Shorepower Technologies.

References

1. IHS Global Insight. Public and Private Sector Interdependence in Freight Transportation Markets. NCFRP-01, National Cooperative Freight Research Program, TRB, Washington, D.C., 2009. https://www.nap.edu/read/14285/chapter/3
2. Bureau of Transportation Statistics. *Freight in America*, U.S. Department of Transportation, Washington, D.C., 2006. https://www.rita.dot.gov/bts/sites/rita.dot.gov.bts/files/publications/freight_in_america/index.html
3. Gaines, L., and P. Weikersheimer. *Status and Issues for Idling Reduction in the United States: Alternative Fuel and Advanced Vehicle Technology Market Trend.* U.S. Department of Energy, Washington, D.C., 2015. https://cleancities.energy.gov/files/u/news_events/document/document_url/93/2015_strategic_planning_idling_reduction.pdf.
4. Stodolsky, F., L. Gaines, and A. Vyas. Analysis of Technology Options to Reduce the Fuel Consumption of Idling Trucks. ANL/ESD-43. Argonne National Lab., IL, 2000. https://www.osti.gov/scitech/biblio/771201
5. Kheirbek, et al. *New York City Trends in Air Pollution and Its Health Consequences*, New York City Department of Health and Mental Hygiene, New York, NY, 2013. https://www1.nyc.gov/assets/doh/downloads/pdf/environmental/air-quality-report-2013.pdf
6. Rahman, S. M., H. H. Ashrafur, M. A. Masjuki, M. J. Kalam, A. Abedin, A. Sanjid, and H. Sajjad. Impact of Idling on Fuel Consumption and Exhaust Emissions and Available Idle-Reduction Technologies for Diesel Vehicles – A Review. *Energy Conversion and Management*, Vol. 74, 2013, pp. 171–182.
7. American Transportation Research Institute. *Idle Reduction Technology: Fleet Preferences Survey*. American Transportation Research Institute, Arlington, VA, 2005. http://www.atri-online.org/research/results/Idle%20Survey%20One%20Page%20Summary.pdf
8. United States Environmental Protection Agency. *Compilation of State, County, and Local Anti-Idling Regulations.* U.S. Environmental Protection Agency, Washington, D.C., 2006. https://www.epa.gov/sites/production/files/documents/CompilationofStateIdlingRegulations.pdf
9. Electric Power Research Institute. Market and Technology Assessment of Electric Transport Refrigeration Units. Electric Power Research Institute, Palo Alto, CA, 2015.
10. Manson, C. *NYSERDA Transportation Program Case Study: Electric Refrigeration Transportation Network*. Industrial Economics Incorporated (IEc), Cambridge, MA, 2016.
11. Electric Power Research Institute. *Electrified Transport Refrigeration Units Technical Assessment: Refrigerated Fleet Technical Assessments for Electrification.* Electric Power Research Institute, Palo Alto, CA, 2017.
12. United States Environmental Protection Agency. *Fiscal Year 2015 Pollution Prevention Grant Summaries.* Collections and Lists. U.S. Environmental Protection Agency, Office of Chemical Safety and Pollution Prevention, Washington, D.C., 2016. https://www.epa.gov/p2/fiscal-year-2015-pollution-prevention-grant-summaries
13. Burgess, E., M. Peffers, and I. Silverman. *The Health, Environmental and Economic Impacts of Engine Idling In New York City*. Environmental Defense Fund, New York, NY, 2009. https://www.edf.org/sites/default/files/9236_Idling_Nowhere_2009.pdf
14. RMI. *Idle Reduction Confidence Report*, Boulder, CO, 2015. http://carbonwarroom.com/sites/default/files/reports/Idle-Reduction_Confidence_Report.pdf
15. Lewis, P., M. Leming, and W. Rasdorf. Impact of Engine Idling on Fuel Use and CO_2 Emissions of Nonroad Diesel Construction Equipment. *Journal of Management in Engineering*, Vol. 28, No. 1, 2012, pp. 31–38.
16. United States Environmental Protection Agency. *Greenhouse Gas Versus Smog Forming Emissions. Overviews and Factsheets.* U.S. Environmental Protection Agency, Office of Air and Radiation, Washington, D.C., 2015. https://www.epa.gov/greenvehicles/greenhouse-gas-versus-smog-forming-emissions
17. Bailey, D., M. Perrella, and C. Hanning. *NRDC: Clean Cargo - A Guide To Reducing Diesel Air Pollution From The Freight Industry In Your Community*, New York, NY, 2013. https://www.nrdc.org/sites/default/files/clean-cargo-toolkit.pdf
18. Janssen, N., G. Hoek, M. Simic-Lawson, P. Fischer, L. van Bree, H. ten Brink, and M. Keuken. Black Carbon as an Additional Indicator of the Adverse Health Effects of Airborne Particles Compared with PM10 and PM2.5. *Environmental Health Perspectives*, Vol. 119, No. 12, 2011, pp. 1691–1699.
19. Oregon Department of Environmental Quality. *Proposed Elements for the VW Environmental Mitigation Plan for the State of Oregon.* Oregon Department of Environmental Quality, Salem, OR, 2017. https://www.oregon.gov/deq/FilterDocs/orElementsVWplan.pdf
20. Doraiswamy, P., W. Davis, T. Miller, N. Lam, and P. Bubbosh. Air Quality Measurements Inside Diesel Truck Cabs During Long-Term Idling. *Transportation Research Record: Journal of the Transportation Research Board*, 2006. 1987: 82–91.
21. ICF International. *California Transportation Electrification Assessment Phase 3-Part A: Commercial and Non-Road Grid Impacts.* Final Report, Davis, CA, 2015. http://steps.ucdavis.edu/files/01-07-2016-Miller_Electrification-of-Off-Road-Vehicles.pdf
22. Electric Power Research Institute. *Market Assessment and Technology Review of Truck Stop Electrification and*

Electrified Transport Refrigeration Units. Electrification Initiative: Commonwealth Edison Case Study Report. Electric Power Research Institute, Palo Alto, CA, 2017. http://www.epri.com/abstracts/pages/productabstract.aspx?prod uctId=000000003002006036
23. Clean Cities. *Idlebase Database*. U.S. Department of Energy, Washington, D.C., 2015. https://cleancities.energy.gov/files/docs/idlebox_idlebase_database.xlsx
24. American Transportation Research Institute. *Idling Ordinances Adopted Since 9/2004 – ATRI Idling Compendium*, Arlington, VA, 2015. http://www.atri-online.org/research/idling/ATRI_Idling_Compendium
25. California Air Resource Board. *Airborne Toxic Control Measure for In-Use Diesel-Fueled Transport Refrigeration Units (TRU) and TRU Generator Sets, and Facilities Where TRUs Operate*. California Air Resource Board, Sacramento, CA, 2004. https://www.arb.ca.gov/regact/trude03/fro1.pdf
26. California Air Resource Board. *Sustainable Freight: Pathways to Zero and Near-Zero Emissions Discussion Document*. California Air Resource Board, Sacramento, CA, 2015. http://www.arb.ca.gov/gmp/sfti/sustainable-freight-pathways-to-zero-and-near-zero-emissions-discussion-document.pdf.
27. California Air Resource Board. *California Sustainable Freight Action Plan*. California Air Resource Board, Sacramento, CA, 2016. http://www.casustainablefreight.org/theplan.html. Accessed June 19, 2017.
28. California Air Resource Board. *Mobile Source Strategy*. California Air Resource Board, Sacramento, CA, 2016. https://www.arb.ca.gov/planning/sip/2016sip/2016mobsrc.pdf
29. Perrot, T., and J. Tario. *eTRU Refrigerated Warehouse Technology Demonstrations*, European Council for an Energy Efficient Economy, Stockholm, Sweden, 2007. http://www.eceee.org/library/conference_proceedings/ACEEE_industry/2007/Panel_3/p3_10/
30. Energy Information Administration. *Gasoline and Diesel Fuel Update - Energy Information Administration*. U.S. Energy Information Administration, Washington, D.C., 2017. https://www.eia.gov/petroleum/gasdiesel/
31. Energy Information Administration. *Short-Term Energy Outlook - U.S. Energy Information Administration (EIA)*. U. S. Energy Information Administration, Washington, D.C., 2017. https://www.eia.gov/outlooks/steo/report/electricity.cfm
32. Electric Power Research Institute. *Electrified Transport Refrigeration Units Technical Assessment: Refrigerated Fleet Technical Assessments for Electrification*. Electric Power Research Institute, Palo Alto, CA, 2017.
33. Alternative Fuels Data Center. *Alternative Fuels Data Center: Truck Stop Electrification for Heavy-Duty Trucks*, Washington, D.C., 2017. https://www.afdc.energy.gov/conserve/idle_reduction_electrification.html. Accessed July 1, 2017.
34. California Air Resource Board. *Diesel PM Control Technology Options for Complying with TRU ATCM*. California Air Resource Board, Sacramento, CA, 2015. https://www.arb.ca.gov/diesel/tru/documents/control_option_matrix.pdf
35. Carrier Transicold. *Vector 8500*, Carrier Transicold, Redmond, CA, 2017. http://www.carrier.com/truck-trailer/en/north-america/products/na-truck-trailer/single-temperature-trailer/vector-8500/
36. Thermo King. *SmartPower™ Electric Standby*, Thermo King Global, Minneapolis, MN, 2010. http://www.tkpitt.com/smartpowerelectricstandby.htm
37. Carrier Transicold. *Vector 8100*, Carrier Transicold, Redmond, CA, 2017. http://www.carrier.com/truck-trailer/en/north-america/products/na-truck-trailer/special-products/vector-8100/
38. Tario, J. *Electric-Powered Trailer Refrigeration Unit Market Study and Technology Assessment*. The New York State Energy Research and Development Authority, Albany, NY, 2005. http://shorepower.com/ElectricPoweredTrailerRefrigeration.pdf
39. CACWNY. *Electrified Parking Spaces (EPS)*. The Clean Air Coalition of W.N.Y., Buffalo, NY, 2011. http://www.cacwny.org/docs/TSE%27s.pdf
40. Miller, M. *California Off-Road Transportation Electrification Demand Forecast*. CEC-200-2016-012. Institute of Transportation and Aspen Environmental Group, California Energy Commission, Sacramento, CA, 2016.
41. Gartner. *Telematics*. IT Glossary, Garnter, Stamford, CT, 2012. http://www.gartner.com/it-glossary/telematics/. Accessed June 7, 2017.
42. Coretex. *Intelligence that Puts You in Control of Your Cold Chain*, Coretex, San Diego, CA, 2017. http://www.coretex.com/cold-chain/Accessed July 21, 2017.
43. International Telematics. *Ibright Benefits*, Coretex, San Diego, CA, 2017. http://www.internationaltelematics.com/ibright-benefits/Accessed July 21, 2017.
44. Blue Tree Systems. *Technology - Blue Tree Systems - Telematics Fleet Technology*. Blue Tree Systems, Galway, Ireland, 2017. http://www.bluetreesystems.com/technology/Accessed July 21, 2017.
45. Orbcomm. *IoT Devices for Satellite, Cellular and Dual-Mode Asset Tracking*, Orbcomm, Paris, France, 2017. https://www.orbcomm.com/en/hardware/devices. Accessed July 21, 2017.
46. U.S. Environmental Protection Agency. *Nonroad Compression-Ignition Engines: Exhaust Emission Standards*. EPA-420-B-16-022, EPA, Office of Transportation and Air Quality, Washington, D.C., 2016. https://nepis.epa.gov/Exe/ZyPDF.cgi?Dockey=P100OA05.pdf.
47. Energy Information Administration. *Frequently Asked Questions: How Much Carbon Dioxide is Produced from Burning Gasoline and Diesel Fuel?* U.S. Energy Information Administration, Washington, D.C., 2017. https://www.eia.gov/tools/faqs/faq.php?id=307&t=11
48. Barham, H. Achieving Competitive Advantage through Big Data: A Literature Review. *2017 Portland International Conference on Management of Engineering and Technology (PICMET)*, Portland StateUniversity, Portland, OR, July 8–9, 2017.

The Standing Committee on Alternative Transportation Fuels and Technologies (ADC80) peer-reviewed this paper (18-05275).

Article

TRR
JOURNAL OF THE TRANSPORTATION RESEARCH BOARD

Transportation Research Record
2018, Vol. 2672(24) 134–143

DOI: 10.1177/0361198118788428
journals.sagepub.com/home/trr
SAGE

Mapping Heavy Vehicle Noise Source Heights for Highway Noise

Carrie Janello[1] **and Paul R. Donavan**[1]

Abstract

The NCHRP Project 25-45 was initiated in 2013 to measure noise sources during heavy truck pass-bys using the acoustic beamforming method. Phase I testing included measurements from four relatively flat sites in Northern California, where heavy trucks traveled under moderate vehicle speeds. Sixteen additional sites in North Carolina were tested as part of Phase II. These sites had varying pavement grades, faster/slower posted speed limits, and varying operating conditions. The source mapping results from both phases were used to develop noise versus height profiles. The profiles were not significantly dependent on site and operating conditions. Average profiles were developed for each site to demonstrate that the predominant noise source was tire/pavement noise, with engine/powertrain being the secondary source; elevated sources occurred rarely and were equal to or greater than tire/pavement for only 0.5% of the total 1,289 trucks measured, while only 1.8% were within 5 dB(A) of the pavement source and 4.7% were within 10 dB(A).

Two previous studies used the acoustic beamforming method to identify the position and strength of heavy truck noise sources under highway operating conditions: NCHRP Project 08-56, as stated in NCHRP Report 635 (*1*) and a similar project performed by California Department of Transportation (Caltrans) (*2*). The scopes in both of these projects were limited. Fifty-seven trucks were measured at a single site in the NCHRP 08-56 study, and 125 trucks were measured at three sites along the same highway in the Caltrans study. Tire/pavement noise was identified as the dominant source in both projects, and noise from the exhaust stack outlets contributed little to the total noise emission; however, the data were minimal, and the beamforming technology provided only source localization. The technology has improved since these projects, and in 2013, NCHRP initiated Project 25-45 with objectives to determine height distributions for heavy vehicle noise sources and to establish a heavy vehicle noise source database for incorporation into traffic noise models. Full results of this research are documented in more detail in NCHRP Report 842 (*3*). The NCHRP Report 842 also included on-board sound intensity and statistical pass-by measurements, which were compared with the Reference Energy Mean Emission Levels (REMELs) database. While these testing methods are not discussed in this paper, the final report summarizes the findings.

The noise contours generated by the acoustic beamforming array for 20 total sites were used to develop vertical noise profiles that show the vertical distribution of noise concentrations. Both the contours and profiles were used to determine the noise sources for each truck and the percentage of trucks at each site that demonstrated noise sources at elevated heights of 12 ft (3.7 m), which would indicate exhaust noise contribution. This paper summarizes the results found for heavy trucks at these sites. For more detailed information, see NCHRP Report 842 (*3*).

Measurement Methods and Sites

The beamforming measurements were made using one of the more recently developed arrays supplied by B&K, designated WL9x6D250 Foldable Wheel Array. This array features 54 microphones, with nine arms and a diameter of 8.2 ft (2.5 m). The performance of the array is sufficient to cover the analysis frequency range from 315 Hz to 4,000 Hz. NCHRP Report 842 (*3*) provides further analysis of the array and comparisons with older technology used in previous studies. A photograph of the array installed at one of the measurement sites is shown in Figure 1. This new B&K array is quite portable and easy to set up, as the arms fold into each other. This allowed measurements to be made at two sites per day, capturing

[1]Illingworth & Rodkin, Inc, Petaluma, CA

Corresponding Author:
Address correspondence to Carrie Janello: cjanello@illingworthrodkin.com

Table 1. Summary of All Measurement Site Locations

Test site	Highway	Vicinity	Approximate grade	Posted speed limit (mph)	Pavement type
Lakeville	Lakeville Road	Petaluma, CA	0.0%	55	Older DGAC
505 SB 1	I-505 SB	Winters, CA	0.0%	55	Newer OGAC
505 SB 2	I-505 SB	Winters, CA	0.0%	55	Older OGAC
505 NB 1	I-505 NB	Winters, CA	0.0%	55	Ground PCC
NC1	I-77 NB	Elkin, NC	+ 3.0%	70	Transverse tine PCC
NC2	I-77 NB	Elkin, NC	−2.6%	70	Transverse tine PCC
NC3	I-77 SB	Elkin, NC	−0.3%	70	Transverse tine PCC
NC4	I-77 NB	Red Basin, NC	−2.9%	70	Fine aggregate DGAC
NC5	I-77 NB	Red Basin, NC	+ 2.9%	70	Fine aggregate DGAC
NC6	I-74 WB	Mount Airy, NC	+ 3.0%	65	Larger aggregate DGAC
NC7	I-74 EB	Mount Airy, NC	−2.5%	65	Larger aggregate DGAC
NC8	I-77 SB	Pine Ridge, NC	+ 1.1%	On-Ramp	Transverse tine PCC
NC9	I-77 NB	Mulberry, NC	+ 1.1%	70	Fine aggregate DGAC
NC10	I-77 SB	Longtown, NC	−2.8%	70	Ground PCC
NC11	I-77 NB	Longtown, NC	+ 2.5%	70	Ground PCC
NC12	I-73 SB/ I-74 EB	Norman, NC	−0.9%	65	DGAC
NC13	I-73 SB/ I-74 EB	Norman, NC	+ 0.3%	65	DGAC
NC14	SR 211 WB	Seven Lakes, NC	+ 0.5%	45	DGAC
NC15	I-73 SB/ I-74 EB	Norman, NC	0.0%	65	DGAC
NC16	SR 211 EB	Seven Lakes, NC	+ 0.8%	55	DGAC

Figure 1. B&K acoustic beamforming array set up on I-505 in North California.

70–80 heavy truck pass-by events at each site. The array was positioned along the roadside at a distance of 25 ft (7.6 m) from the center of the near lane of travel and 20 ft (6.1 m) from the assumed source plane. Statistical isolated pass-by measurements were also obtained for each truck pass-by event. These data were examined when lower frequency noise (below 315 Hz) was perceptible.

Truck noise source mapping measurements were made at 20 total sites. Pavement types included dense grade asphalt concrete (DGAC), open grade asphalt concrete (OGAC), and Portland cement concrete (PCC). At four relatively flat sites in Northern California, 223 heavy trucks operating at cruising conditions were measured. The first site in California was a two-lane highway, and the other three were made along Interstate 505 (I-505), which has four lanes with a median divider. At each of these sites, the maximum speed limit was 55 mph. Sixteen additional sites were measured in North Carolina, for a total of 1,066 heavy truck measurements. Eleven sites were made in the hilly western part of the state, near the Virginia border. At these sites, trucks operating on upgrades, downgrades, and on level surfaces were measured. One of the sites in this region was an on-ramp to capture slower speed, accelerating trucks, while the other sites in this region had posted speed limits of 65–70 mph. The final five sites were measured in the central southern region of North Carolina. Posted speed limits were 65 mph at three sites along I-73/I-74 and ranged from 45 mph to 55 mph on two sites along State Route 211 (SR 211). SR 211 was a two-lane highway, and while one of the sites consisted of vehicles traveling mostly under cruise conditions, the other was located just after a stop light, which caused slower speeds. Details regarding each test site are summarized in Table 1.

Acoustic Beamforming Measurements

Noise Contour Maps

The contour figures for all test sites were constructed for each pass-by event using the B&K data processing

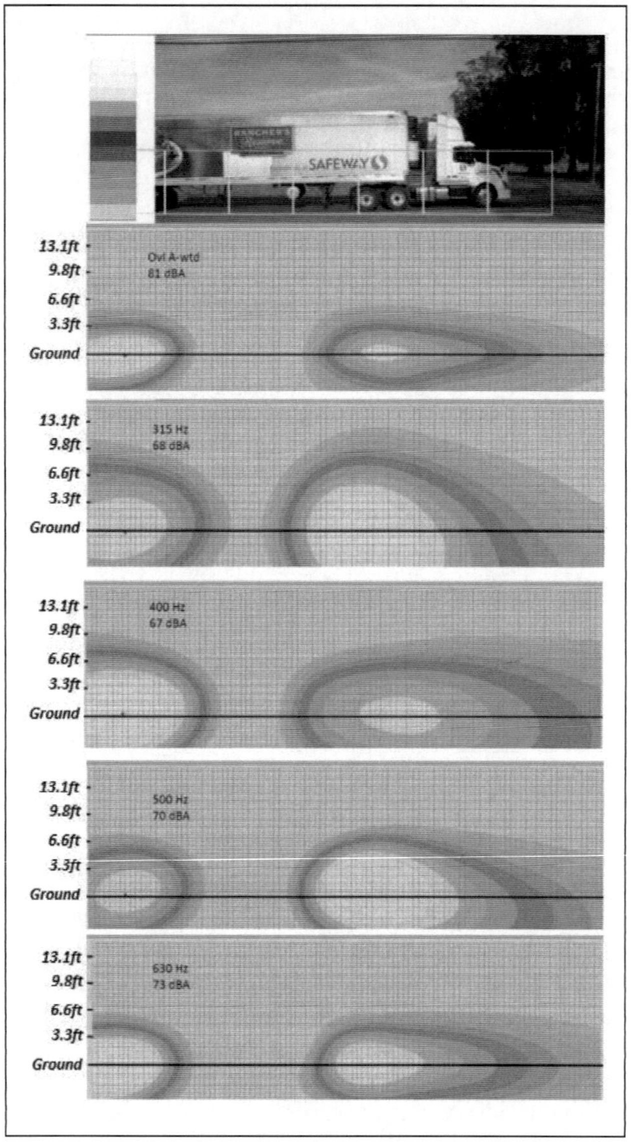

Figure 2. Sound level contours for a typical truck—overall A-weighted and 1/3 octave bands from 315 Hz to 630 Hz.

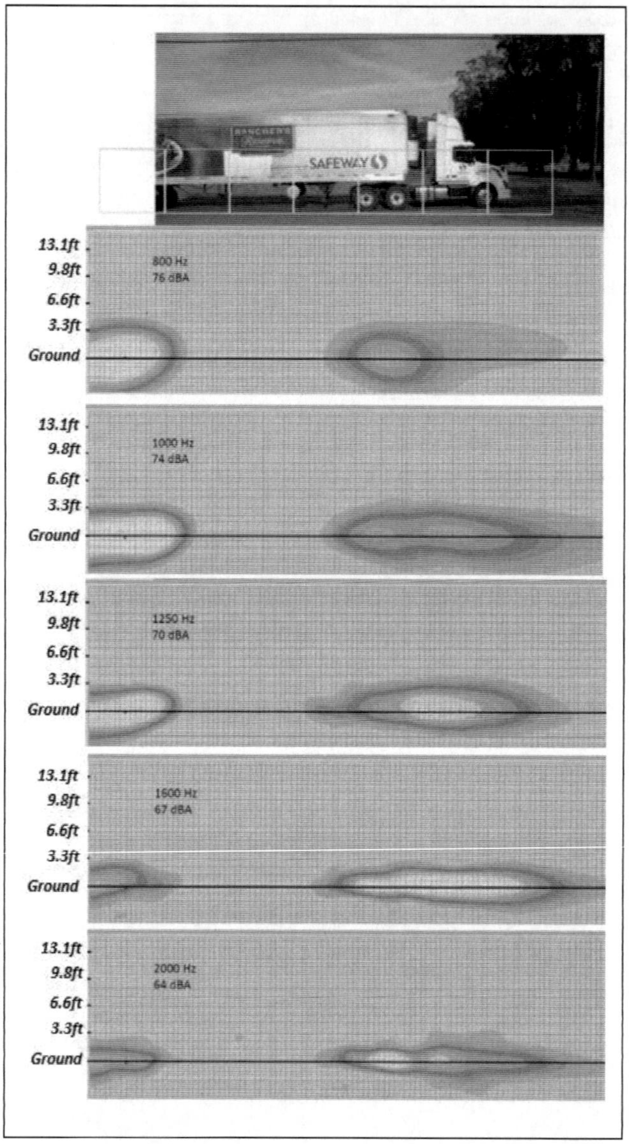

Figure 3. Sound level contours for a typical truck—1/3 octave bands from 800 Hz to 2,000 Hz.

software. For each measurement, data were collected for the frequency range of 315–4,000 Hz, and an example of the noise contours for a heavy truck are shown in Figures 2–4 for each frequency band, as well as the overall contour. The photograph of the truck is scaled and aligned with the grid. The yellow boxes in the photograph are 8 ft by 8 ft (2.44 m by 2.44 m), and the noise level scale of the contours is shown in the upper left of the figure, with the pale-yellow color representing the maximum level for any one plot. The colors correspond to 1 dB increments, and the maximum level is provided on each contour plot. The approximate location of the pavement ground plane is shown with each contour, and sound reflected by the pavement is indicated below the ground plane. In each contour, the grid spacing is 0.33 ft (0.1 m) in the vertical direction and 0.66 ft (0.2 m) in the horizontal direction, with references indicated in feet.

The example shown in the figures was taken at the Lakeville site in Northern California; however, most of the tractor/trailer pass-by events at each of the sites produced contours similar to this example. The contours of the overall level indicate primary noise regions that are low and split by the ground plane. The regions are concentrated at the drive axles of the tractor and at the axles of the trailer, indicating that tire noise is the dominant noise source. The contours for the tractor are also stretched toward the very front of the truck, indicating tire noise for the front axle and/or powertrain noise coming from beneath the truck. At the lower frequency bands (315 Hz and 400 Hz), the noise sources are extended in

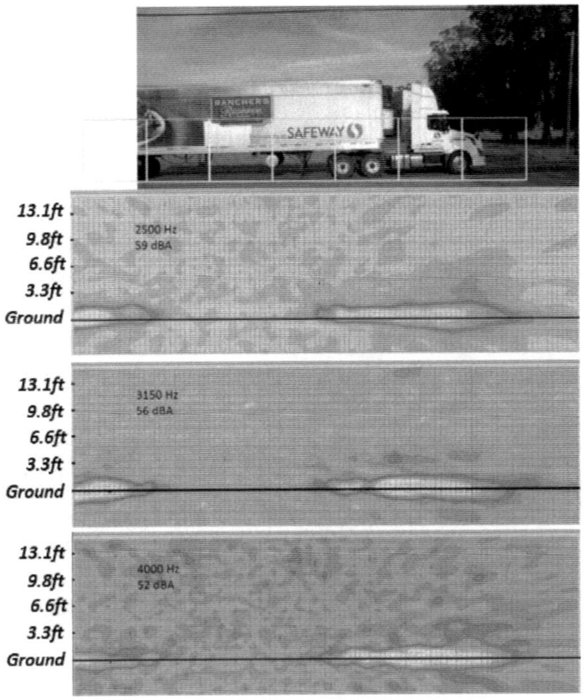

Figure 4. Sound level contours for a typical truck—1/3 octave bands from 2,500 Hz to 4,000 Hz.

the vertical direction to heights up to 13 ft (4 m); however, the cause for these low frequency distortions cannot be determined from the plots. This low frequency content does not appear to be affecting the measured overall contour plot. At the higher frequencies, powertrain noise is further indicated, as the source region extends the entire length of the tractor. Source regions at elevated heights above 3.3 ft (1 m) are not indicated in Figures 2–4.

An example of a truck with exhaust noise is shown in Figure 5. This truck was measured at the same site as the truck shown in Figures 2–4. The truck in Figure 5 shows a more unique influence from exhaust system noise. For overall A-weighted contours, the level at the top of the stack at a height of 13.8 ft (4.2 m) above the ground is only 6 dB lower than the maximum level of 79 dB(A) at the pavement surface. Traces of exhaust noise can be found in nearly every frequency band of the spectra, including the 500 Hz and 1,000 Hz bands. In the 2,500–4,000 Hz band contours, the exhaust outlet also produces levels equal to the maximum levels produced by sources at pavement level.

Analysis of Overall A-Weighted Profiles

The vertical profiles were calculated by energy summations of the noise levels in each horizontal row of the contour figures, yielding vertical distributions of the sound pressure levels for each vehicle tested. As shown in each of the contours, reflected sound levels are indicated below the pavement surface. To demonstrate the above-ground distribution of noise levels, the below-ground noise levels are cut off in each of the vertical profiles. Examples of the overall vertical profiles are shown in Figures 6–10 for the following categories: (1) in Northern California, which is relatively flat with a speed limit of 55 mph; (2) at an upgrade site in North Carolina; (3) at a downgrade site; (4) at a level site; and (5) at a slower speed site. For each site, the profiles were normalized to the maximum profile level at each respective site.

The contours shown above are two of the 44 total heavy trucks measured at the Lakeville site, which is relatively flat with a speed limit of 55 mph. The vertical distributions at this site for all trucks are shown in Figure 6, which plots source height on the vertical axis and sound level on the horizontal axis. The highest overall levels for each truck were located at the ground level, indicating that the dominant noise source was tire noise, whether from the front or rear axles or a combination of both. Other noise sources found below 9.8 ft (3 m) include engine noise, which could radiate from under the front of the vehicle, from the front wheel well, or possibly fan noise radiating from the front grille of the truck. The typical profile at the Lakeville site reduces from the maximum profile level by 10 dB(A) at heights ranging from 5.3 ft to 8 ft (1.6–2.4 m). This includes the typical vehicle shown in the contours of Figures 2–4 (identified as Run 38). In some cases, the profiles start to reduce like the typical truck runs and show a jump in height at a particular noise level. Run 2, identified in the figure, "bumps-up" at 88 dB(A). This may be indicative of a small contribution of exhaust outlet noise. Two profiles in Figure 6 produced levels at 12 ft (3.7 m) that were within 10 dB(A) of the ground level maximum level. Run 55, which is the truck identified in the contours of Figure 5, has levels that remain within 10 dB(A) for the entire height of the truck. This is due to exhaust noise. For Run 17, the cause is high levels of low frequency noise distorting the overall level. The cause of this low frequency noise is not apparent. Out of 44 trucks, 4.5% had elevated source levels within 10 dB(A) of the ground level maximum. The percentage of heavy trucks at the other California sites with elevated source levels within 10 dB(A) of the ground level maximum range from 1.8% to 6.1% of all trucks tested, and only one truck had levels within 5 dB(A) of the maximum.

Sixty-six trucks were measured at NC13 (Figure 7) in North Carolina. Of all the upgrade sites, this site had the most number of heavy trucks with high levels at heights above 11.5 ft (3.5 m). Three trucks (Runs 66, 40, and 58) resulted in elevated noise levels within 10 dB(A) of the ground level maximum, and three more trucks (Runs 22,

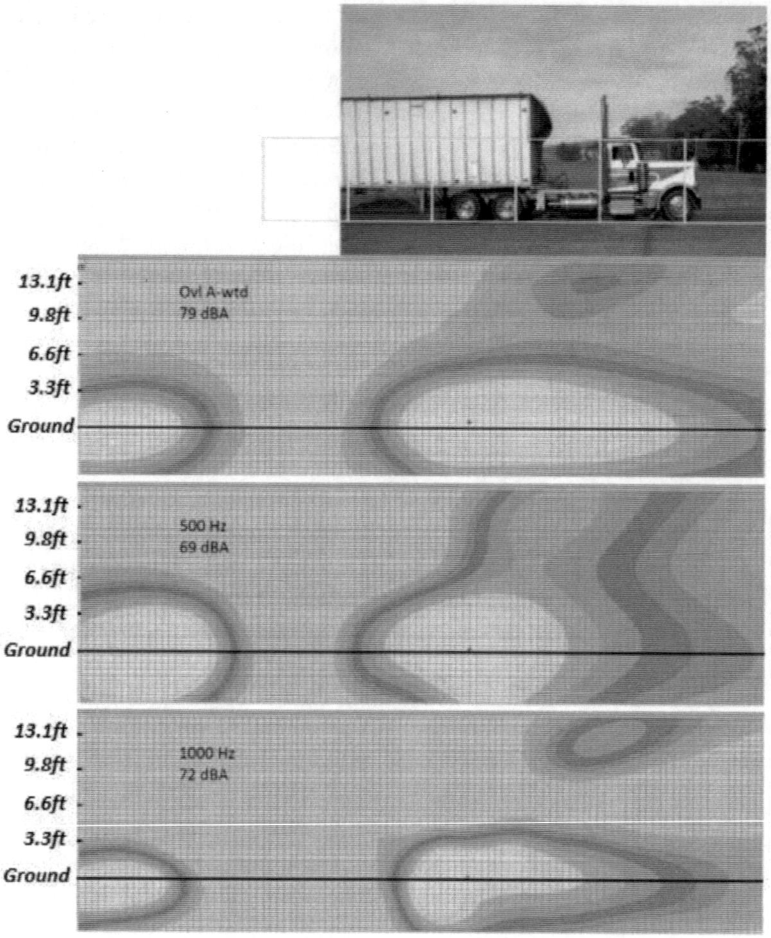

Figure 5. Sound level contours for a truck with exhaust noise—overall A-weighted and 1/3 octave bands at 500 Hz and 1,000 Hz.

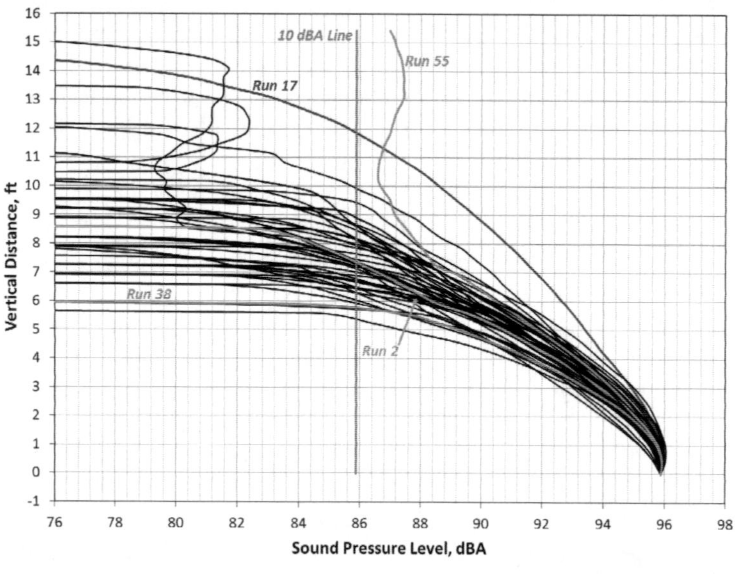

Figure 6. Overall A-weighted levels for trucks measured at Lakeville (California), normalized to 95.9 dB(A).

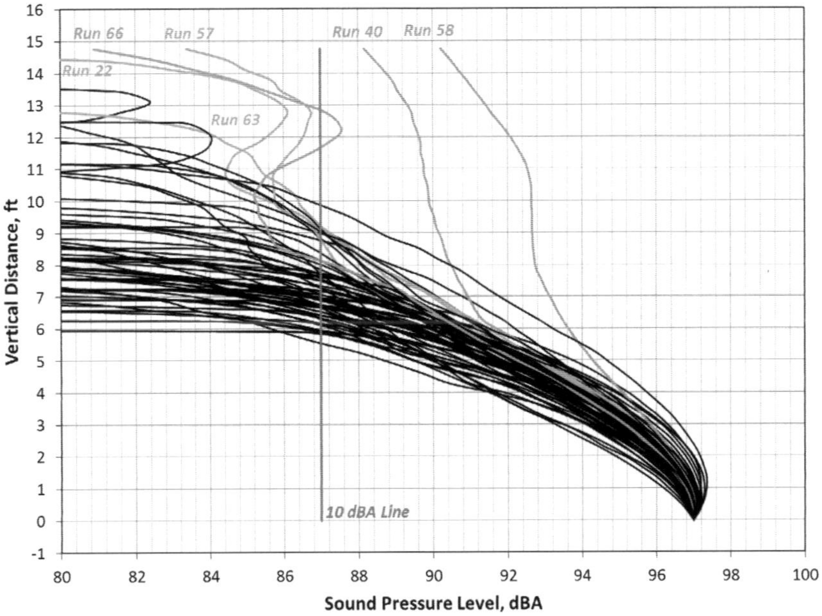

Figure 7. Overall A-weighted levels for trucks measured at NC13 (+ 0.3% upgrade on DGAC), normalized to 97.0 dB(A).

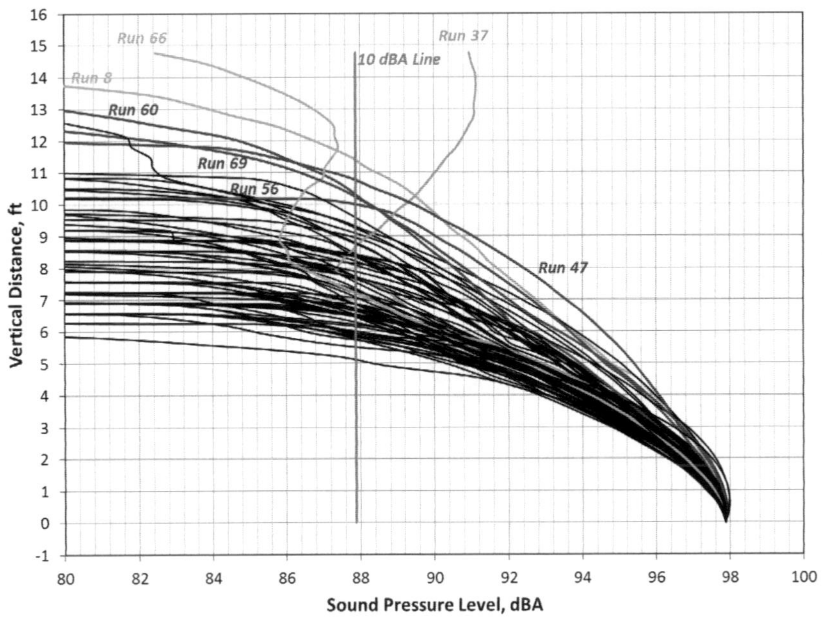

Figure 8. Overall A-weighted levels for trucks measured at NC12 (–0.9% downgrade on DGAC), normalized to 97.9 dB(A).

63, and 57) resulted in elevated noise levels within 13 dB(A) of the maximum. These levels, for all six trucks, were due to exhaust noise. Five additional trucks indicated some noise within 15 dB(A) of the ground level maximum at heights above 11 ft (3.4 m), and for three of these trucks, exhaust noise was present. Of the 66 total heavy trucks measured at NC13, 4.5% had levels within 10 dB(A) of the ground level maximum, and 1.5% had levels within 5 dB(A). In addition to trucks with exhaust noise, two trucks indicated some low frequency content.

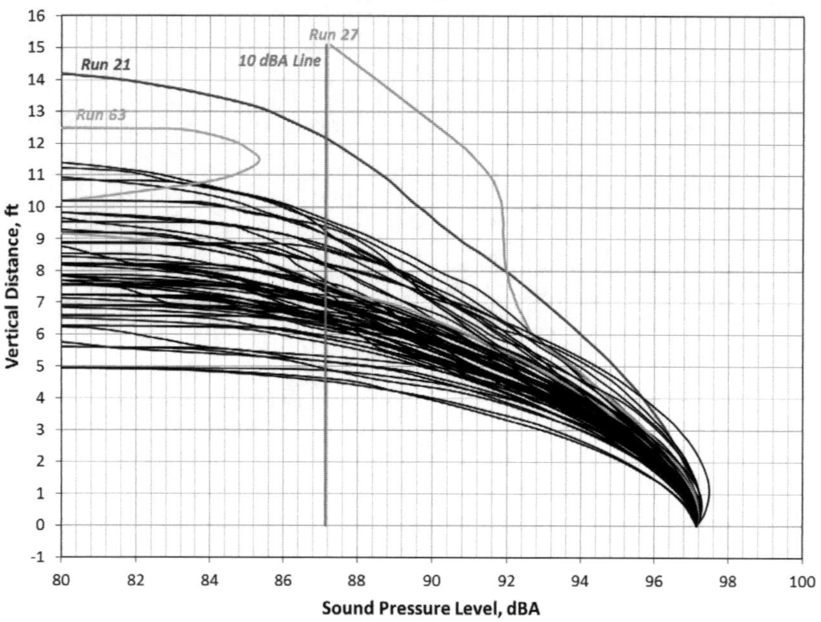

Figure 9. Overall A-weighted levels for trucks measured at NC15 (+ 0.0% level on DGAC), normalized to 97.1 dB(A).

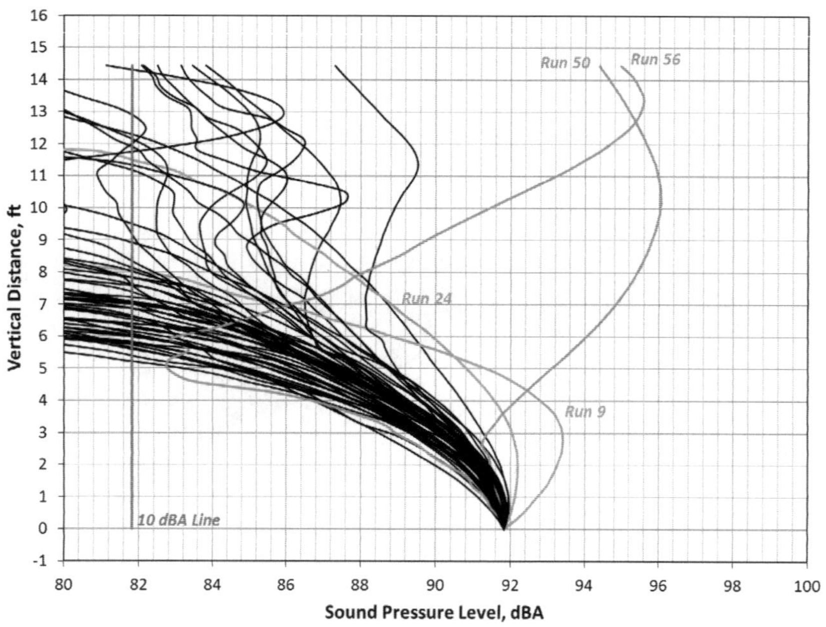

Figure 10. Overall A-weighted levels for trucks measured at NC8 (slower speed on transverse tine PCC), normalized to 91.8 dB(A).

The percentage of trucks at all uphill sites with elevated source levels within 10 dB(A) of the ground level ranged from 3.9% to 8.5%, and within 5 dB(A) of the ground level ranged from 0 to 3%. Two trucks at uphill sites had elevated noise sources equal to or greater than the ground level maximum.

Site NC12 (Figure 8) had the same pavement type and was located along the same roadway as NC13 but

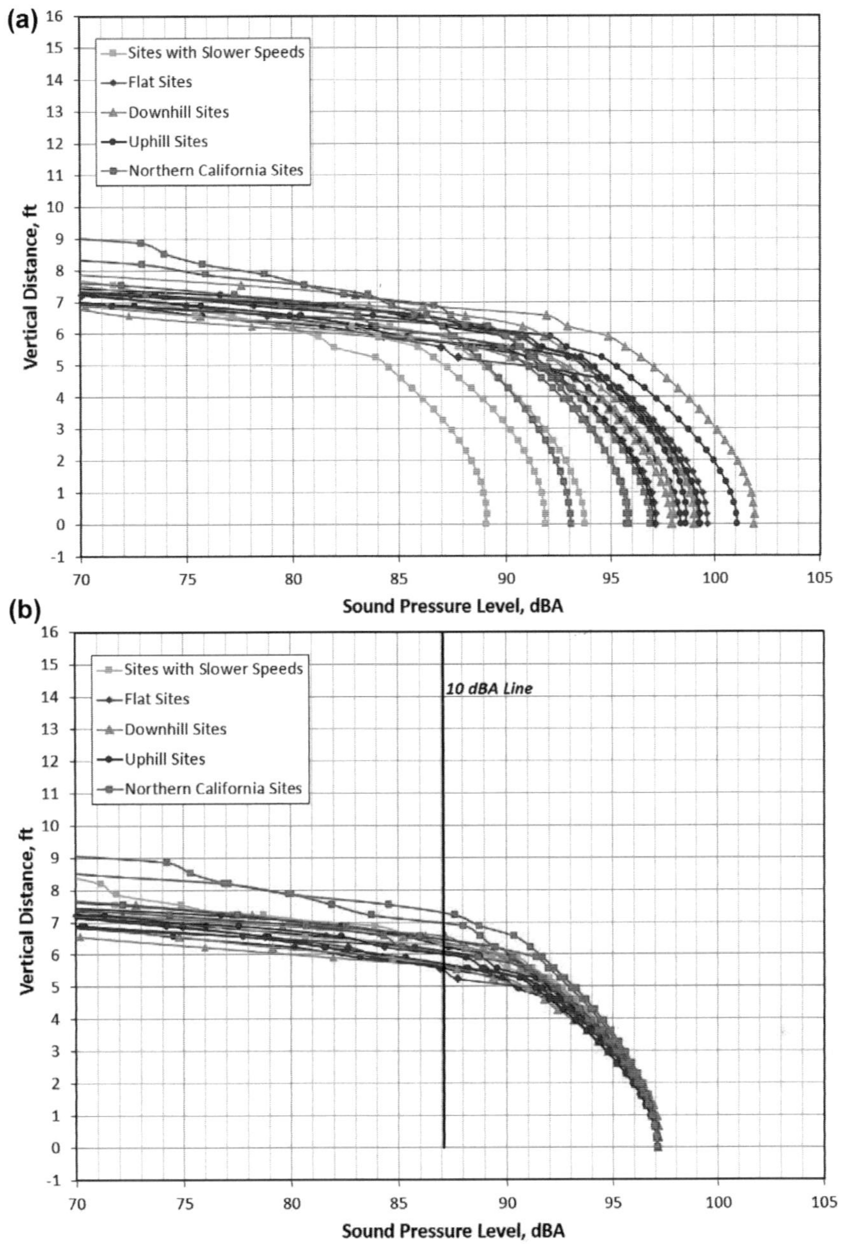

Figure 11. Average overall A-weighted levels for all trucks: (*a*) as measured and (*b*) normalized to 97.1 dB(A).

had a downgrade. Three trucks (Runs 8, 66, and 37) at this site, out of the 65 measured, had noise levels within 11 dB(A) of the ground level maximum at heights above 11.5 ft (3.5 m), and it was determined that all three trucks had exhaust noise. Of these, only one (Run 37) had exhaust noise levels within 10 dB(A) of the ground maximum, which is 1.5% of the heavy trucks measured at NC12. Four additional trucks (Runs 60, 69, 56, and 47) at this site had low frequency content. Overall, elevated noise sources were identified at only three of the five downgrade sites. Elevated sources within 10 dB(A) of the ground maximum ranged from 0 to 4.2% at the downgrade sites, and only one truck at the downgrade sites had elevated levels within 5 dB(A) of the ground maximum.

While trucks at three of the five downgrade sites did not have noise levels above 90 dB(A) at heights above 13 ft (4 m), NC12 and one other site had one truck each with high levels at these elevated heights. In contrast, five out of the six uphill sites had multiple heavy trucks with noise levels above 90 dB(A) at these heights. There was only one heavy truck at the flat, 55 mph limit California sites with levels exceeding 90 dB(A) at these heights.

Along the same roadway as NC13 and NC12, NC15 also had DGAC pavement but with a 0% grade. All but three heavy trucks at this site (Figure 9) had levels at least 12 dB(A) lower than the maximum level at the ground level [97.1 dB(A)] at heights of 10.2 ft (3.1 m) or below. Of these trucks, two had exhaust noise (Runs 27 and 63), and the other (Run 21) had low frequency content. Out of 62 trucks tested at NC15, only one (1.6%) had exhaust noise within 10 dB(A) of the ground level maximum (Run 27), which is similar to the downgrade site with this same pavement but less than the upgrade site. The other level site in North Carolina did not have any trucks indicating elevated noise sources.

NC8, as shown in Figure 10, was located at an on-ramp along I-77 and consisted of transverse tine PCC. The speed measured for trucks at this site was averaged to 29 mph, with a range from 18 mph to 39 mph, since the trucks were accelerating from stop. The majority of the heavy trucks had levels at least 10 dB(A) lower than the maximum noise level at the pavement surface for heights below 10 ft (3 m). Approximately 17 trucks indicated noise levels 10 dB(A) or greater than the pavement surface maximum above 10 ft (3 m). It was determined that 16 out of the total 58 trucks measured at NC8 had exhaust noise, while two additional trucks had low frequency content that could not be assigned to the exhaust stack. Out of the trucks with exhaust noise, 11 of the 58 total (19%) produced levels within 10 dB(A) of the ground level maximum, and two (Runs 56 and 50) had levels greater than ground level at heights of 10 ft (3 m) and greater. All vehicles at this site showed varying degrees of engine noise. This is clearly apparent in the two profiles (Runs 24 and 9) where the levels around 1.6 ft (0.5 m) bulge out to be greater than the ground level maximum.

NCHRP Report 842 (*3*) provides the profiles for all 20 test sites in Northern California and North Carolina.

Test Site Averages

Truck profiles at each site were averaged to develop an overall profile for each site (Figure 11*a*). These average profiles were normalized to the same maximum profile level of 97.1 dB(A), which is the average of all the sites (Figure 11*b*). The curvatures for each of the profiles were similar. However, the North Carolina test sites (varying grades and speeds) seemed to decrease from the maximum profile level slightly faster than the Northern California sites (flat with speed limits of 55 mph). Overall, there is no indication of a noise source at heights above 9 ft (2.7 m), which at heights of 10 ft (3 m) and above, the average profile levels for each site are effectively zero. NCHRP Report 842 (*3*) provides a comparison of these overall results to the profiles generated in previous studies.

Table 2. Summary of Heavy Trucks with Elevated Exhaust Noise Source Regions

Test site	Heavy truck count	No. with exhaust noise	% with exhaust noise
Lakeville	44	4	9.1
505 SB 1	57	7	12.3
505 SB 2	49	9	18.4
505 NB 1	73	9	12.3
NC1	69	6	8.7
NC2	61	0	0
NC3	66	0	0
NC4	73	1	1.4
NC5	76	5	6.6
NC6	71	6	8.5
NC7	72	6	8.3
NC8	58	16	27.6
NC9	80	3	3.8
NC10	66	1	1.5
NC11	67	5	7.5
NC12	65	3	4.6
NC13	66	9	13.6
NC14	46	11	23.9
NC15	62	2	3.2
NC16	68	11	16.2
Total No.	***1,289***	***114***	***8.8***

Note: Bold indicates the summary of all previous lines (the total).

Summary and Conclusion

Of the 20 sites tested, two were slow speed sites (posted speed limits were 45 mph), five were medium speed sites (posted speed limits were 55 mph), and 13 were high speed sites (posted speed limits were 65 mph and up). While the average profiles shown in Figure 11 do not indicate elevated sources above 9 ft (2.7 m), a few heavy trucks were found to have some noise located in the vicinity of the exhaust stack outlet, at heights around 11 ft (3.4 m) and above. Some traces of elevated noise were identified at higher frequencies but did not result in a significant noise contribution in the overall noise profile compared with the ground-level noise source. This noise could be aerodynamic and produced by "jetting" of the exhaust gases through the outlet. The occurrence and source strength of elevated noise sources is a primary concern of this research. Table 2 summarizes the number of heavy trucks at each site that showed some indication of elevated noise sources, which was 8.8% of the total 1,289 trucks in this study.

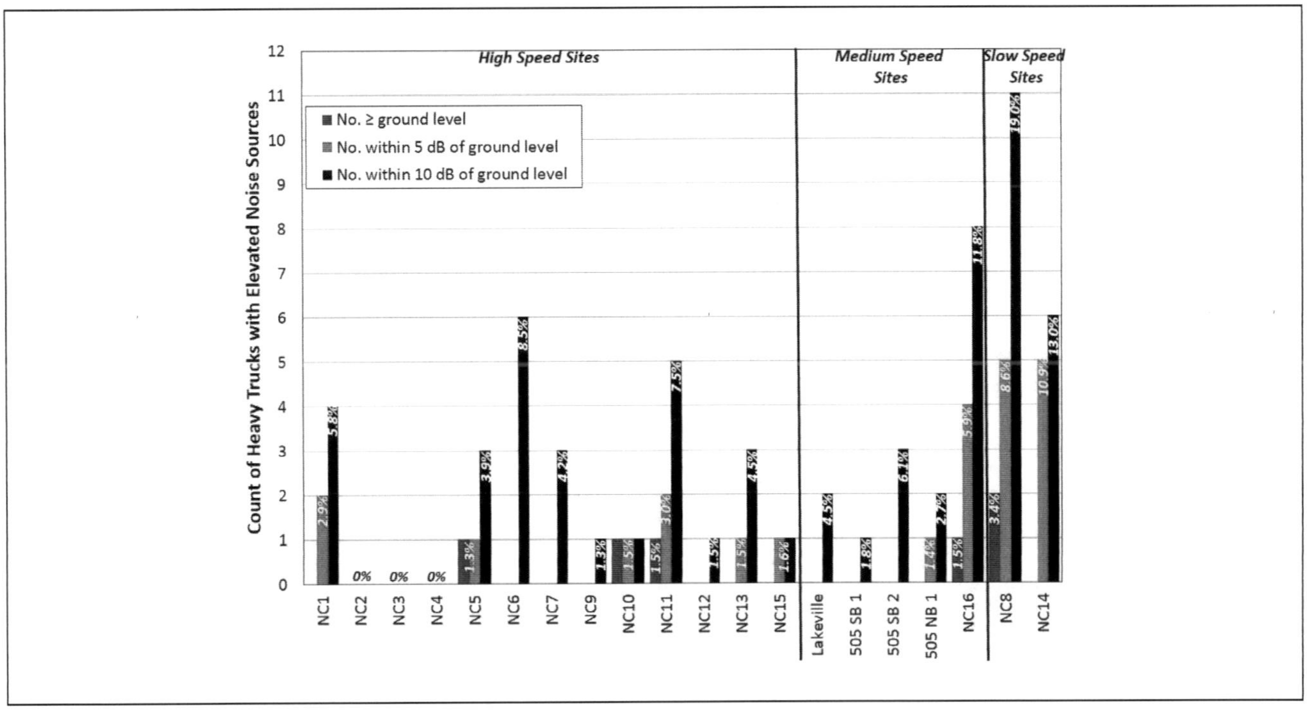

Figure 12. Truck counts at each site with noise source strength at 12 ft or above.

Table 2 shows that 8.8% of all heavy trucks tested resulted in traces of exhaust noise regardless of the magnitude compared with the level of the ground height source. Figure 12 shows the number of trucks at each test site that resulted in noise levels at 12 ft (3.7 m) within 10 dB(A) of the profile level at the ground (shown in black), noise levels at 12 ft (3.7 m) within 5 dB(A) of the profile level at the ground (shown in red), and noise levels at 12 ft (3.7 m) equal to or greater than the profile level at the ground (shown in blue). Only six trucks out of the total 1,289 resulted in noise levels at 12 ft (3.7 m) that were equal to or greater than noise levels at 0 ft. This comes to 0.5% of all heavy trucks. It should be noted that for the trucks with noise levels at 12 ft (3.7 m) that are within 10 dB(A) of the ground level maximum not all these levels are due to exhaust noise. Of the 61 cases in Figure 12, only 56 (4.3%) have identifiable exhaust noise. The others are due to high levels of low frequency noise that extend up to 12 ft (3.7 m).

Acknowledgments

The research in this paper was sponsored by NCHRP. Information on measurement site locations and pavement types was provided by Bruce Rymer, Caltrans Division of Environmental Analysis, and Gregory Smith, Human Environment Unit, North Carolina Department of Transportation. Additional technical assistance with the beamforming technology was provided by Tony Frazer, Brüel & Kjær.

Author Contributions

The authors confirm contributing to the paper as follows: study conception and design: Paul R. Donavan; data collection: Paul R. Donavan and Carrie J. Janello; analysis and interpretation of results: Paul R. Donavan and Carrie J. Janello; draft manuscript preparation: Carrie J. Janello. All authors reviewed the results and approved the final version of the manuscript.

References

1. Gurovich, Y., K. Plotkin, D. Robinson, W. Blake, and P. Donavan. *NCHRP Report 635: Acoustic Beamforming: Mapping Sources of Truck Noise*. Transportation Research Board of the National Academies, Washington, D.C., 2009.
2. Donavan, P., and B. Rymer. Measurement of the Vertical Distribution of Truck Noise Sources during Highway Cruise Pass-bys Using Acoustic Beam-Forming. *Transportation Research Record: Journal of the Transportation Research Board*, 2009. 2123: 145–152.
3. Donavan, P., and C. Janello. *NCHRP Report 842: Mapping Heavy Vehicle Noise Source Heights for Highway Noise Analysis*. Transportation Research Board of the National Academies, Washington, D.C., 2017.

The Standing Committee on Transportation-Related Noise and Vibration (ADC40) peer-reviewed this paper (18-03102).

Article

TRR
JOURNAL OF THE TRANSPORTATION RESEARCH BOARD

Transportation Research Record
2018, Vol. 2672(24) 144–153
© National Academy of Sciences:
Transportation Research Board 2018
Article reuse guidelines:
sagepub.com/journals-permissions
DOI: 10.1177/0361198118797188
journals.sagepub.com/home/trr
SAGE

New Hampshire Department of Transportation Statewide Noise Barrier Study

Jason C. Ross[1], Mark Arnoldy[1], and Jonathan Evans[2]

Abstract

This paper presents the results of a statewide noise barrier study for the New Hampshire Department of Transportation (NHDOT). The feasibility and reasonableness of noise barriers throughout the state has been evaluated according to NHDOT acoustical and cost-effectiveness criteria. The study has provided the DOT with an estimate of the potential noise barrier material costs associated with a newly implemented Type II noise barrier program. The study also has identified municipalities that the DOT can coordinate with for enacting noise-compatible planning regulations. The paper describes the methodology used to develop the screening-level noise barrier evaluation. Highway noise levels throughout the state and the distances from the highway where noise levels exceed the noise abatement criteria were determined. A method to predict the noise reduction of barriers (insertion loss) based on a simple geometrical relationship of receptors and potential noise barriers was established and has been shown to correlate well to detailed traffic noise models using actual roadway and terrain geometry. The simplified prediction method was used to evaluate over 300 miles of highway, 300 candidate noise barrier study areas, and over 30,000 receptors. This screening-level modeling approach is critical to the practicality of using the Traffic Noise Model across the entire state given the significant number of receptors and noise barrier study areas. The results of the study including information about all 30,000 receptors and 300 noise barrier study areas were collected in Google Earth™ data files for easy access when the DOT responds to noise complaints from the public.

The New Hampshire Department of Transportation (NHDOT) has recently initiated a Type II noise barrier program to address growing concerns from the public with highway noise and the growing trend of new noise-sensitive uses being developed near the state's highways. Type II noise barriers are those constructed on existing highways when it is not part of a highway construction project that substantially alters the existing roadway geometry and/or increases capacity. The Type II Noise Barrier Program applies to residences and is documented in NHDOT's *Policy and Procedural Guidelines for the Assessment and Abatement of Highway Traffic Noise for Type I & II Highway Projects* (November 2016) which is the DOT's noise policy for implementing the FHWA regulation *Procedures for Abatement of Highway Traffic Noise and Construction Noise* (23 CFR 772).

In considering the Type II noise barrier program, the DOT administrators were interested in getting a preliminary cost estimate of noise barriers that may be considered feasible and reasonable according to the NHDOT noise policy and the length, height, and location of the barriers. Understanding the approximate barrier costs will help in identifying local, state, and federal funding for Type II noise barriers. Additionally, understanding where Type II noise barriers may be feasible and reasonable will help the DOT coordinate with specific municipalities that need to enact planning and development regulations which require avoidance, minimization, or mitigation of exterior noise impact for new noise-sensitive developments.

This statewide noise barrier study included the development of a screening procedure to estimate highway noise emissions from all Tier 1 highways across the state and evaluate distances exceeding the NHDOT noise abatement criteria (NAC). This study also included an investigation of fundamental noise barrier design principles using the FHWA's Traffic Noise Model (TNM). Noise barrier principles that were analyzed included

[1]Vanasse Hangen Brustlin Inc., Watertown, MA
[2]New Hampshire Department of Transportation, Bureau of Environment, Concord, NH

Corresponding Author:
Address correspondence to Jason C. Ross: jross@vhb.com

Table 1. DEI Adjustments Based on Land Use

Percentage	Adjustment for percentage of receptors that were permitted or constructed before opening date of highway	Adjustment for percentage of receptors that were permitted after November 28, 1995
1 to 25%	+ 100	−100
25 to 50%	+ 200	−200
50 to 75%	+ 300	−300
75 to 100%	+ 400	−400

Source: NHDOT, 2016.

determining how far back from a barrier benefits can be found, how the horizontal shielding angle of a noise barrier relates to insertion loss, and how building rows, barriers, and terrain affect insertion loss. The study also provided easy access to the wealth of information on all 30,000 receptors and 300 noise barrier study areas that the DOT can use when responding to noise complaints from the public.

This paper includes a summary of the NHDOT criteria for Type II noise barriers to be considered feasible and reasonable; the method to identify and locate the large database of noise receptors adjacent to highways throughout the state; results of the analysis to determine the distances to highway noise levels approaching the NAC; and the sensitivity of these distances to the range of traffic volumes, speeds, truck percentages, and highway geometry; and the relationships of barrier shielding angle to insertion loss. The overall results of the statewide noise barrier feasibility and reasonableness evaluation are provided in regard to the total number of barriers, lengths, and costs that were found to likely be feasible and reasonable. A detailed TNM analysis would be conducted for any potential Type II barrier being considered.

Type II Barrier Eligibility Criteria

The NHDOT noise policy includes criteria to determine whether a noise barrier is eligible for the Type II program. The barrier shall be along an existing Tier I highway. This study evaluated approximately 300 miles of highways including the FE Everett Turnpike, Spaulding Turnpike, Blue Star Turnpike (I-95), Interstate 93, Route 101, Interstate 89, and Interstate 393. Type II noise barriers shall not be along a section of highway for which a Type I eligible project is programmed in the 10-year Transportation Improvement Plan. This study included Type I barriers in the evaluation for planning purposes. Prior Type I studies were reviewed to identify noise barriers that were previously determined not to be feasible and reasonable as these areas are not eligible for Type II barriers under the DOT policy.

The cost-effectiveness criteria are modified based on how recent residences were developed in relation to the existence of the highway. If a greater percentage of residences pre-date the highway, it is more reasonable to provide noise abatement. The criteria require that at least one benefiting receptor must have been permitted for development before the original opening date of the highway or before November 28, 1995, as required by FHWA regulation. The NHDOT uses a dimensional effectiveness index (DEI) criterion to evaluate the cost-effectiveness of potential noise barriers. If the actual DEI of the potential barrier is below the DEI criterion (base criterion of 1,500 square feet per receptor), then the barrier is considered to be cost-effective. The base DEI criterion is adjusted negatively based on the percentage of receptors that have been developed in a study area after November 28, 1995, and it is adjusted positively based on the percentage of receptors that existed in a study area before the opening date of the highway according to Table 1.

Methodology

The following section describes the methodology used to identify residential receptors within 1,500 feet of the highway median, compute highway noise levels at all receptors, compute the insertion loss of 10, 15, 20, or 25-foot noise barriers and evaluate the DEI to determine eligibility of barriers in accordance with the NHDOT Type II Noise Barrier Program.

Identifying Receptors

Statewide parcel land use data have been used to identify receptors according to residential state land use codes. For land use codes indicating two to four dwelling units, the higher number of dwelling units was assumed to provide a slightly conservative estimate of the DEI. Land use information was reviewed along with aerial photography to further identify residential land uses. In some areas with multi-family buildings and condominiums, it was necessary to estimate the number of dwelling units

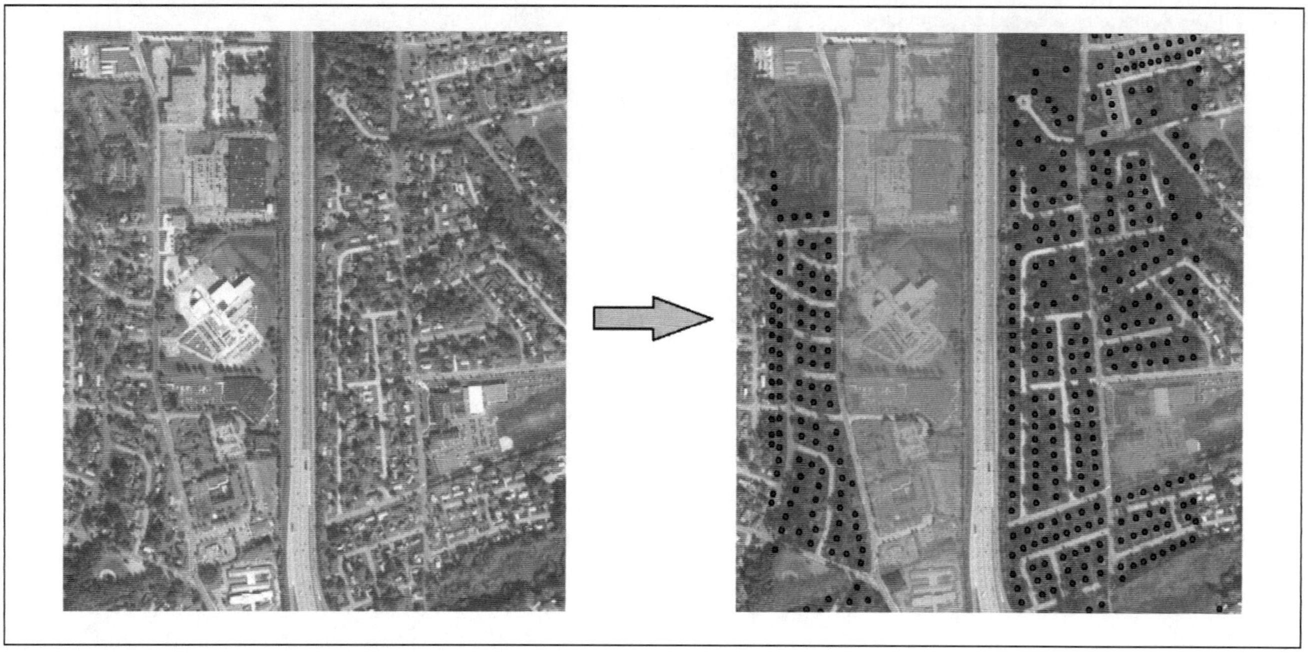

Figure 1. Parcel data, land use, and receptors.

based on reviewing Google Streetview™ images of buildings and/or parking lots (assuming that each dwelling unit would have two to three parking spaces). Receptors were located in the geometric center of the parcels. These locations were generally representative of outdoor areas with frequent human use.

U.S. Census data for housing units within "census tracts" were analyzed between 1950 and 2015. The percentage of housing units built after 1995 was calculated for each census tract. Similarly, the percentage of housing units built before the highway opening date (typically late-1950s to mid-1960s) was calculated. The results were applied to each barrier study area. On average, 25% of the housing units were built after 1995 and 35% were built before the highway opening. Therefore, the average DEI criteria among all barrier areas evaluated was 1,575. Figure 1 shows a typical aerial location (left) and the identification of residential land use and receptors (right). Each colored block indicates a tax parcel and its respective land use code, while each dot is the receptor location used for sound level analysis.

Highway Noise Calculations

Highway noise levels were calculated using TNM version 2.5 using a straight and flat highway geometry accounting for the number of travel lanes and traffic data for each segment of highway. The most recent three to five years of traffic data available from the NHDOT Bureau of Traffic were used in the analysis. These data included average annual daily traffic volumes (AADT), peak-hour factors (K-factor), posted and measured speeds, and percentages of medium and heavy trucks. For highway segments where one or more of these traffic variables were not available, the nearest segment with the required data was assigned. Highway noise levels at distances every 10 feet up to 1,500 feet of the median were computed for all 137 exit-to-exit highway segments. Additionally, a sensitivity analysis of how the different traffic conditions relate to different highway noise levels was conducted. The following summarizes the range of traffic data and highway noise levels throughout the state:

- Highways ranged from one to five lanes per direction and were typically composed of 12-foot travel lanes with approximately a 45-foot median and 5 to 10-foot shoulders.
- AADT ranged from 3,100 to 120,828 vehicles per day across all Tier 1 highway segments. This large variation in AADT corresponds to a 16 dBA difference between the minimum and maximum highway noise levels.
- K-Factors across all the roadways varied from 8.5% to 15.9%. Multiplying the AADT by the K-Factor results in the peak hour volume (PHV). The minimum PHV of 491 vehicles per hour (vph) occurs on I-93 between Exits 37 and 38 and the maximum of 12,445 vph is on the Everett Turnpike between Exits 5 and 6.
- Medium Truck percentages ranged from 2% to 6% of total volumes and Heavy Truck percentages ranged from 2% to 10%. The range of

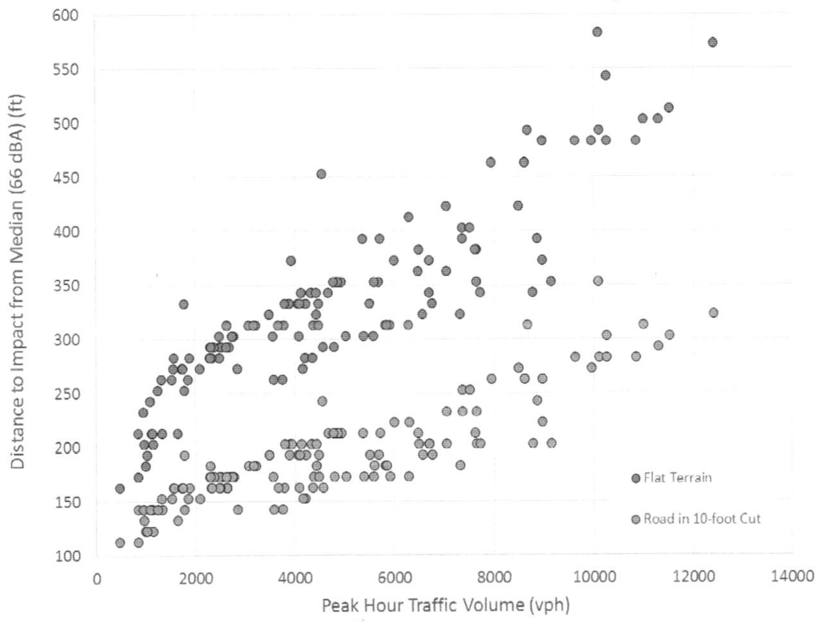

Figure 2. Distance to noise impact for all exit-to-exit highway segments.

Medium Truck percentages correspond to only a 0.3-dBA variation in noise levels. The range of Heavy Truck percentages correspond to a 2-dBA variation in noise levels.

- Average speeds ranged from 59 to 73 mph which correspond to a 4-dBA variation in noise level.

Highway noise levels were calculated at every receptor (approximately 30,000) within 1,500 feet of the highway median. The distance to impact from the highway median (where noise levels approach or exceed the NHDOT NAC for FHWA Activity Category B [residential] land use) was determined for each exit-to-exit highway segment. Figure 2 presents the distance to impact for all highway segments for both flat ground and for when the highway is in a 10-foot cut. This figure shows that the distance to impact is typically 300 to 400 feet from the median for flat ground and 150 to 250 feet when the highway is in a 10-foot cut.

It should be noted that building rows were not included in the highway noise calculations. While this overestimates highway noise levels for receptors with intervening building rows, it does not substantially affect the DEI calculations or the determination of Type II eligibility. This is because the criteria for eligibility only requires one or more receptors to exceed the NAC (which typically occurs at first row receptors with no intervening buildings) and the barrier must benefit a sufficient number of receptors. As discussed in the next section, the presence of building rows has been found to not substantially affect barrier insertion loss.

Detailed TNM models were developed in two locations: the Pannaway Manor Neighborhood south of the Spaulding Turnpike adjacent to I-95 in Portsmouth, NH, and the Keating Avenue neighborhood in Dover along the Spaulding Turnpike. The detailed TNM models include terrain lines, building rows, and the actual highway geometry along with site-specific traffic conditions. Additionally, results from previous detailed TNM studies conducted for I-93 Salem to Manchester and Spaulding Turnpike Newington to Dover have been included. Figure 3 compares the highway noise predictions of the detailed TNM models and the method described here for receptors in relatively flat sections of study areas. This figure shows that noise level predictions for most receptors are within 3 dB for both computation methods.

Insertion Loss Calculations

Type II noise barrier study areas were defined by reviewing all highway segments and grouping locations of relatively dense receptors. A total of approximately 300 noise barrier study areas were evaluated. The acoustical effectiveness or insertion loss of noise barriers was evaluated based on a grid analysis which included a matrix of receptors behind a noise barrier along a straight segment of highway. The grid analysis computed the insertion loss of receptors behind 5, 10, 15, 20, and 25-foot

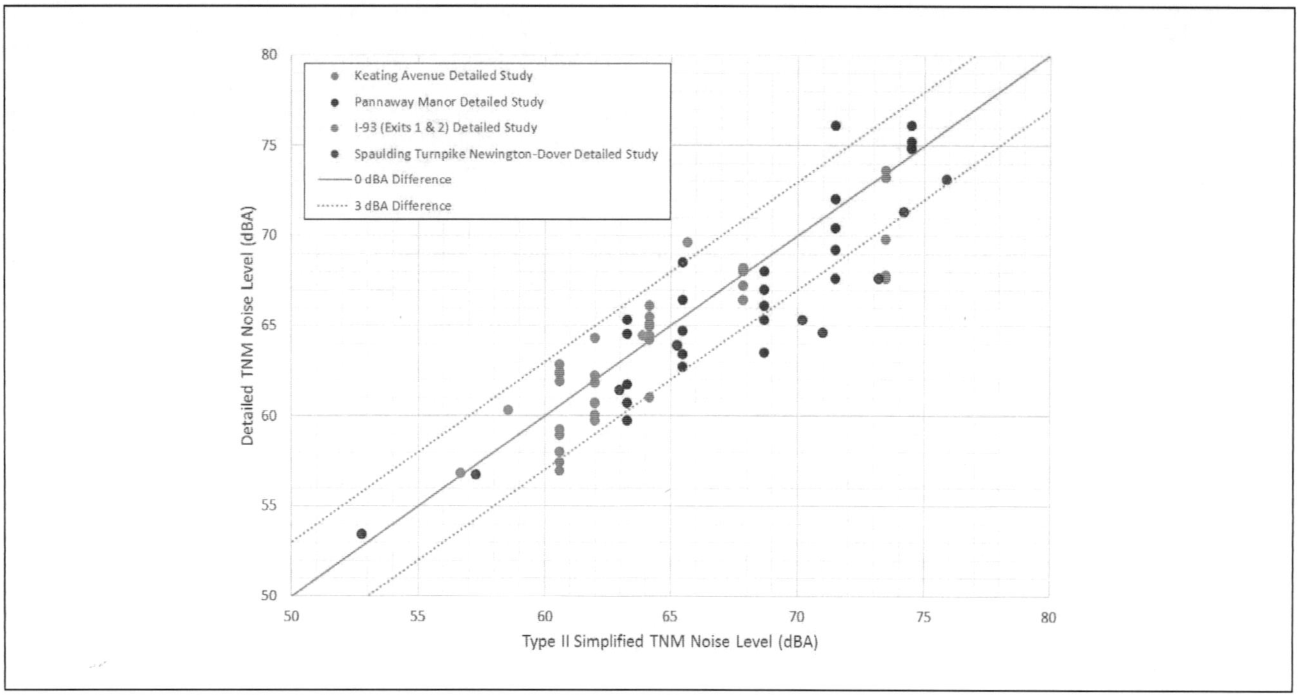

Figure 3. Detailed TNM vs. simplified TNM noise levels.

barriers at distances out to 1,500 feet. This analysis showed that the insertion loss provided by a certain height barrier can be predicted with high accuracy based solely on the angle of shielding that the barrier would provide. This relationship allows us to estimate the insertion loss of various height barriers based only on the geometry of the receptors and the barriers which can be calculated in the Geographic Information System (GIS) program. This screening-level modeling approach is critical to the practicality of using TNM across the entire state given the significant number of receptors and noise barrier study areas.

As shown in Figure 4, the insertion loss of a 1,200-feet-long and 20-feet-tall barrier depends on the proximity of the receptor to the barrier and where along the barrier the receptor is. This figure shows that 5 dB of insertion loss, which is the minimum needed to be considered a benefited receptor, extends out up to approximately 500 feet from the highway median in the center of the barrier.

Figure 5 shows the same results as in Figure 4, although the insertion loss of the barrier is plotted against the angle of shielding which the barrier provides. This curve shows that there is a high correlation between the barrier shielding angle and insertion loss and, in fact, that this relationship is not dependent on the distance of the receptor to the barrier. This relationship was found to be the same for shorter and longer barriers from 300 to 3,600 feet. This figure also demonstrates how much farther past the last receptor a noise barrier should extend. A common rule of thumb is to maintain a 4 to 1 ratio; for example, if a receptor is 100 feet from a barrier, the barrier should extend 400 feet past it. This rule of thumb is equal to a shielding angle of 165 degrees which does maintain a high level of insertion loss. This figure also shows that if a noise barrier is extended less than the rule of thumb, such as only a 1 to 1 ratio (135 degrees of shielding), the insertion loss may be reduced 5 dB or more, but the barrier can still benefit receptors.

Barrier Height Effects on Insertion Loss

The grid analysis was conducted for 1,200-feet-long barriers with heights of 5 to 25 feet. Figure 6 shows the relationships of insertion loss to barrier angle for 5, 10, 15, 20, and 25-foot barriers assuming flat ground. This figure shows that there is relatively little difference in insertion loss as a function of barrier height among barriers of 15, 20, and 25 feet height and that in some cases extending the length of the barrier can provide greater increases in insertion loss than increasing the barrier height.

Building Row and Building Barrier Effects on Insertion Loss

The relationship of insertion loss versus shielding angle is not substantially affected by intervening building rows of 25% or 50% building. Figure 7 shows the change in insertion loss with a 20-foot building row or 20-foot barrier located within 100 feet of the highway. This figure

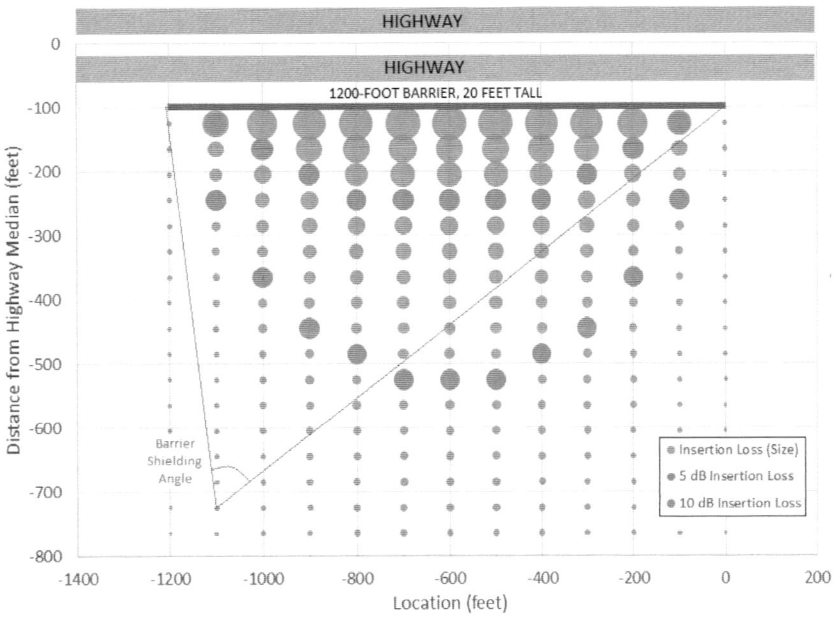

Figure 4. Noise barrier insertion loss grid analysis.

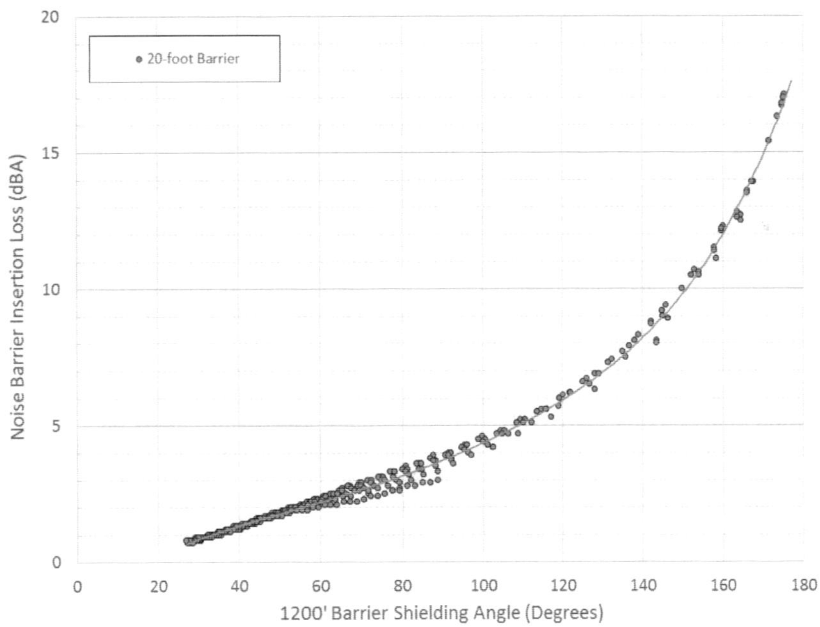

Figure 5. Noise barrier insertion loss vs. shielding angle.

shows the building rows reduce insertion loss only up to 1 dBA at certain shielding angles. The 20-foot barrier, on the other hand, has a more substantial effect on insertion loss.

Terrain Effects on Insertion Loss

Another important factor that has been shown to substantially affect insertion loss is whether there is intervening terrain (i.e., if the highway is in a cut). When the

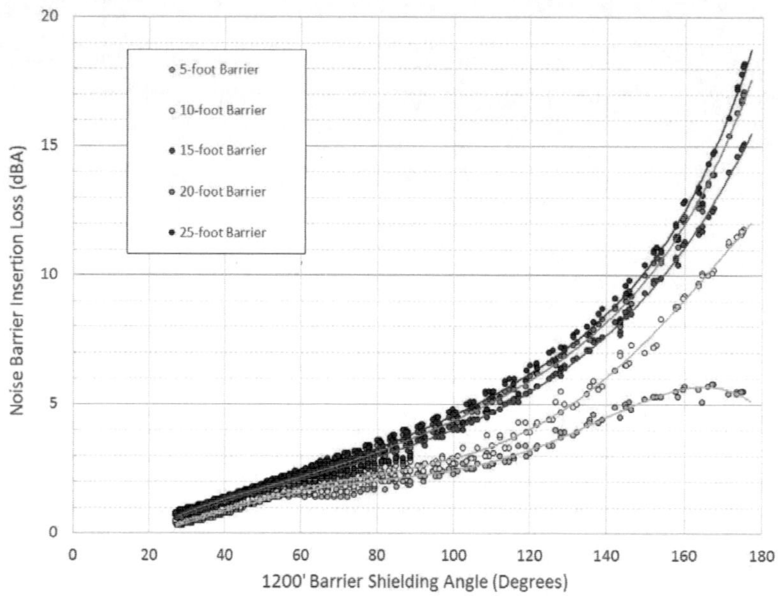

Figure 6. Noise barrier insertion loss vs. shielding angle (barrier heights 5 to 25 feet).

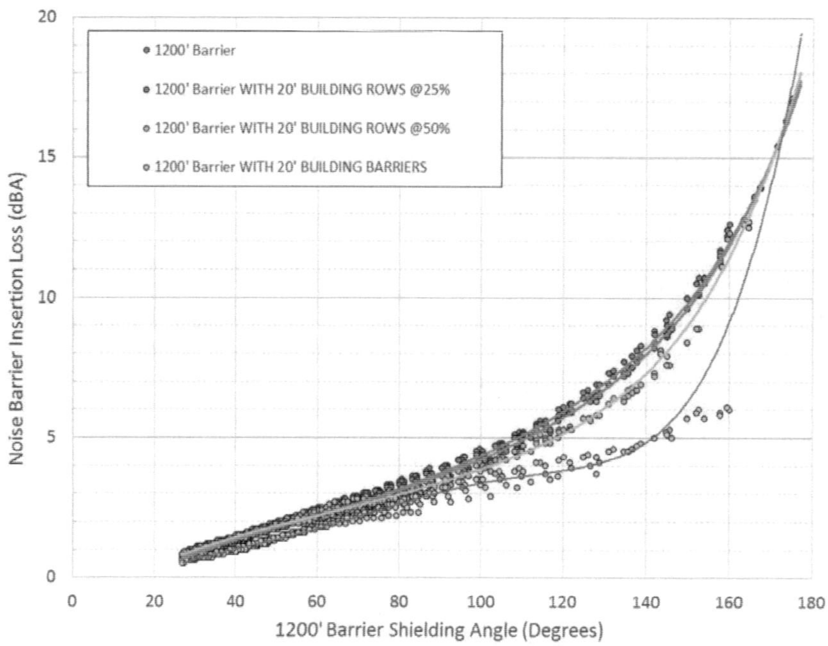

Figure 7. Insertion loss vs. shielding angle, building rows, and buildings as barriers.

highway is in a cut, as shown in Figure 8, the existing terrain can break the line of sight to certain noise sources which both reduce highway noise and reduce the effectiveness of the incremental increase in acoustic shielding provided by a noise barrier.

Figure 9 shows the significant degradation of insertion loss when the highway is in a 5 to 20-foot cut. To provide a conservative estimate of potentially eligible barriers, this evaluation assumes that the terrain of each study area is relatively flat ground and the highway is not in a

Figure 8. Existing terrain breaking the line of sight and reducing barrier effectiveness.

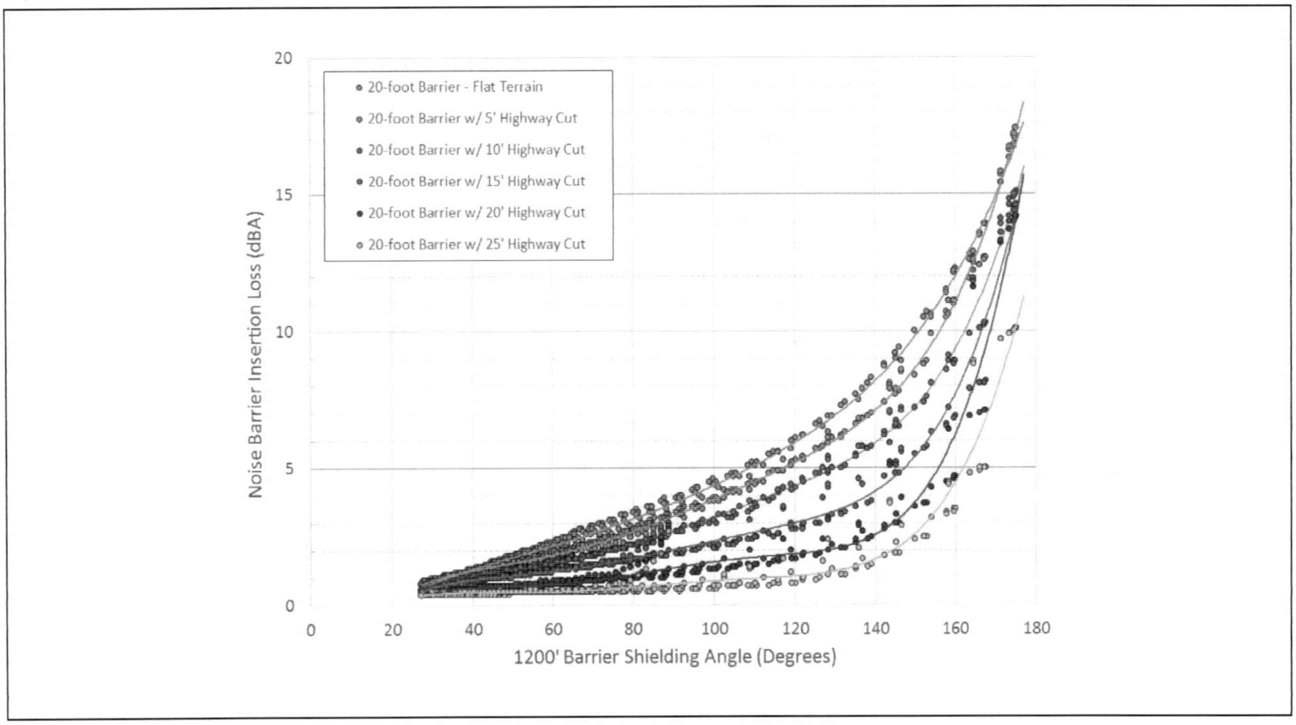

Figure 9. Noise barrier insertion loss vs. shielding angle (highway in cut 5 to 25 feet).

significant cut or on an embankment. The distance to noise impact, as shown in Figure 2, and insertion loss, as shown in Figure 9, are reduced when the terrain already breaks the line of sight between the receptors and the pavement, so the acoustical effectiveness and noise barrier DEI is substantially reduced. Therefore, detailed TNM noise modeling to further evaluate the eligibility of Type II barriers should include these terrain effects. Without any terrain effects, the results of this study conservatively assume there is a greater potential for feasible and reasonable barriers.

Comparison of Screening-Level and Detailed Noise Barrier Evaluations

Figures 10 and 11 compare the insertion loss calculated in the detailed TNM studies at Pannaway Manor, Keating Avenue, and I-93 with the grid analysis calculations. This figure shows that the insertion loss results are typically within 3 dB for receptors in a flat terrain.

It should be noted that the extent of the study areas—how far away from the highway receptors are included—has a significant effect on the number of benefits a barrier can provide and consequentially on the DEI of the barrier. If receptors only out to 750 feet from the highway median are included, the total number of eligible barriers is reduced substantially. Limiting the distance of receptors included in the analysis may provide a more realistic determination of barrier insertion loss and potential benefit because background ambient noise becomes a more important factor at distances farther from the highway. Some DOTs such as North Carolina DOT actually limit the distance of receptors to 800 feet for Tier 1 highways and do not allow additional receptors to be included for the purposes of determining benefit. All receptors that would be impacted must be included in traffic noise studies regardless of distance from the highway.

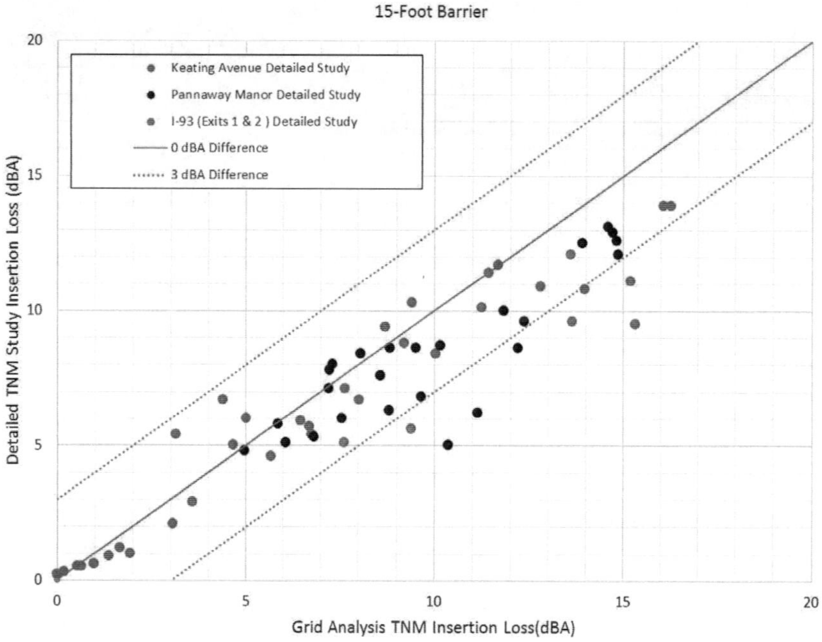

Figure 10. Insertion loss for detailed TNM study vs. grid analysis (15-foot barrier)—flat terrain.

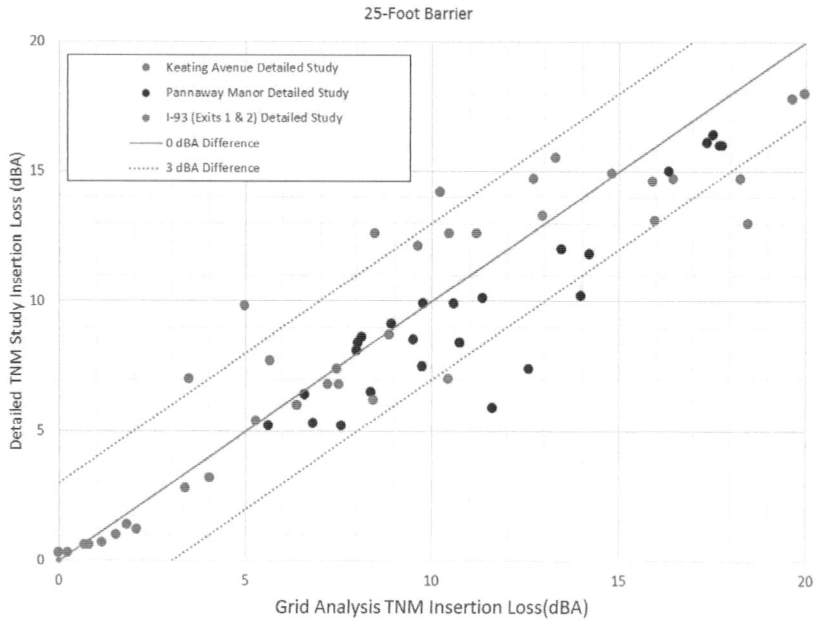

Figure 11. Insertion loss for detailed TNM study vs. grid analysis (25-foot barrier)—flat terrain.

Feasibility and Reasonableness Evaluation

For all study areas, noise barriers have been evaluated for feasibility and reasonableness according to the following criteria:

- Highway noise levels during existing loudest-hour conditions meet or exceed the NAC (66 dBA Leq).
- The barrier must provide a minimum of 7 dB insertion loss for at least one benefited receptor.

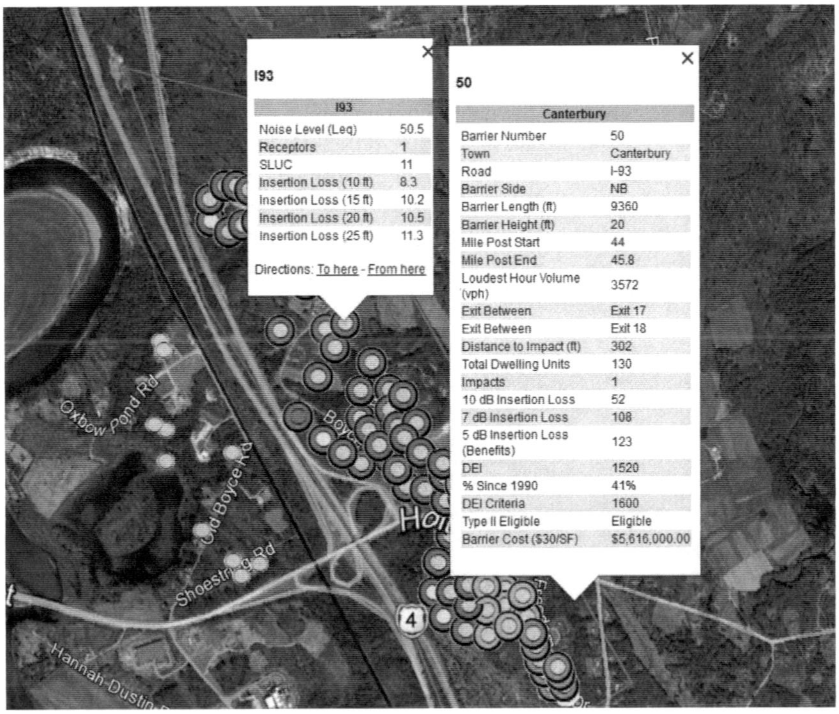

Figure 12. Google Earth™ noise barrier and receptor information.

- The DEI of the barrier (square footage of barrier per benefited receptor) shall be less than the criteria which is nominally 1,500 plus adjustments based on the date of developments.
- Constructability and safety shall be considered.
- It is also a goal (not a requirement) that the barrier should provide 10 dBA or greater insertion loss at first row receptors.

Based on highway noise level predictions and the grid analysis insertion loss calculations, the DEI for 10, 15, 20, and 25-foot barriers were computed. Study areas were evaluated according to NHDOT Type II eligibility, feasibility, and reasonableness criteria. For all eligible barriers, the tallest barrier that meets all criteria was assumed. For ineligible barriers, the barrier height resulting in the lowest DEI that can be achieved has been reported.

Summary of Results

The simplified prediction method was used to evaluate over 300 miles of highway, 300 candidate noise barrier study areas, and over 30,000 receptors. The statewide noise barrier study showed that a total of 49 barriers over 37 miles at an estimated cost of approximately $125,000,000 are likely to be eligible for the Type II program.

This screening-level modeling approach is critical to the practicality of using TNM across the entire state given the significant number of receptors and noise barrier study areas. The results of the study including information about all 30,000 receptors and 300 noise barrier study areas were collected in Google Earth™ data files for easy access when the DOT responds to noise complaints from the public. The files allow the user to view all study areas and their associated results (i.e., town/city, highway, distance to noise impact, number of impacts, number of benefits, DEI, and eligibility) and receptors and their associated results (i.e., number of dwelling units, noise level, and insertion loss for barriers 10 to 25 feet tall). Figure 12 shows a typical noise barrier study area and receptor information that is available in the Google Earth™ files.

Author Contributions

The authors confirm contributing to the paper as follows: study conception and design: Jason Ross, Mark Arnoldy, and Jon Evans; data collection: Jason Ross, Mark Arnoldy, and Jon Evans; analysis and interpretation of results: Jason Ross, Mark Arnoldy, and Jon Evans; draft manuscript preparation: Jason Ross, Mark Arnoldy, and Jon Evans. All authors reviewed the results and approved the final version of the manuscript.

The Standing Committee on Transportation-Related Noise and Vibration (ADC40) peer-reviewed this paper (18-04698).

Sage Publications USA
sagepub.com

ISBN 978-1-5108-7828-0

9 781510 878280

CURRAN ASSOCIATES INC.
proceedings.com

Traffic Monitoring: Automobiles, Trucks, Bicycles, and Pedestrians 2018

Transportation Research Record: Journal of the Transportation Research Board Volume 2672, Issue 43

SAGE Publishing